FERTIGATION

FERTIGATION

Authors:
C. Burt
K. O'Connor
T. Ruehr

Funding Provided By:

California Energy Commission
Agricultural Energy Assistance Program

Published by:

Irrigation Training and Research Center
California Polytechnic State University
San Luis Obispo, CA 93407

Written and published in the United States of America.

Copies may be ordered from
The Irrigation Training & Research Center
California Polytechnic State University (Cal Poly)
San Luis Obispo, CA 93407
(805) 756-2434

Fertigation
C. Burt, K. O'Connor, and T. Ruehr
320 pages

Funding for development of this book was provided by the California Energy Commission, Agricultural Energy Assistance Program.

LEGAL NOTICE:
This report was prepared as a result of work sponsored by the California Energy Commission. It does not necessarily represent the views of the Energy Commission, its employees, or the State of California. The Commission, the State of California, its employees, contractors, and subcontractors make no warranty, express or implied, and assume no legal liability for the information in this report; nor does any party represent that the use of this information will not infringe upon privately owned rights.

Library of Congress Catalog Card No.: 94-75957
ISBN 0-9643634-1-0

Cover Photograph: Fertigation Containment, T&A, Salinas, CA; Charles Burt, photographer
Printed on 60# Recycled Offset paper with 10pt coated recycled cover stock
Typeset in Adobe Garamond with Helvetica and Palatino

Book Design and Production by Lindsay Roy
Printing by Central Coast Printing, Grover Beach, CA

Acknowledgments

By the time any book on a technical topic is completed, the authors fully realize that they could not have completed the book without the help of many others. Such is the case for this book, and our acknowledgments will suffer because we simply cannot list all the people who provided valuable insight. However, some special names stand out.

Bill Dickinson, a consultant from Port Hueneme, provided valuable review on some chapters regarding fertilizer compounds.

Mike Wharton, a graduate student at Cal Poly, spent a summer interviewing farmers and consultants regarding fertigation practices. He then assembled the first parts of the information.

Numerous growers, equipment manufacturers, and consultants throughout California willingly shared their knowledge. Special names include Gideon Cohen (Three Flags Ranch, Brawley), Angelo Mazzei, Skip Purdee (Crop Production Services, Stockton), Gary Tanimura (T&A, Salinas), McFarland Associates (Clovis), Gary McKenzie (Calgene Fresh), Gallo, M. Caratan Farms, Steve Beck (Crown Packing), Steve Flanagan, Robert Viets (Double B Farms), Jerry Rivers (Growers Testing Service, Visalia), Hugh Rathbun (Dellavalle Labs, Fresno), Larry Ferrini (Bonita Packing, Santa Maria), Craig Reade (Betteravia Farms), Brock Taylor (Vaquero Farms) and Steve Moss (Christopher Farms, Gilroy).

Staff members of the USEPA, the California EPA (Department of Pesticide Regulation), and various Agricultural Commissioners provided information on backflow prevention regulations.

The authors would like to express their sincere appreciation to the California Energy Commission for financial support. In particular, staff members Ricardo Amon and Henry Traylor recognized the importance of examining the overall energy efficiency of farming, including the concept of how much yield is obtained per unit of energy consumed. Their understanding of the importance of good fertilizer management, and how fertigation will play an increasingly important role in energy/resource management, allowed this book to be developed.

Finally, Lindsay Roy must be thanked for the many long days (and some weekends) spent on graphics and editing.

TABLE OF CONTENTS

CHAPTER 1. INTRODUCTION
General 1
Extent of Chemigation Use 1
Energy Conservation 3

CHAPTER 2. SAFETY
Highlights 5
Regulations 5
  General 5
  Chemigation Safety Hardware 6
  Pesticide Labels 10
  Pesticide Notification or Posting 11
Proper Materials for Hardware 11
  Hoses 11
  Fittings 12
  Tanks 12
Containment Structures 12
Neatness 13
Chemical Safety 13

CHAPTER 3. CHEMICAL INJECTORS
Highlights 15
Injector Flow Rates 15
Limiting the Amount of Injected Chemical 15
Single vs. Multiple Units 17
Injector Flushing 19
Drip Systems – Injection Upstream or Downstream of the Filter? 20
Proportional Control 20
Obtaining a Pressure Differential 21
  In-line Pressure Differential 21
  Large Venturi Bypass 22

Plumbing across a Booster Pump 23
Bypass Pumps 23
Injector Designs 24
Venturi 24
Float Valves 26
Differential Pressure Tank 28
N_2 Gas-Powered Pumps 28
N_2 Pressurized Tank 30
Chicken Feeders 31
Water Powered Pumps 31
Diaphragm Pumps 34
Piston Pumps 35
Gypsum Injectors 36
SO_2 Generators 37
Injector Calibration Accuracy 38

CHAPTER 4. INJECTION TECHNIQUES FOR VARIOUS IRRIGATION METHODS

Highlights 43
California Considerations 43
Moving vs. Stationary Irrigation Systems 44
Continuous Move Irrigation Systems – General 44
Moving Systems – Surface Irrigation 44
Stationary Irrigation Systems 47
Variations in Chemical Injection Concentrations with Time 48
Centralized vs. Mobile Injection Units 49
Continuous vs. Non-Continuous Injection 50
Chemical Travel Time in Pipelines 51

CHAPTER 5. IRRIGATION PRINCIPLES, LEACHING, AND FERTILIZER UNIFORMITY

Highlights 53
Irrigation Distribution Uniformity and Scheduling 53
Chemical Leaching 58
Nitrogen 58
Other Nutrients 58
Pesticides 59
Preferential Flow 59
Broadcasting vs. Injecting Nitrogen 60
Leaching 60
Soil Nitrogen Uniformity 60

CHAPTER 6. NITROGEN TRANSFORMATIONS AND PROCESSES

Highlights 63
Nitrogen Cycle 64
Nitrogen Transformations 65
Nitrogen Fixation 65
Mineralization 65

Nitrification ... 65
Immobilization ... 65
Denitrification ... 66
Volatilization ... 66
Ammonium ... 66
General ... 66
Soil Acidification with NH_4^+ and NH_3 Fertilizers ... 67
Correcting Acidity ... 67
Nitrification Inhibitors ... 67
Nitrate ... 68
General ... 68
Leaching ... 68
Ammonia ... 68
General ... 68
Volatilization ... 68
Avoiding Volatilization Losses ... 70
Organic Nitrogen Fertilizers ... 71
Urea ... 71
General ... 71
Urea Hydrolysis ... 71

Chapter 7. Nitrogen Uptake

Highlights ... 73
Cation-Anion Balance ... 73
Nitrogen Source and Effect on Soil pH ... 74
Nitrogen Movement in the Plant ... 75
Ammonium vs. Nitrate Nutrition ... 75

Chapter 8. Other Nutrient Processes

Highlights ... 77
Mechanisms for Nutrient Uptake ... 77
Nutrient Interactions ... 78
Phosphorus ... 78
General ... 78
Phosphate Movement ... 79
Phosphorus Application ... 80
Phosphorus Uptake ... 81
Potassium ... 81
General ... 81
Potassium Movement ... 82
Potassium Uptake ... 83
Secondary Nutrients ... 83
Calcium ... 84
Magnesium ... 84
Sulfur ... 84

Micronutrients 84
General 84
Chloride 85
Boron 85
Molybdenum 85
Metal Micronutrients 86
Copper 87
Iron 87
Manganese 87
Zinc 87
Metal Chelates 88

CHAPTER 9. SOLUBILITY AND COMPATIBILITY
Highlights 91
Fertilizer Solubility 91
General 91
Dry Fertilizer Conditioners 91
Cooling Effect with Mixing 92
Fertilizer Compatibility 94
General 94
The Jar Test 94
Basic Mixing Rules 96
Nitrogen 97
General 97
Liming Effect with Ammonia Fertigation 97
Phosphorus Solubility 98
Potassium 100
Calcium 100
Micronutrients 100
Corrosion 102

CHAPTER 10. SPECIFIC FERTILIZERS AND AGRICULTURAL MINERALS
Highlights 103
Trends in Fertilizer Use 104
Anhydrous Ammonia (82-0-0) 106
Aqua Ammonia (20-0-0) 107
Ammonium Nitrate Solution or AN-20 (20-0-0) 107
Urea Ammonium Nitrate Solution or UAN-32 (32-0-0) 107
Calcium Ammonium Nitrate or CAN-17 (17-0-0-8.8Ca) 107
Ammonium Phosphate (8-24-0) 108
Ammonium Polyphosphate (9-30-0, 10-34-0 and 11-37-0) 108
Ammonium Polysulfide (20-0-0-45) 109
Ammonium Thiosulfate (12-0-0-26) 109
Calcium Polysulfide or Lime Sulfur 110
Metal Chelates 110

Phosphoric Acid (0-54-0 "White" & 0-52-0 "Green" Acids) ... 113
Potassium Chloride ... 113
Potassium Nitrate ... 113
Potassium Phosphate (0-52-34) ... 114
Potassium Sulfate ... 114
Potassium Thiosulfate (0-0-25-17 and 0-0-22-23) ... 114
Sulfuric Acid ... 114
Urea Solid (46-0-0) and Urea Solution (23-0-0) ... 115
Urea Phosphate (17-44-0) ... 116
Urea Sulfuric Acid ... 116
    General ... 116
    Safety Considerations ... 117
    Urea Sulfuric Acid as a Fertilizer ... 118
    Urea Sulfuric Acid as a Maintenance Strategy ... 119

CHAPTER 11. PLANT AND SOIL TESTING

Highlights ... 121
New Attitudes about Nutrient Management ... 121
Stages in Improving Nutrient Management ... 122
Categories of Tests ... 123
Describing Amounts of Nutrients ... 124
    General Guidelines ... 124
    Milliequivalent Conversions ... 125
Soil Sample Testing ... 126
    Highlights ... 126
    Standard Laboratory Procedures ... 126
    On-Farm Quick Tests ... 127
    Soil Sampling ... 128
    Soil Sample Nitrogen ... 129
    Soil Sample Phosphorus ... 130
    Soil Sample Potassium ... 130
    Soil Secondary and Micronutrients ... 131
    Interpretations from Soil Sample Tests ... 132
Soil Solution Testing ... 133
    Highlights ... 133
    General ... 133
    Soil Solution Testing – Procedures ... 134
    Soil Solution Testing – Interpretations ... 135
Plant Tissue Analysis ... 136
    Highlights ... 136
    General ... 136
    Plant Tissue Sampling Procedures ... 138
    Tissue Test Interpretation – Critical Level Approach ... 139
    Tissue Test Interpretation – Sufficiency Range Approach ... 139
    Tissue Test Interpretation – Nutrient Ratio/Product Approach ... 147

Tissue Test Interpretation –
Diagnosis and Recommendation Integrated System (DRIS) Approach 149
Plant Sap Testing 152
Irrigation Water Testing 154
How Much Nutrient to Apply 154
Conclusions 160

CHAPTER 12. DRIP SYSTEM MAINTENANCE
Introduction 161
Chemicals Used for Plugging Prevention 162
Chlorine 162
Acids 165
Overview of Synthetic Compounds 167
Polyphosphates 168
Phosphonates 171
Polyelectrolytes 171
Bromide Materials 172
Copper Sulfate 172
Specific Plugging Problems and Their Solutions 172
Small Slimy Bacteria 172
Iron and Manganese Bacteria 173
Iron and Manganese Sulfides 174
Calcium and Magnesium Carbonate Precipitation 175
Root Intrusion 175

CHAPTER 13. INFILTRATION PROBLEMS
High Adjusted Sodium Adsorption Ratio 177
Pure Irrigation Water 180
Gypsum Injection 181
High Magnesium/Calcium Ratios 183
Fertigation with Monovalent Cations 184
Polymers 184
Wetting Agents 184
Polyelectrolytes 184
Others 184

CHAPTER 14. SOIL pH MODIFICATION
Soil Acidification 185
Benefits of Liming 186
Neutralizing Materials 186
Determining the Liming Schedule 187
Liming Material Injection Techniques 189

CHAPTER 15. INSECTICIDE, FUNGICIDE, AND HERBICIDE INJECTION (CHEMIGATION)
General 191
Chemigation Advantages 191

Chemigation Disadvantages 192
Herbigation 192
Fungigation 193
Insectigation 194
Nemagation 194
Specific Chemicals 195
Vapam® 195
Nemacur® 196
Vydate® 196
Furadan 4F® 196

CHAPTER 16. GROWER DRIP FERTIGATION EXPERIENCES
Bell Peppers 197
Lettuce 198
Plant and Soil Testing 198
Fertilizers for Lettuce 199
Special Practices on Lettuce 199
Tomatoes 200
Plant and Soil Testing 200
Fertilizers for Tomatoes 200
Tomato Summary 202
Grapes 203
Plant and Soil Testing 203
Fertilizers for Grapes 204
Other Chemicals for Vines 206
Summary for Grapes 207

CHAPTER 17. SAMPLE FERTIGATION CALCULATIONS
Cost Comparisons of Fertilizers 209
Injection Rate Calibration 209
Relevant Conversions 209
Density of Various Liquid Fertilizers 210
Calculation of Fertilizer Injection Rate 210

APPENDIX A. UNITS OF SALINITY MEASUREMENT
Electronical Conductivity (EC) 215
Parts Per Million (PPM) 216
Milliequivalents per Liter (meq/L) 217

APPENDIX B. QUICK TEST EQUIPMENT 219

APPENDIX C. QUICK TEST PROCEDURES
Soil Nitrate Test Strip Procedure 223
Soil Nitrate Colorimetric Procedure 225
Sap Nitrate Test with Cardy Meter® 226

APPENDIX D. EPA LABEL IMPROVEMENT PROGRAM 229

Appendix E. California Regulations 245

Appendix F. Sample Pesticide Labels 253

References 281

Glossary 287

Index 291

LIST OF TABLES

Table 1. Extent of fertigation used on specific crops for 1984 3
Table 2. Descriptions of required safety devices 7
Table 3. Approved alternative devices for chemigation equipment 8
Table 4. Manufacturer's information regarding venturi efficiencies and injection capacities. 25
Table 5. Manufacturer's information on fertigation equipment 39
Table 6. The effect of pH on ammonia volatilization 69
Table 7. Plant composition as influenced by the type of nitrogen nutrition 74
Table 8. Documented nutrient interactions 78
Table 9. Solubility information on various fertilizer compounds 93
Table 10. Relative corrosion of various metals 102
Table 11. A summary of the fertilizers sold specifically for fertigation in California 104
Table 12. Liquid fertilizer and agricultural minerals used in California 106
Table 13. Densities of various ammonium polyphosphate fertilizers 108
Table 14. Summary and constituents of N-pHURIC® products 117
Table 15. Examples of safe mixtures of N-pHURIC® 118
Table 16. Differences between nutrient test categories 123
Table 17. Conversions of relevant nutrient units of measurement 124
Table 18. Common constituents of fertilizer, soil, and water 125
Table 19. General rules for interpreting soil test results 132
Table 20. Manufacturer's information regarding soil solution access tubes 134
Table 21. General interpretations of nutrients in the soil solution 135
Table 22. Plant analysis guide 140
Table 23. Total nutrient analyses for diagnosis of the total nutrient level of vegetable crops 144
Table 24. Sample laboratory petiole analysis for citrus 147
Table 25. Criteria for selecting a nutrient ratio or product 148
Table 26. Nutrient ratios derived from actual laboratory petiole analyses results 149

Table 27. Comparison of successes and failures of fertilizer recommendations 150
Table 28. Citations for published DRIS norms 151
Table 29. Tissue and sap petiole NO_3-N sufficiency value ranges 153
Table 30. Conversion of water analysis of ions to pounds material per acre-foot of water 154
Table 31. Quantity of nutrients contained in the harvested portion of various crops 155
Table 32. Nitrogen uptake (net) by growth stage for various crops 156
Table 33. Weekly nitrogen fertigation estimate of (gross) requirements of vegetable crops ... 159
Table 34. Guidelines for potential water plugging 161
Table 35. Quantities of N-pHURIC® required 166
Table 36. Equivalent rosestone injection to counter the "liming effect" 169
Table 37. Ca_x values in meq/L for use in calculating R_{Na} of irrigation water 178
Table 38. Permeability hazard of irrigation water 178
Table 39. Traditional liming materials and their calcium carbonate equivalent 187
Table 40. One farmer's general fertigation strategy on bell peppers 198
Table 41. One farmer's general fertigation schedule for tomatoes 202
Table 42. Guidelines for calibrating injector flow rates 210
Table 43. Densities and grades of various liquid fertilizers 210
Table 44. Common constituents of irrigation waters 217
Table 45. Manufacturer information on various quick test equipment 219
Table 46. Relationship of NO_3-N concentration of petiole sap and dry petiole tissue 228

LIST OF FIGURES

Figure 1. Chemigation station layout 6

Figure 2. Combination spring-loaded check valve and injection point 9

Figure 3. Gooseneck pipe loop approved by USEPA in lieu of irrigation line check valve 10

Figure 4. Containment structure built of concrete block walls 12

Figure 5. Containment structure made from a large poly tank 13

Figure 6. Limiting the amount to be injected by the depth of the suction hose 16

Figure 7. Diaphragm pump with four independent pump heads 17

Figure 8. Schematic of diaphragm pump with four independent pump heads 18

Figure 9. Neat configuration of multiple venturi injectors 19

Figure 10. Multiple venturi injectors on a permanent drip installation 19

Figure 11. Some injectors utilize a pressure differential in the irrigation pipeline 21

Figure 12. Use of a large venturi to create adequate pressure differential 22

Figure 13. Using the pressure differential across a booster pump to power an injector 23

Figure 14. The use of a bypass pump to provide motive flow for a venturi injector 24

Figure 15. A simple venturi injector 24

Figure 16. A typical commercial unit incorporating a bypass pump and a venturi 26

Figure 17. A typical float valve assembly 26

Figure 18. Utilizing a float valve to discharge into a canal or open pipe 27

Figure 19. Float valves located on a canal upstream of a drip system booster pump 27

Figure 20. A differential pressure tank 28

Figure 21. A typical nitrogen gas (N_2) powered pump installation 29

Figure 22. Use of N_2 gas and pressurized tank with a simple orifice for flow adjustment 29

Figure 23. On-site N_2 pressurized tank with a float valve box on a standpipe 30

Figure 24. A typical N_2 pressurized tank with a float valve box on a standpipe 30

Figure 25. Homemade chicken feeder assembly used for low rates of injection 31

Figure 26. Water-powered pump using a linear hydraulic motor 32

Figure 27. Water-powered pump using diaphragm action 33
Figure 28. A typical diaphragm pump 34
Figure 29. An electric powered piston pump 35
Figure 30. A traditional electric powered piston pump design 36
Figure 31. A typical commercial gypsum injector 37
Figure 32. A typical sulfur dioxide gas generator 38
Figure 33. Distance vs. advance time along a hypothetical furrow 45
Figure 34. Infiltration rates decrease over time at a point along a furrow or border strip 46
Figure 35. Depths of water infiltrated at various distances along a furrow 46
Figure 36. An injector pump and portable tank and stand 49
Figure 37. The concept of non-uniformity 54
Figure 38. The concept of adequate irrigation 55
Figure 39. The concept of improved uniformity 56
Figure 40. The concept of poor timing 57
Figure 41. The nitrogen cycle 64
Figure 42. Plant leaves can be damaged if exposed to ammonia 70
Figure 43. The relationship between soil pH and phosphorus availability 79
Figure 44. The potassium environment 82
Figure 45. Relationship between soil pH and nutrient availability 83
Figure 46. Compatibility chart for common fertilizers and agricultural minerals 95
Figure 47. Fertigation must often supplement the slow natural release of potassium 131
Figure 48. Nitrate nitrogen readings from the Cardy Meter® 153
Figure 49. Hypochlorous acid concentration as a function of water pH 163
Figure 50. Acid titration curves for two different water samples 166
Figure 51. A gypsum injection machine 183
Figure 52. Approximate tons of limestone required to raise the soil pH 188

PREFACE

This book is provided as a reference for agricultural students, consultants, fertilizer specialists, and growers. Both the novice and the expert will find new information.

Some excellent Chemigation references are available for the Pacific Northwest, Texas, and other states through the extension services there. We (the authors) did not try to duplicate the information found in those references, although they are quoted where appropriate. We did decide to only marginally cover the topics of insecticide, fungicide, and herbicide injection, since those are covered very well in other references.

This book does address many new ideas and consolidates others. For example:

- we cover many types of chemical injection hardware which are widely used, but are not referenced in most literature,
- special considerations which must be made depending on whether one plans to apply chemicals through a furrow, sprinkler, or drip system,
- and, the importance of understanding irrigation system uniformity.

Of particular interest to us at the start was the whole arena of what types/mixes of fertilizers should be applied, and when. Along those lines, we tried to consolidate concepts such as maintaining proper cation/ion balances, DRIS, and the proper ratio of ammonium/urea/nitrate forms of nitrogen fertilizers.

Finally, in California we have important questions regarding water penetration problems (infiltration problems) and the maintenance of drip/micro irrigation systems. Each topic was given a separate chapter.

Charles M. Burt, P.E., Ph.D., Professor of Agricultural Engineering and ITRC Director
Kris O'Connor, ITRC Special Problems Investigator
Thomas Ruehr, Ph.D., Professor of Soil Science

June 1995

CHAPTER 1. INTRODUCTION

GENERAL

"Chemigation" is the application of any chemical through irrigation water. This may include insecticides, fumigants, nematicides, fertilizers, soil amendments, and other compounds. By far, the most common form of *Chemigation* is *"Fertigation,"* which refers to fertilizer application in the irrigation water. This book emphasizes *Fertigation.*

Fertigation offers several distinct advantages in comparison to conventional application methods.

1. Soil compaction is avoided because heavy equipment never enters the field.
2. The crop is not damaged by root pruning, breakage of leaves, or bending over, as occurs with conventional chemical field application techniques.
3. Less equipment may be required to apply the chemical.
4. Less energy is expended in applying the chemical.
5. Usually less labor is needed to supervise the application.
6. The supply of nutrients can be more carefully regulated and monitored.
7. The nutrients can be distributed more evenly throughout the entire root zone or soil profile.
8. The nutrients can be supplied incrementally throughout the season to meet the actual nutritional requirements of the crop.
9. Nutrients can be applied to the soil when crop or soil conditions would otherwise prohibit entry into the field with conventional equipment.

EXTENT OF CHEMIGATION USE

Fertilizers have been applied through a wide range of irrigation systems for many years. Because most of the older traditional methods had significant worker contact with the water (such as with furrow irrigation and hand move sprinklers), fertilizers were the most commonly injected chemicals.
With the introduction of center pivots and linear moves, the application of various forms of pesticides became more widespread, especially in the Midwestern U.S. Significant advancements have been made to the design of special components of these machines to enhance chemigation,

including the development of commercial under-canopy spray heads to apply insecticides to the undersides of leaves, and high speed gearboxes for the drive units, enabling fast movement of the machines across the field for light applications of chemicals. The center pivots and linear moves have some peculiar traits for pesticide application which are not common to some other irrigation methods—they do not require the presence of people in the field during irrigation, and they are capable of quick, small, and very uniform applications of water. Furthermore, they can wet the leaves of the plants, which surface irrigation methods cannot do.

Fertigation, through center pivots in the Midwest, has become fairly sophisticated in some areas. This sophistication has been assisted by the fact that most center pivots are found on a relatively small number of crops—principally grains, soybeans, and hay. The fertility requirements of crops are fairly well established based on decades of research.

Drip and microirrigation development, especially in California, have stimulated a parallel growth in chemigation. An increasingly wide range of fungicides, herbicides, and insecticides are injected through drip and microirrigation systems. The extent of this type of chemigation appears to be largely dependent on the crop type, and is more prevalent and sophisticated on row crops (vegetables and strawberries) than on trees and vines.

Drip and microirrigation have a characteristic not shared by other irrigation methods—fertigation is not optional, but is actually necessary. Fertigation provides the only good way to apply fertilizers physically to the crop root zone for permanent crops. On high value drip irrigated crops, such as lettuce, tomatoes, and peppers, the level of fertigation management for achieving high yields and crop qualities appears to exceed what is found with other irrigation methods and crops.

Results from a 1984 Farm and Ranch Irrigation Survey by the USDA Economic Research Service indicate that fertigation in the United States has been prevalent for some time. That survey revealed that fertigation was used on all of the major crops in the country to varying degrees (Table 1).

Table 1. Extent of fertigation used on specific crops for 1984.
(U.S. Department of Commerce, 1986)

Crop	# of Farms Using Fertigation (%)	Total
Alfalfa	13	7,900
Barley	18	3,300
Cotton	21	2,200
Dry edible beans	19	1,200
Hay (other)	1	2,000
Grain corn	27	11,200
Grain sorghum	14	1,800
Orchard crops	23	11,500
Pasture	11	5,800
Potatoes	45	2,200
Rice	31	2,100
Small grains (others)	9	400
Silage corn	22	2,900
Soybeans	4	600
Sugar beets	22	600
Tobacco	17	700
Wheat	25	6,100
Vegetables	22	3,400

ENERGY CONSERVATION

Chemigation offers a considerable number of benefits in terms of energy conservation. The most obvious savings occurs because vehicles do not need to traverse a field to apply pesticides or fertilizers.

More significant energy savings occur due to the following:

- Less chemical is generally applied by chemigation than with conventional application techniques. There are significant energy requirements in the manufacture of nitrogen fertilizer (approximately 24,600 BTU per pound of nitrogen).

  The ITRC, working with the CEC, has documented on a pepper field with row crop drip that the nitrogen fertilizer can represent 44% of the total annual energy consumption for a row crop drip system (9.3 MBTU/acre out of 20.9 MBTU/acre total) and 51% for a solid set sprinkler system on the same field. The total energy consumption included the energy for manufacturing plastic, filters, tractor travel, etc. In the documented field, the fertilizer application stayed approximately the same for the first year of drip irrigation, but the yield increased by 50%. Effectively, this means that the fertilizer requirement was reduced per ton of produce.

Many row crop growers report typical decreases in fertilizer application to be in the 25% range once they convert to drip irrigation and employ good fertigation practices.

The MBTU/acre saved will obviously depend on the crop. For example, vines traditionally have low fertilizer inputs, so the MBTU/acre saved would be less.

- Fertigation allows growers to manage crop nutrients at a level which is unprecedented, and impossible to achieve with conventional fertilizer practices. The result can be much higher yields and crop quality. Assuming a conservative estimate of a 10% increase in production due to better nutrient management, this represents a 10% increase in energy efficiency for all farming operations.

When discussing energy efficiency, the following definition is often used:

$$\text{Energy efficiency}_{\text{(old concept)}} = \frac{\text{Energy consumption}}{\text{acre}} \times 100$$

Our understanding of energy efficiency has improved, and now we discuss agricultural energy efficiency as:

$$\text{Energy efficiency}_{\text{(new concept)}} = \frac{\text{Energy consumption}}{\text{amount of crop sold}} \times 100$$

By increasing the amount of crop produced or sold through fertigation, the energy efficiency is increased.

- Chemigation with pesticides also allows growers to increase yields or crop quality. With some forms of irrigation, notably drip and center pivots, systemic insecticides can be applied to the crop immediately after an insect infestation is noticed. The soil does not have to dry out for tractor access after a rainfall, for example. Pesticides can be applied quickly before the blight or population spreads. This minimizes damage, and also reduces the amount of pesticide which must be applied.

In summary, the implications of using good irrigation/fertigation practices are tremendous in terms of improving agricultural energy efficiency. For many crops, good fertigation/irrigation practices provide more energy conservation than other practices which have previously been promoted.

CHAPTER 2. SAFETY

HIGHLIGHTS

- Safety is an essential consideration for sound chemigation practices.
- This section handles both the mechanical and chemical aspects of chemigation safety.
- Mechanical components suggested for a safe chemigation system prevent chemical backflow into the water supply, chemical spilling, and chemical injection into empty irrigation lines.
- The materials for hoses, fittings, and tanks must be appropriate for handling specific chemicals.
- Maintaining a neat injection area promotes safer handling of chemicals and simplifies identifying leaks and spills.
- Precautions should be taken when mixing chemicals: follow manufacturer's guidelines, mix a test batch prior to injection, and consult a professional when in doubt.

REGULATIONS

GENERAL

The regulations specific to chemigation involve several issues:

1. Chemigation hardware
2. Pesticide labeling
3. Pesticide notification or posting

The laws have been created to protect the environment, including groundwater, and for worker safety. Since chemigation is relatively new, there is little legislation *specific* to chemigation. In addition, the codes are not uniform between federal, state, and local laws. Prior regulations pertaining to general pesticide use and application still apply to pesticide chemigation. In addition, the Worker Protection Standards (WPS) are being revised, so laws concerning chemical handling, posting, and field reentry are changing. Growers should contact the Pesticide Regulation Department of the California Environmental Protection Agency for more details.

CHEMIGATION SAFETY HARDWARE

Most safety sketches of safety hardware have been developed around injection devices which are commonly used in the Midwestern states. Those systems typically use piston or diaphragm pumps that are powered by electric motors, or are belt-driven from a drive shaft between the engine and the irrigation pump. Figure 1 shows the sketch of such a chemigation layout with the U.S. Environmental Protection Agency (USEPA) required safety devices.

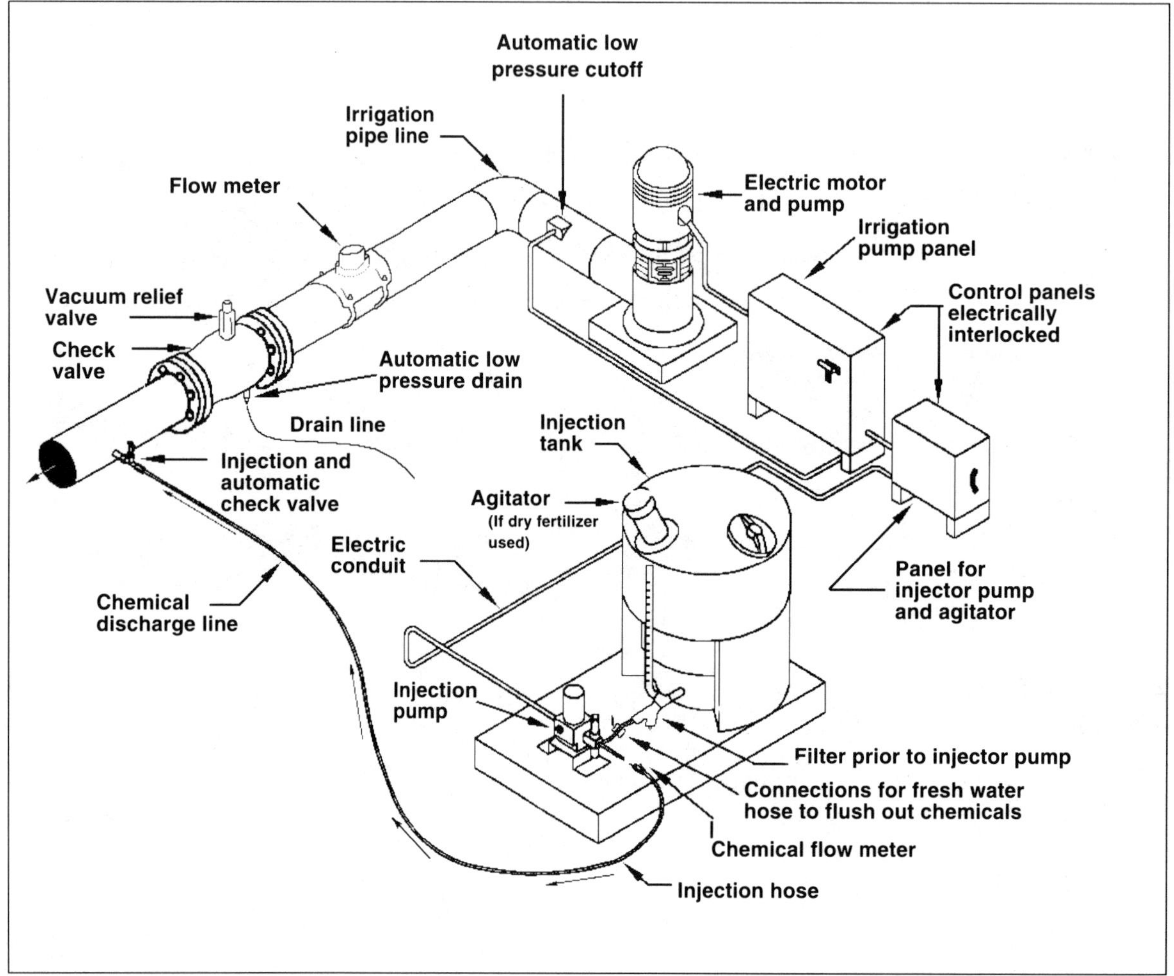

Figure 1. Chemigation station layout.

Table 2. Descriptions of required safety devices.

Device	Description/Location	Purpose
Irrigation check valve *	Between well and injection points	Prevents chemicals from flowing backwards and entering the water source
Injection line check valve	At the injection point and is a one way valve with a 10 psi spring which closes when not under pressure	Prevents water from flowing backwards into the chemical tank, which would cause the tank to overflow and spill
Vacuum relief valve	Between check valve and well	Prevents vacuum when pump shuts off; reduces chance of backflow
Low pressure cutoff	On irrigation pipeline	Turns off injector power when irrigation water pressure is low
Low pressure drain *	Between well and irrigation line check valve	Discharges any water which might leak through the check valve after irrigation pump is shut off
Normally closed solenoid valve	Between injection pump and chemical tank	Prevents tank from emptying unless injector is working
Interlock	Between injection pump and irrigation pump panels/power	Prevents injection if irrigation pump stops

Note: Devices are marked with an asterisk (*) can be substituted **only** by an approved alternative device which became effective in March 1989 by the USEPA (Table 3). Appendix E includes the specific text from the USEPA regarding alternative devices.

The above mentioned devices are required according to the United States Environmental Protection Agency (USEPA) Label Improvement Program (LIP) which became effective in April 1988 (Appendix E). They are required for ground water protection. If there is no possibility of water source contamination (e.g., the injection point is downstream of an air gap), many of these devices may not be required. Regardless, many of the devices incorporate common-sense principles to minimize spills and operator hazards.

Table 3. Approved alternative devices for chemigation equipment.

Original Device	Approved Alternative Device[1]
Functional normally closed, solenoid-operated valve located on the intake side of the injection pump	1) Functional spring-loaded check valve with a minimum of 10 psi cracking pressure (Figure 2) 2) Functional normally closed hydraulically operated check valve 3) Functional vacuum relief valve located in the pesticide injection line between the positive displacement pesticide injection pump and the check valve
Functional main water line check valve and main water line low pressure drain.	1) Gooseneck pipe loop located in the main water line immediately downstream of the irrigation water pump (Figure 3)
Positive displacement pesticide injection pump	1) Venturi systems including those inserted directly into the main water line, those installed in a bypass system, and those bypass systems boosted with an auxiliary water pump
Vacuum relief valve	1) Combination air and vacuum relief valve

1 See details in Appendix D for alternative device specifications.

The spring loaded check valve with a minimum of 10 psi cracking pressure (Figure 2) is generally preferred by growers over the normally closed, solenoid-operated valve located on the intake side of the injection pump. Several manufacturers sell a combination check valve/ injection port device which is located at the discharge end of the chemical hose. This combination device provides the safety feature required by USEPA, and also places the chemical *into the midstream of the irrigation water flow*, providing better chemical mixing. Such midstream placement is especially important when injecting acids; the walls of the pipe may corrode if the acid is injected next to the pipe walls.

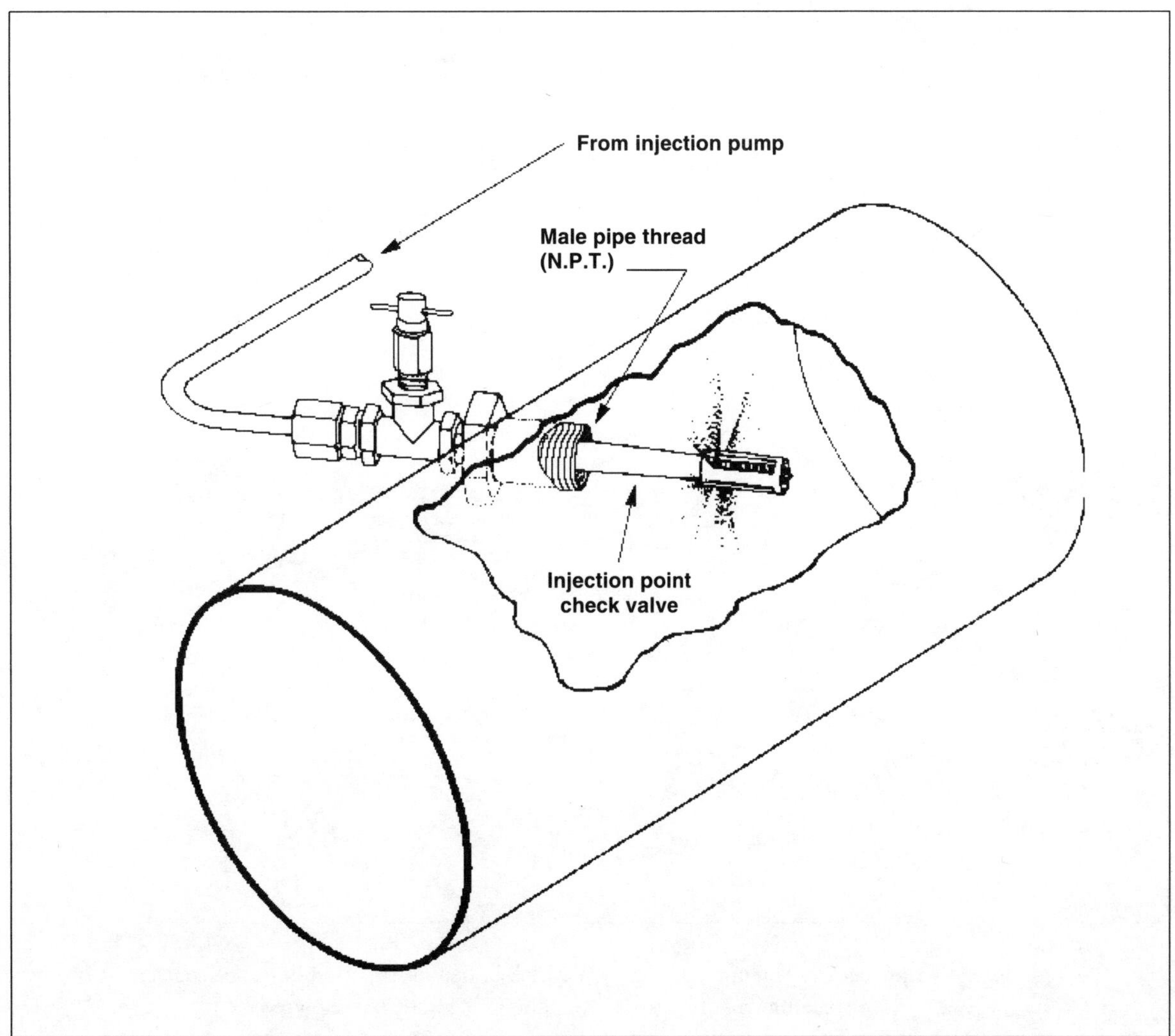

Figure 2. Combination spring-loaded check valve and injection point.

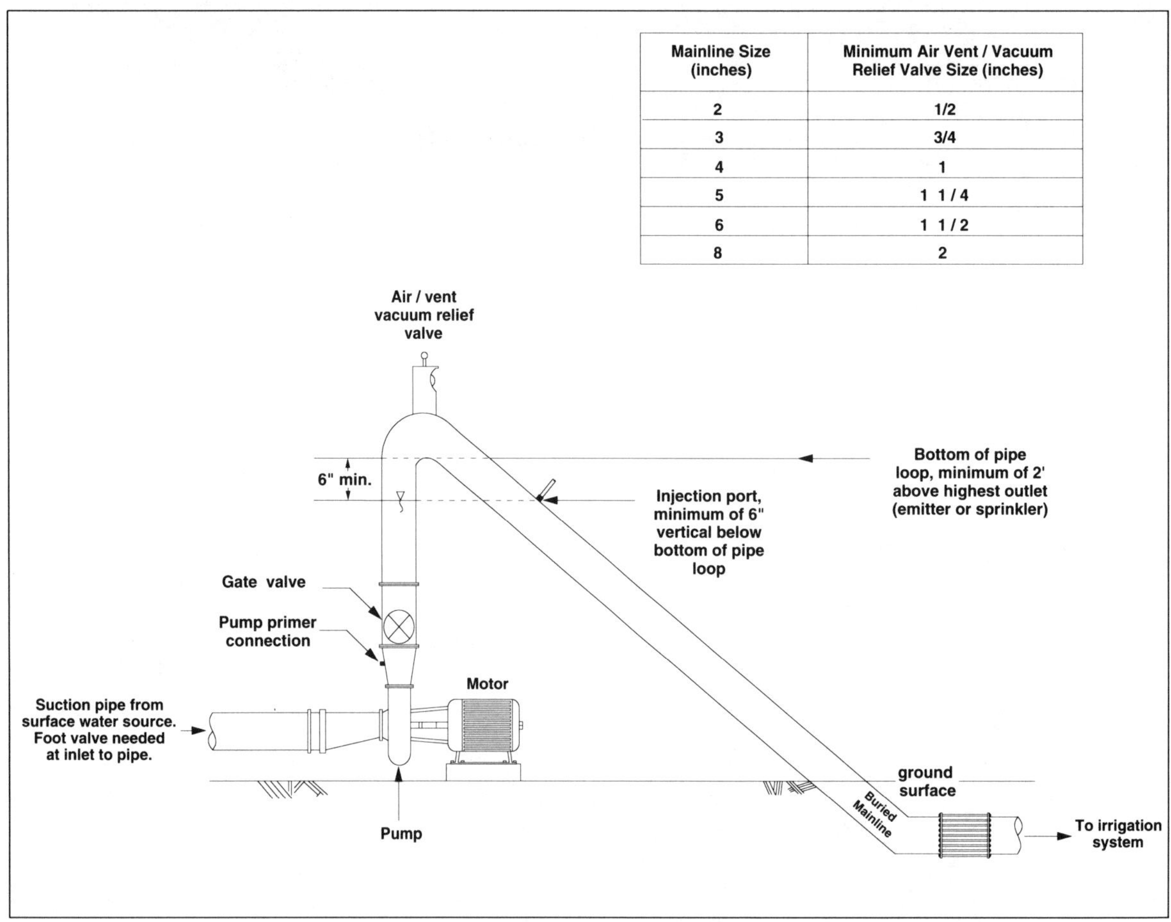

Figure 3. Gooseneck pipe loop approved by USEPA in lieu of irrigation line check valve. Not recommended by the ITRC, due to possible backflow.

The state public health law (Title 17) and agricultural law (Title 3) require that irrigation systems which inject fertilizers, herbicides, or pesticides have one of three backflow prevention devices: air-gap separation, reduced pressure principle backflow prevention device, or double check valve assembly (Appendix E). Such devices must be acceptable to the water purveyor and the local health department. At the time of this publication, there were conflicting requirements for check valve assemblies in California. Some local water agencies have been able to implement regulations which only require a single, spring-loaded check valve. The Environmental Hazard Department of local health departments will provide lists with County approved devices. County agricultural commissioners enforce the pesticide laws and should be contacted regarding local regulations for backflow prevention.

PESTICIDE LABELS

According to the USEPA Label Improvement Program (Appendix D), which deals specifically with restricted products which are chemigated, labels must provide detailed information regarding application rates, reentry intervals, safety clothes, etc. **The label must state that the product can be chemigated and applicators must adhere to the instructions provided on the container labels.** Chemicals are registered in each state for specific crops and methods of application. Special use permits and temporary use permits are sometimes available for nonregistered applications. There are instances when the labels disagree with the guidelines specified in the state code. In these cases, the stricter instruction should be followed. In California, the Cal EPA Department of Pesticide Regulation should be contacted regarding these discrepancies. Appendix F includes some sample labels.

PESTICIDE NOTIFICATION OR POSTING

Before starting the application of certain pesticides through the irrigation system, California codes require that signs be posted at normal points of field entry and shall not exceed 600 foot intervals. The signs shall be in English and Spanish and state: "Danger, Pesticides are applied in water through the irrigation system. Do not drink the water from the irrigation system." These signs should be readable at 25 feet. The USEPA has different requirements regarding chemigation posting. Both Cal EPA and USEPA are in the process of developing a sign which would satisfy both requirements.

Other than these specific regulations pertaining to chemigation, the general pesticide regulations pertaining to field reentry, equipment cleaning, protective clothing, and notification or posting *still apply* to chemigation practices. For a small fee, local Agricultural Commissioners can provide copies of the State Food and Agricultural Code pertaining to Pesticides and Pest Control Operations.

PROPER MATERIALS FOR HARDWARE

HOSES

In most cases, hoses should be constructed of reinforced-braided Ethyl Vinyl Acetate (EVA). EVA has the following characteristics:

- Flexible at a wide range of temperatures
- Capable of working at up to 200 psi pressure
- Does not deteriorate in sunlight
- Is chemically compatible with fertilizers and herbicides
- Can be thick enough to work under suction without collapsing

Hoses should be inspected regularly for leaks and cracks and flushed at the end of every injection.

Fittings

For fertigation only, plastic fittings (nylon or polypropylene) are usually sufficient. However, when pesticides and herbicides are injected, the material of choice is generally 316 stainless steel. Some pesticides will destroy PVC fittings. Urea-sulfuric fertilizers require special gaskets in injection pumps. The manufacturers of the chemicals and injection equipment should always be contacted regarding compatibility and corrosivity of materials.

Tanks

Tanks should be constructed of poly or fiberglass; mild steel tanks should be avoided because of their potential for corrosion. If stainless steel is used, it should be constructed of 316 stainless. There should always be an on-off valve attached to the tank itself so that the injection mechanism can be removed. An easily cleaned 40 – 80 mesh filter should be attached downstream of the on-off valve.

Although not related to safety, there are advantages and disadvantages to various shaped tanks. For example, a tank with a conical bottom can be completely emptied.

Containment Structures

Some farmers locate chemical tanks within chemical containment structures if the chemicals are potentially hazardous in the event of a spill. Acids, pesticides, and insecticides fall under this category. Two containment structure designs are illustrated in the following figures. The first is constructed of cinder block walls around a concrete pad (Figure 4). The second is a larger poly tank which essentially acts as a "double hulled" unit (Figure 5).

Figure 4. Containment structure built of concrete block walls.

Figure 5. Containment structure made from a large poly tank.

NEATNESS

For safety reasons, it is important to maintain a neat chemigation unit, including the area surrounding the injector and chemicals. Messy installations encourage lax operations, which provide a hazard to both the operator and the environment. With a neat installation, spills and leaks are easy to identify, isolate, and correct.

CHEMICAL SAFETY

It is critical that manufacturer's guidelines are followed regarding mixing chemicals. There is a potential for dangerous reactions, so caution must be taken when a grower makes their own blends. Proper clothing, eye protection, and gloves should be worn when handling and mixing the chemicals. Always test small samples at a time in approximately the same concentrations which would be used during the injection. It is also important to follow directions regarding the *order* of mixing. Adding chemicals in the wrong order can be extremely dangerous. For example, ***always* add acid to water rather than water to acid**. Always follow label instructions and local or USEPA rules for safety.

CHAPTER 3. CHEMICAL INJECTORS

HIGHLIGHTS

There are many ways to inject chemicals into irrigation systems. The choice of method and equipment will depend on the following:

- Injecting liquid versus solid materials. Liquid fertilizers may not need agitation or mixing in the field, whereas solid chemicals need mixing.
- Potential hazard of the chemical. Injection of most nitrogen fertilizers can be done without serious concern to worker safety, whereas injection of acids or pesticides requires special precautions.
- Availability of power. In some sites, electricity is unavailable. An injector must be powered by the water, an internal combustion engine, or other means.
- Portability versus permanent installation. Some injection equipment is too cumbersome for portable usage.

INJECTOR FLOW RATES

In many systems, more than one injector is necessary because of different flow rate requirements for different chemicals. Herbicides, insecticides, and nematicides are generally injected at very low rates, often 10% less than fertilizer injection rates. As a rule of thumb, fertilizer injection should be possible at a rate of 0.1% of the irrigation flow rate. For example, if the irrigation pump delivers 1000 GPM, a fertilizer injector should be capable of delivering at least 1 GPM.

LIMITING THE AMOUNT OF INJECTED CHEMICAL

Limiting the maximum amount of chemical which can be applied during an irrigation event can be important for these reasons:

1. The timing or duration may need to match some rule such as the quarter-half-quarter rule.
2. Equipment or operator errors might cause a gross over-application that could kill the crop.
3. A grower may want to inject a specific volume of fertilizer on each set, even though the set pressures and flow rates may change.

There are several methods of limiting the amount of chemical to be injected. They are:

1. Put the amount of chemical to be injected into a small tank. Inject from that small tank rather from the larger storage tank. This can be used in addition to the other methods listed below. It prevents gross over-application of chemicals, in the case of failure of other protection mechanisms.
2. Insert the suction hose for the chemigation pump into the top of the chemical tank, and only immerse the suction hose deep enough to access one day's supply (see Figure 6). This is not recommended, because once the chemical level is down to the level of the hose suction, air will be pumped into the irrigation system. On hilly and undulating ground, air in pipelines can cause flow rate restrictions and water hammer. Venturi injectors, in particular, can inject air at very high rates once the suction tube entrance is exposed to air.

Figure 6. Limiting the amount to be injected by the depth of the suction hose. Not recommended because of air entrainment problems.

3. Some injection units have feedback control. A controller typically reads the electrical conductivity (EC) of the irrigation water downstream of the injection point, and adjusts the injection rate to maintain a constant EC. This can work if the fertilizer is a salt and if the EC of the water source does not change with time.
4. For water-powered injector pumps, some growers use a simple volumetric limiting device on the water supply to the pump. These fittings can be purchased for about $30, and are designed for home gardening systems. One dials in the volume of water desired, and the valve shuts the water supply off when that volume has been applied.

Since most water-powered injector pumps have a direct proportion of drive water to injected chemical, it is easy to calculate how many gallons to set the valve for in order to inject a certain number of gallons of chemical.

5. With electric-powered pumps, it is relatively simple to incorporate a time clock on the electrical supply.

SINGLE VS. MULTIPLE UNITS

Many growers use more than one injector at an installation. Reasons include:

- It was noted in the prior section that pesticides require lower rates of injection than do fertilizers.
- Fittings do not need to be taken apart when switching tanks. This can minimize health risks and also keep the fittings clean.
- A single injector can be calibrated and dedicated to a chemical with a unique viscosity and specific gravity. It does not need to be recalibrated constantly as the chemicals are changed.
- There are incompatibilities with different chemicals. Isolating various chemicals, and their injection equipment, helps to minimize problems.
- Several chemicals are often injected simultaneously. Each will need a separate rate control.

Multiple pumps or pump heads are standard equipment in the greenhouse industry. However, they are not as common in agricultural field production.

Figure 7. Diaphragm pump with four independent pump heads, allowing injection of up to four chemicals simultaneously at different injection rates. (Courtesy Edward Kessel, Santa Ynez, Ca.)

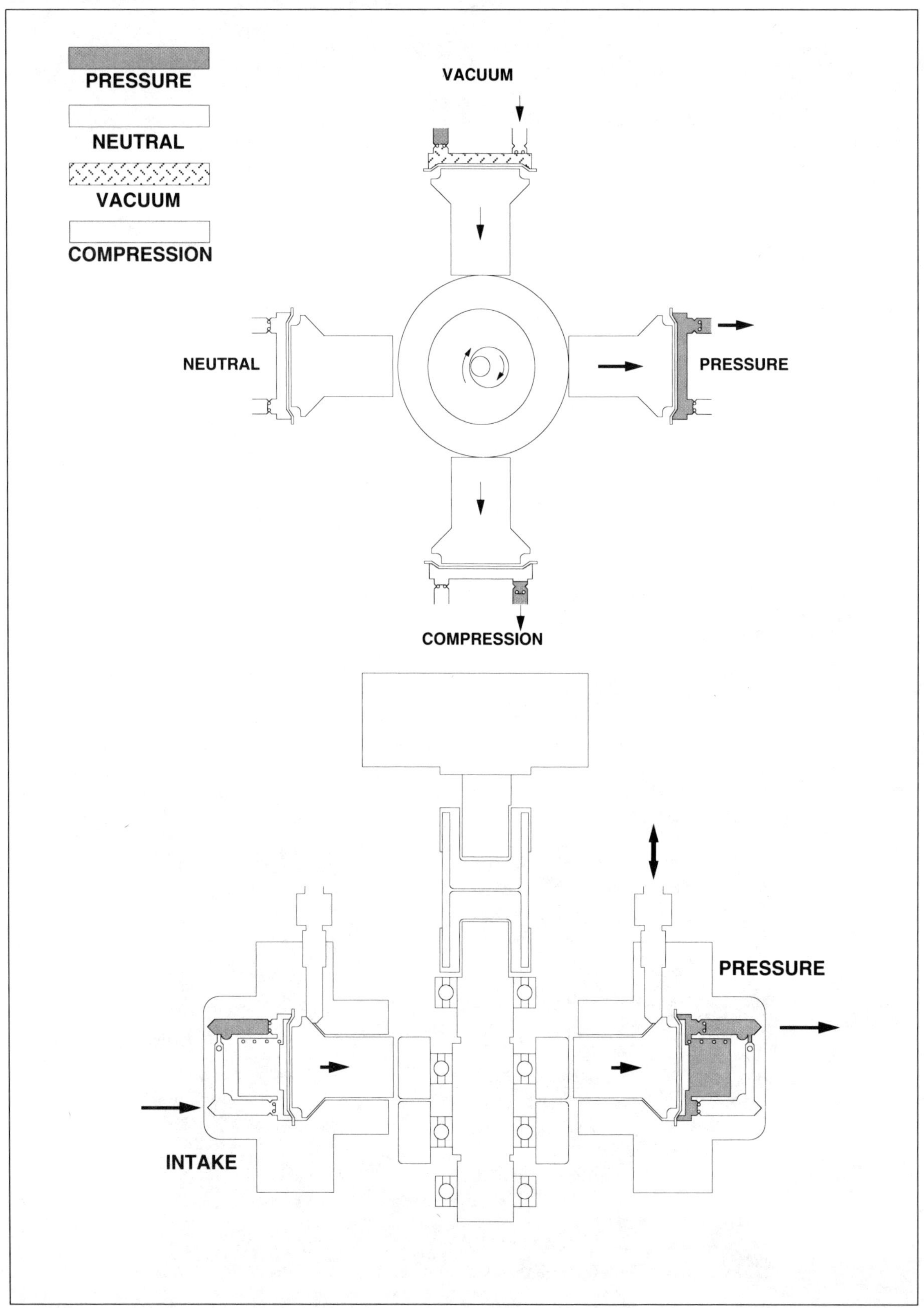

Figure 8. Schematic of diaphragm pump with four independent pump heads. (Courtesy Ozawa R & D, Inc.)

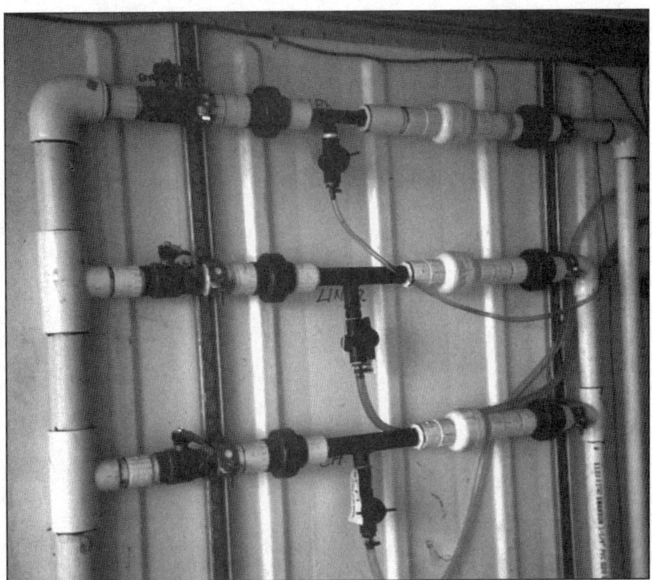

Figure 9. Neat configuration of multiple venturi injectors arranged within a portable trailer for row crop drip. (Courtesy Bill Moresco, Salinas, Ca.)

Figure 10. Multiple venturi injectors on a permanent drip installation. (Courtesy Gideon Cohen, Three Flags Ranch, Brawley, Ca.)

Injector Flushing

Ideally, all injection systems should be plumbed so that one can easily flush fresh water through the injectors and the fittings. Most injectors should be flushed after use, as this increases the life of gaskets and metals, and helps to prevent encrustation problems. Flushing the complete unit also allows one to purge any chemical from the system before connecting with another chemical, thus minimizing the potential for precipitation problems.

Drip Systems – Injection Upstream or Downstream of the Filter?

There are clear arguments for injecting chemicals either upstream or downstream of the filter. The primary reason for injecting upstream of the filters is that any introduced contaminants, or immediate chemical precipitations, resulting from chemigation will be trapped by the filters and thereby the irrigation system will be saved. This is a major advantage, because dirty connections are made and accidents do happen.

The major problems with upstream injection are:

1. When acids are injected, they may damage the filter.
2. The filter flush water will contain chemicals.

In general, the following rules are applicable:

1. Inject acids downstream of the filter.
2. Always conduct a compatibility test of the acids and the irrigation water and any other chemical being injected.
3. Maintain very clean hose connections. Use a separate injector for each chemical and for each tank. Do not connect and disconnect hoses unless absolutely necessary, and have all connection points on clean concrete.
4. As a first priority, inject other chemicals upstream of the filters. However, the chemical injection should be discontinued for a period before and during the filter flushing process. Controllers are available which can control both the flush valves and the chemical injector simultaneously.

Proportional Control

Some of the electrically-driven diaphragm and piston pumps can be automatically controlled with a feedback system to deliver a desired proportion of fertilizer into the irrigation water. Such feedback systems generally monitor the flow rate of the irrigation pipeline and then vary the injection rate accordingly. They can be quite accurate. Typical accuracy problems with such units revolve around sloppy measurements of the irrigation water flow rate:

- Some of these units use inexpensive and inaccurate flow meters.
- The flow meters are often improperly placed with respect to turbulence-causing fittings, thus giving inaccurate flow rate measurements.

OBTAINING A PRESSURE DIFFERENTIAL

Some injectors, notably venturis and differential pressure tanks, use the motive (driving) water to carry the chemical back into the irrigation line. Such injectors generally have a fairly large pressure drop across them.

IN-LINE PRESSURE DIFFERENTIAL

Since the motive water originates and discharges into the irrigation pipeline, many configurations use a pressure differential in the irrigation pipeline (Figure 11).

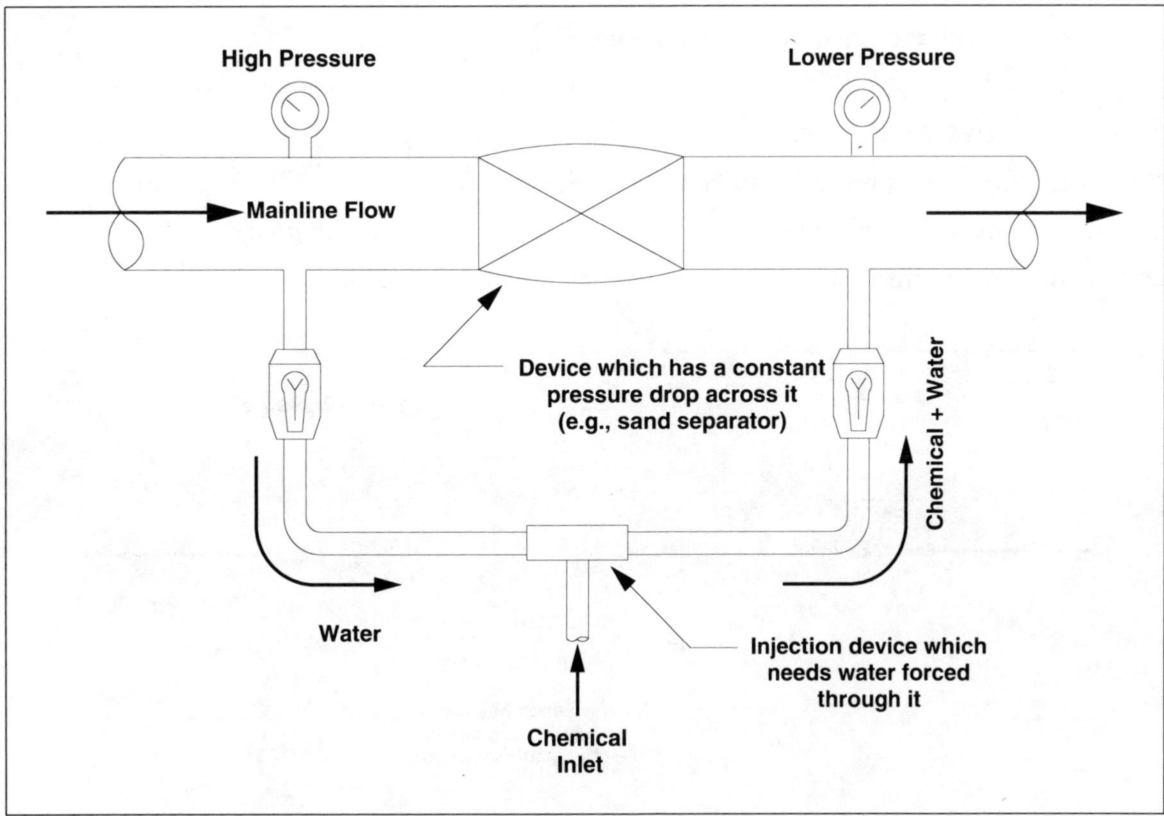

Figure 11. Some injectors utilize a pressure differential in the irrigation pipeline to force water through the injector.

Probably the biggest error with such installations is that valves are used as the pressure differential device, and the valves are temporarily throttled (closed down) during fertigation in order to power the injector.

CAUTION Temporary throttling of a valve will cause a loss of water distribution uniformity throughout the field—the chemical will be injected, but will be distributed unevenly.

If it is necessary to utilize a pressure differential in the irrigation pipeline, it should be done with a device which has a constant pressure drop across it, (e.g., certain types of valves or filters). Generally, a minimum of 10 – 20 psi drop is needed for an injector in a drip system—this drop may be hard to obtain. One should *not* install a device with a permanent pressure drop in a line just to operate an injector—this practice would be a waste of power (unless the irrigation system receives water from a very high pressure distribution system with excess pressure continuously available). If the pressure drop is obtained by plumbing across a pressure regulating valve or pump control valve, there can be problems because these devices automatically modulate with time. Consequently, the pressure differences across them changes with time. That will eventually cause differences in the injection rate.

LARGE VENTURI BYPASS

If there is only a small pressure drop available in the irrigation pipeline, a large venturi can be placed in the mainline or in a bypass line (Figure 12). The pressure difference between the inlet and throat of the venturi can be used to power the actual chemical injector.

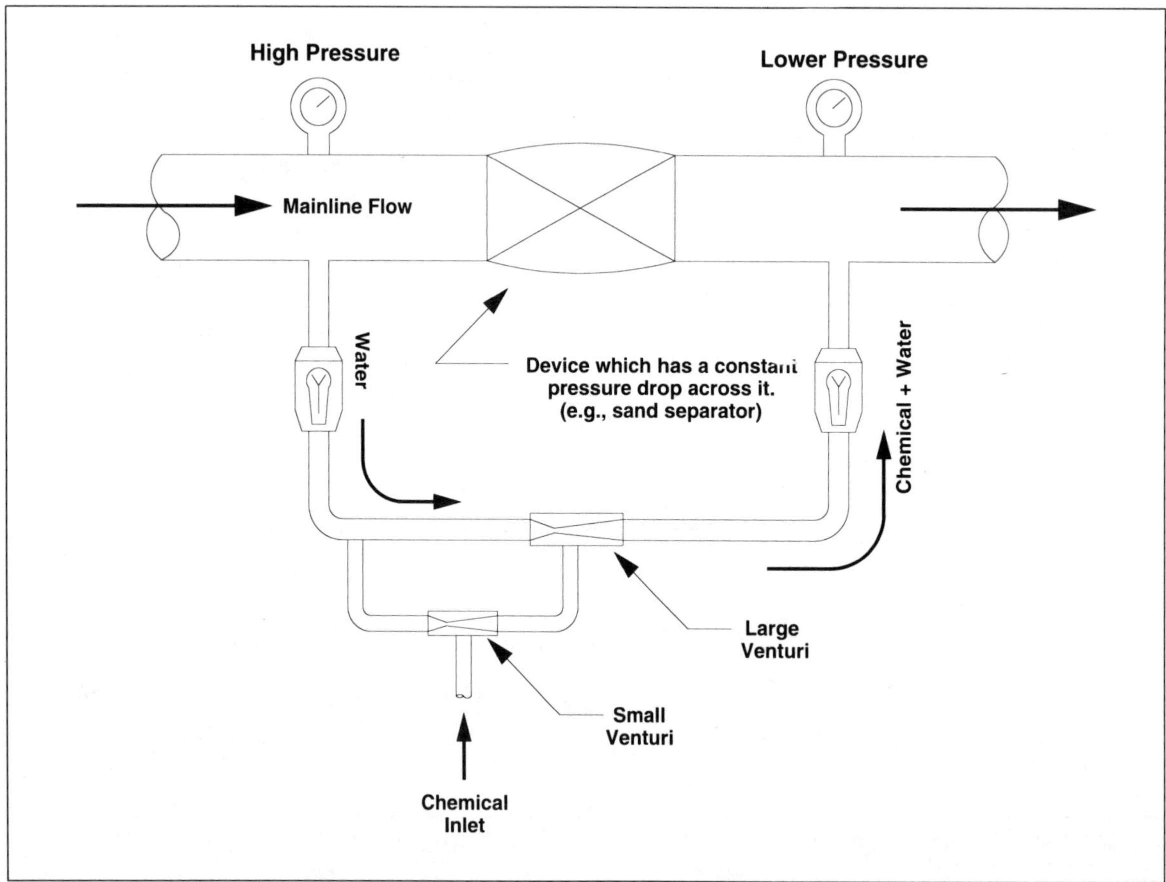

Figure 12. Use of a large venturi to create adequate pressure differential to operate a chemical injector.

PLUMBING ACROSS A BOOSTER PUMP

A booster pump always has a pressure differential between its inlet and outlet (Figure 13). As long as the flow rate through the irrigation water booster pump stays constant, the bypass flow rate will also remain constant (therefore maintaining a constant chemical injection rate).

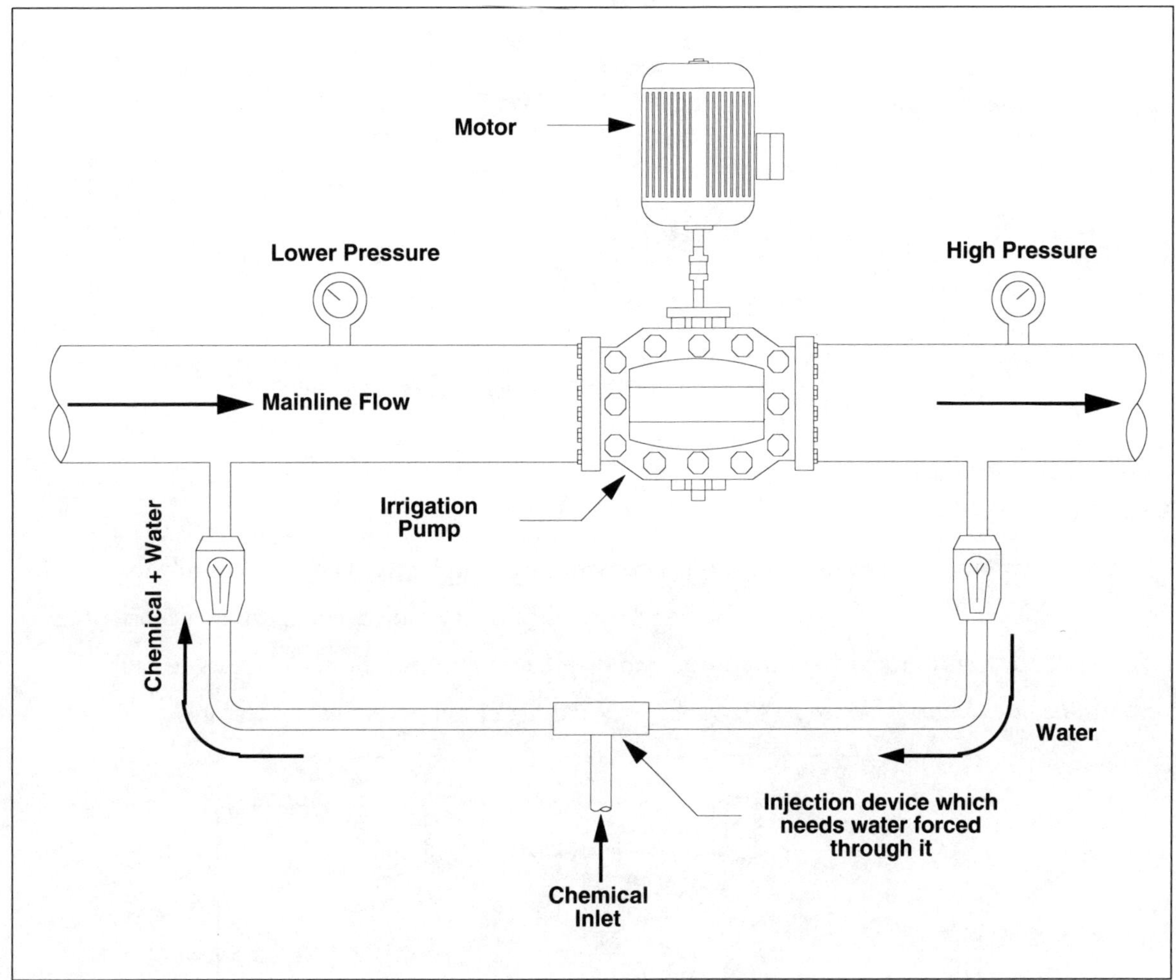

Figure 13. Using the pressure differential across a booster pump to power an injector.

Problems with this configuration include possible damage to the pump impellers if the injector runs out of chemical and begins to inject air, and damage from corrosive chemicals.

BYPASS PUMPS

Bypass pumps are quite popular for providing the motive flow through venturi injectors. These pumps do not come in contact with the chemicals, so they can be inexpensive water pumps. They provide an almost constant pressure differential across the venturi injectors, so the injection rate can remain relatively constant regardless of flow rate variations in the irrigation pipeline itself.

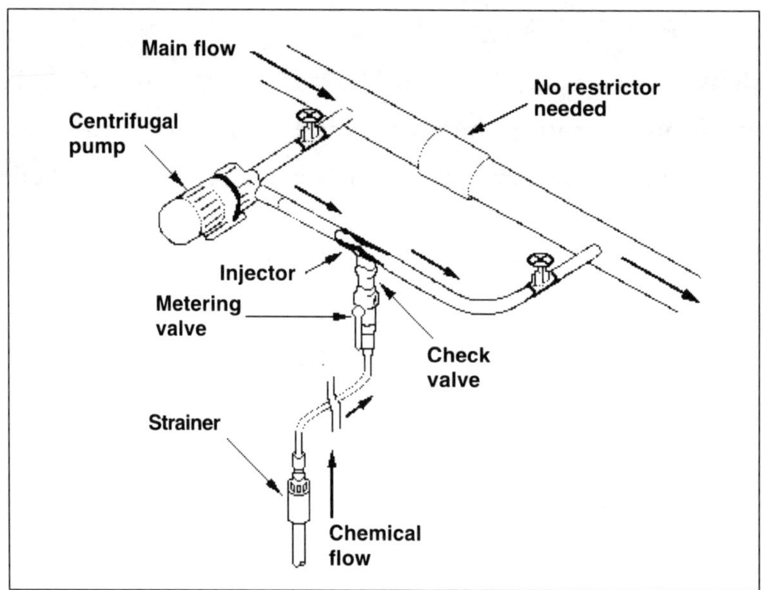

Figure 14. The use of a bypass pump to provide motive flow for a venturi injector.

INJECTOR DESIGNS

VENTURI

Venturis are commonly used for fertigation because of their simplicity. They are typically made from materials which are inert, and have no moving parts. Because they inject chemicals at a constant rate (as opposed to the pulsing found with piston and diaphragm pumps), they can be easy to calibrate with in-line flow meters. Figure 15 shows the basic configuration of a venturi injector.

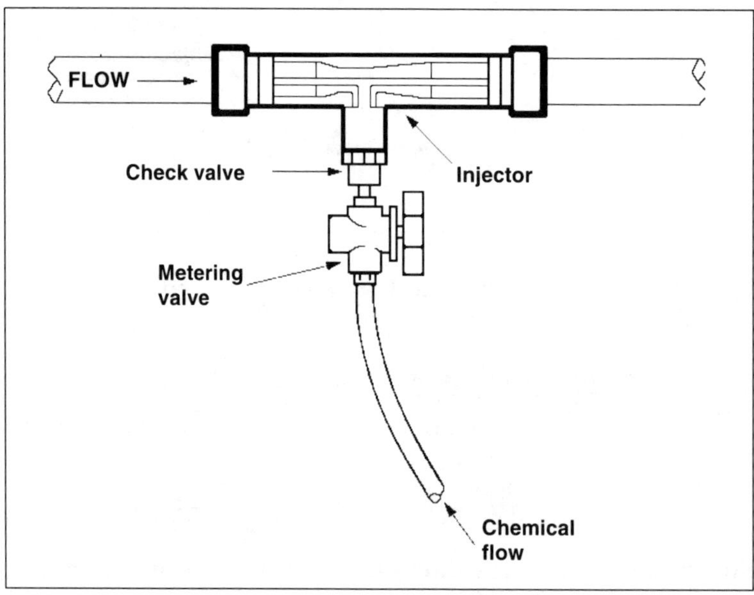

Figure 15. A simple venturi injector.

Table 4 provides an indication of the flow rate requirements, injector capabilities, and pressure drops for one commercial model.

Table 4. Manufacturer's information regarding venturi efficiencies and injection capacities. (Courtesy Mazzei Injector Corp.)

Model No.	Efficiency (%)	Motive Flow (GPM)	Injection Capacity (GPH)
1	26	0.5	6
2	25	2.1	10
3	18	3.4	17
4	18	3.4	17
5	16	6.4	25
6	16	12.0	60
7	18	17.0	75
8	18	34.0	180
9	18	34.0	180
10	18	101.0	500
11	18	101.0	500
12	50	2.1	35
13	32	12.0	140
14	35	36.0	350
15	67	29.0	1130
16	67	29.0	1130

Efficiency Δ P required to initiate suction or minimum pressure differential to create a vacuum.
Motive Flow Actual water flow through injector orifice at 50 psi.

The venturis themselves do not constitute an injection system. They must be configured with all of the required support hardware. Some companies sell complete units which can include a bypass pump, calibration device, flow meter, and check valves, etc. Figure 16 shows such a unit.

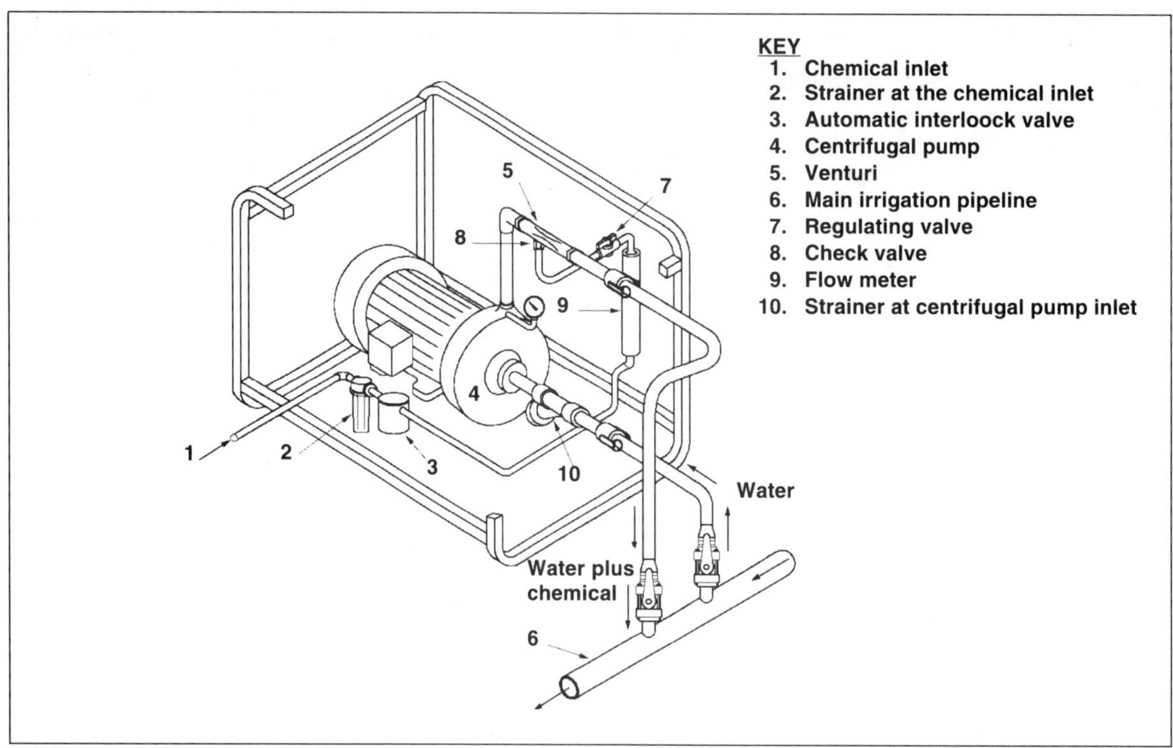

Figure 16. A typical commercial unit incorporating a bypass pump and a venturi. (Adapted from CSI Chem. Sys.)

FLOAT VALVES

Float valves have been used for many years to inject chemicals into canals and open, low pressure pipeline stands without the need for injector pumps. The float, installed within a plastic container, maintains a constant pressure against the adjustment valve—even if the pressure into the float valve changes. As the supply tank empties the discharge flow rate remains constant.

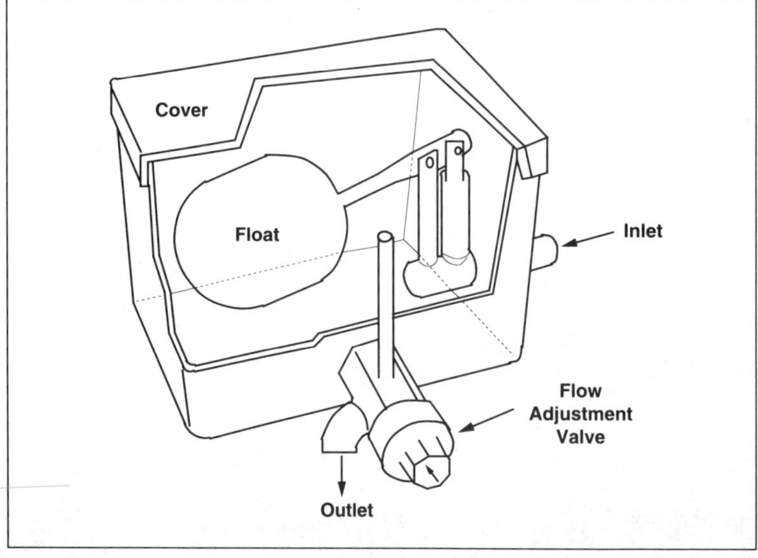

Figure 17. A typical float valve assembly.

Float valves are often used to inject chemicals into a pressurized (sprinkler or drip/micro) irrigation system by locating the assemblies over a canal or open water supply upstream of the irrigation booster pump. This provides a very simple and inexpensive method of injecting chemicals into a pressurized system, as long as the booster pump is the final destination of the surface water.

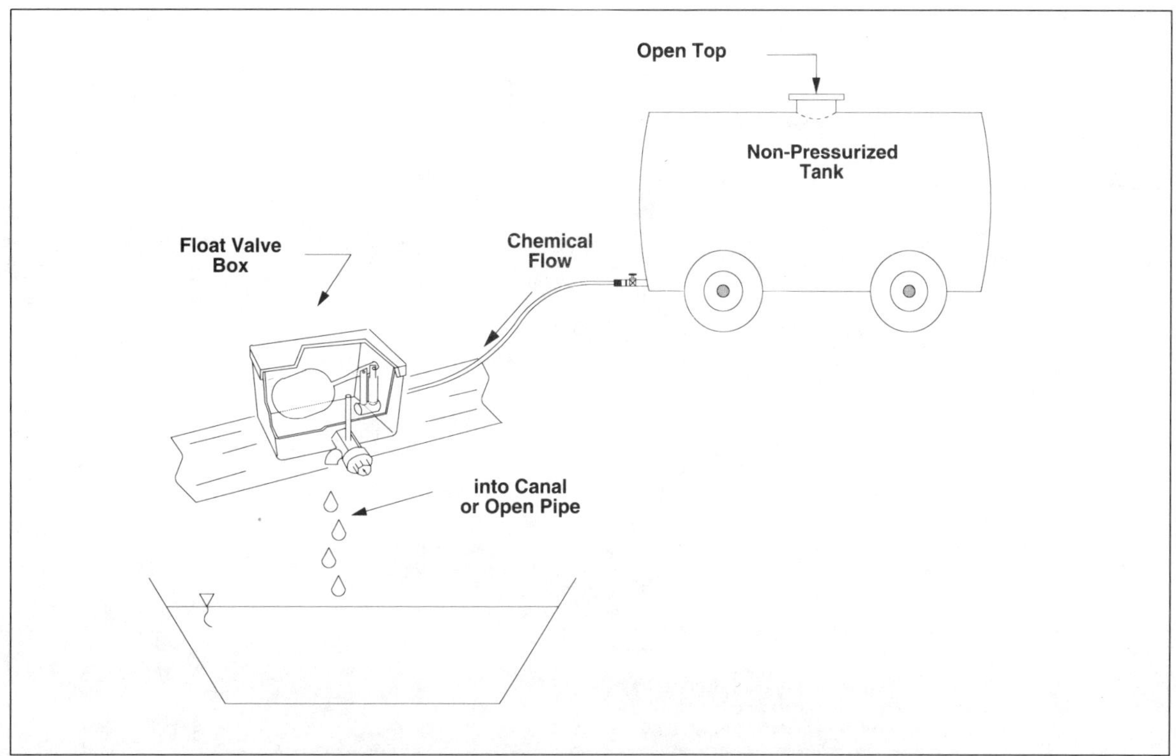

Figure 18. Utilizing a float valve to discharge into a canal or open pipe.

Figure 19. Float valves located on a canal upstream of a drip system booster pump.

Inexpensive commercial float valve assemblies constructed from inert ingredients are available and can tolerate most injected chemicals.

Differential Pressure Tank

A sketch of a differential pressure tank (Figure 20) is found in almost all chemigation books.

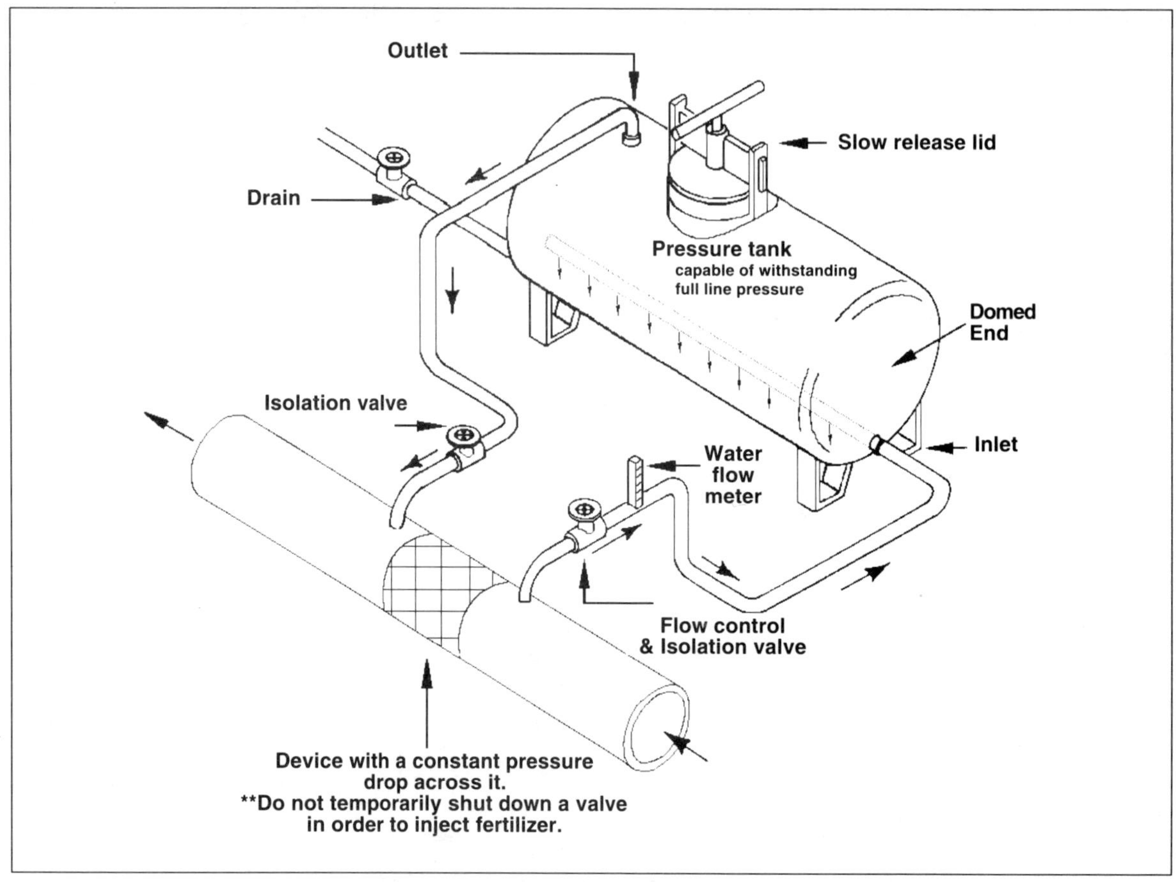

Figure 20. A differential pressure tank.

There are few, if any, good reasons to use differential pressure tanks. They are expensive, potentially dangerous (if a person opens a lid while the tank is under pressure), quite small, and incapable of injecting a chemical at a constant rate. The rate at which the chemical is released from the tank decreases with time, as the tank becomes more diluted with water.

N_2 Gas-Powered Pumps

These pumps are widely used in some areas because of their simplicity and low cost. Nitrogen gas (or compressed air) powers a small diaphragm pump, which can inject liquid chemicals into an irrigation system. The N_2 gas is *not* a fertilizer supply. The chemical is stored in a non-pressurized vessel. It is an ideal injection system for some cases in which a portable, reliable injector is needed in areas with no electricity at the injection site.

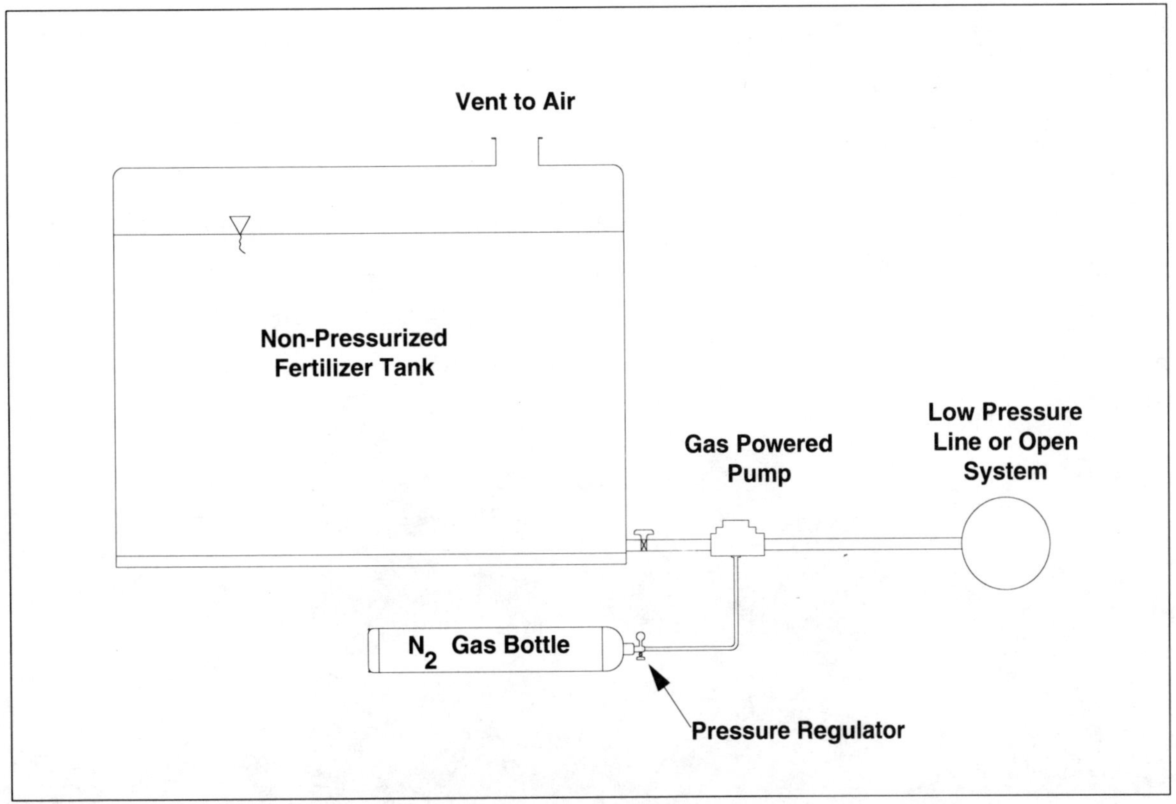

Figure 21. A typical nitrogen gas (N_2) powered pump installation.

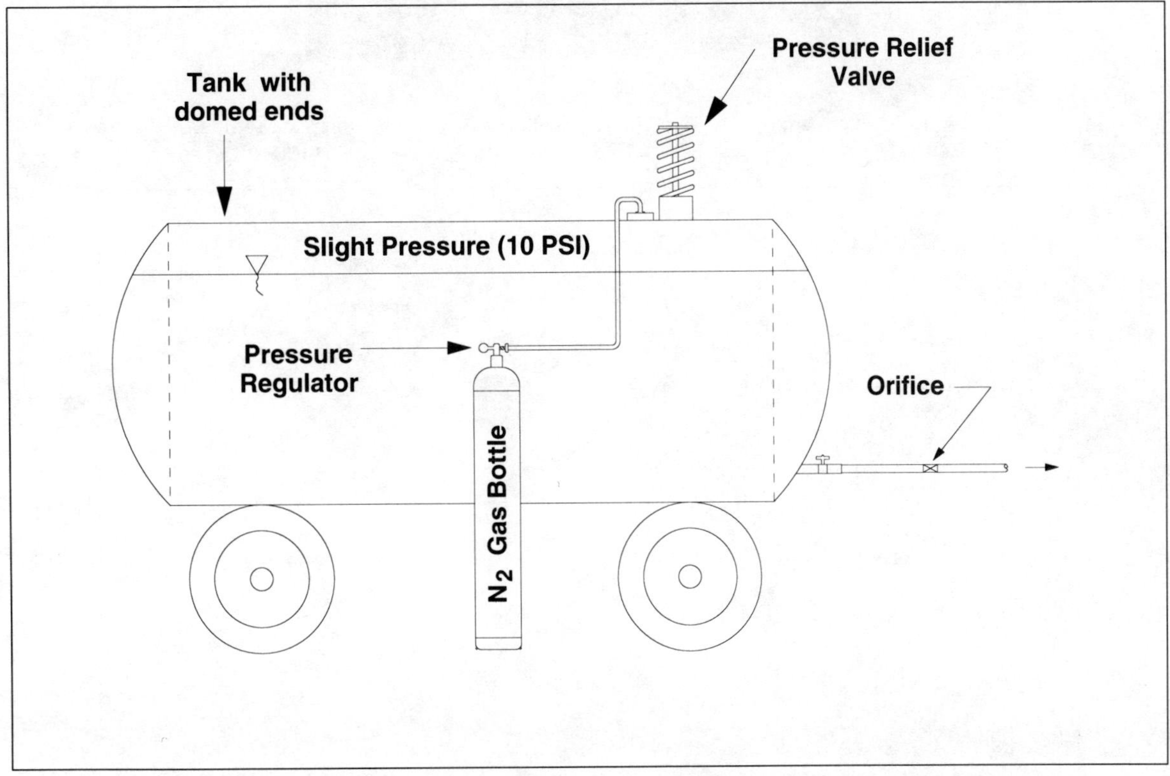

Figure 22. Use of N_2 gas and pressurized tank with a simple orifice for flow adjustment.

N_2 PRESSURIZED TANK

Some growers use a low pressure to force liquid chemicals out of a portable tank. N_2 gas is usually used to provide the pressure. The tank must be capable of safely withstanding the pressure on it (typically less than 10 psi), and it must be equipped with an excellent pressure relief valve to prevent an explosion of the tank due to accidental over-pressurization. Since the pressure on the discharge flow control point will change as the tank empties, growers generally use a float valve at the end of a hose to regulate the fertilizer flow rate. This is a convenient method to inject chemicals into the open top of an irrigation standpipe.

Figure 23. On-site N_2 pressurized tank with a float valve box on a standpipe.

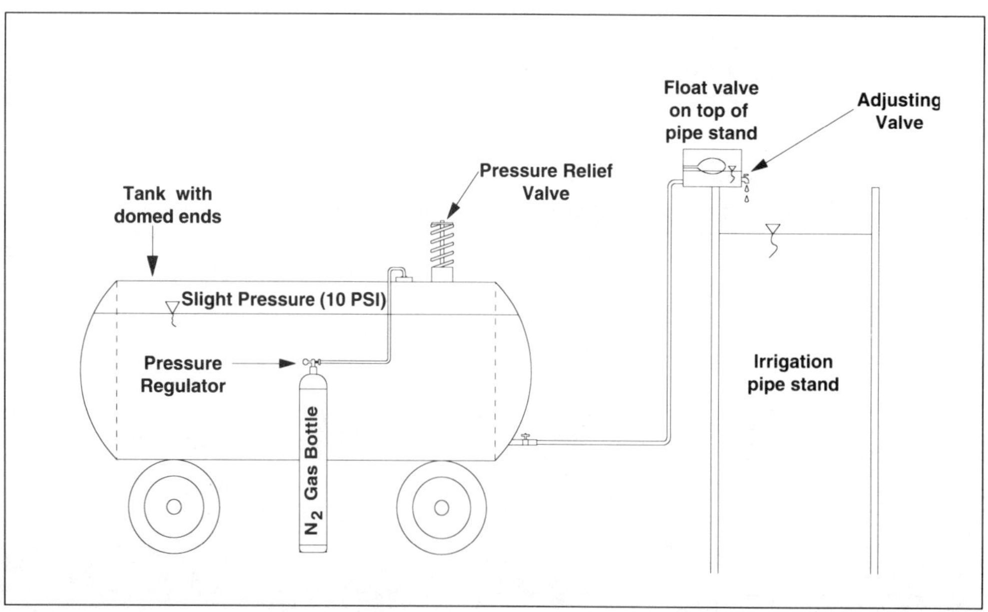

Figure 24. A typical N_2 pressurized tank with a float valve box on a standpipe.

CHICKEN FEEDERS

"Chicken feeder" designs are used to drip very low rates of chemicals into canals or open pipelines. They provide a constant discharge rate regardless of the height of the chemical in the drum. A tee valve gives the gross adjustment of flow rate and the rate can be further modified by pivoting the 'tee' valve/air vent assembly slightly to change the amount of pressure on the valve.

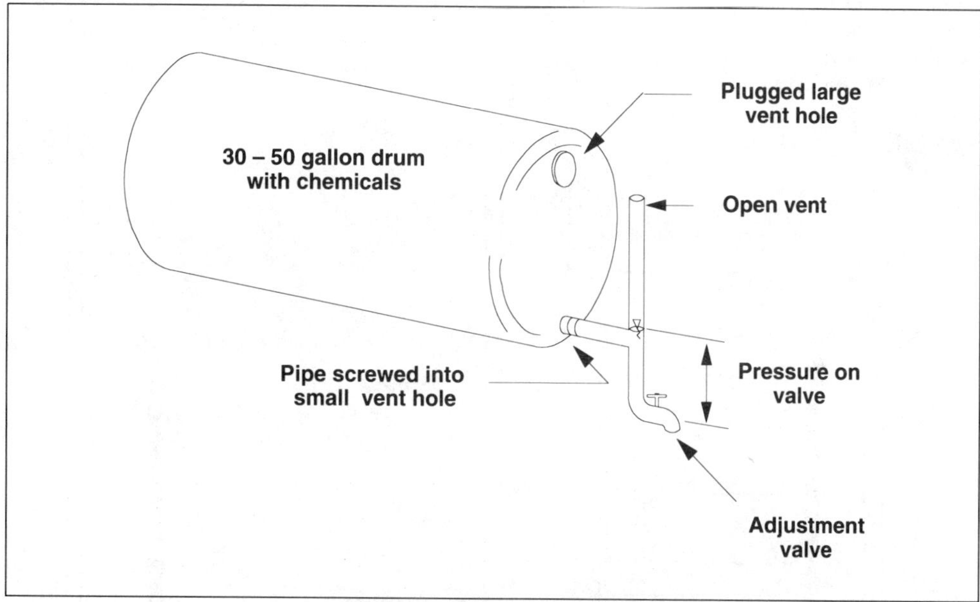

Figure 25. Homemade chicken feeder assembly used for low rates of injection.

WATER POWERED PUMPS

There are several designs of water-powered pumps. Earlier designs sometimes used a propeller (turbine) in the irrigation pipeline to power the pump. Most current designs use a small percentage of the irrigation water to power a piston or diaphragm pump. Basically, a water motor powers the pump instead of an electric motor. The irrigation water which drives the pump action must be discharged into the atmosphere. This creates a disposal problem in some locations.

The chemical being injected never contacts the drive water. The pumps are able to inject the chemical directly into the pressurized irrigation pipeline.

There are major differences in construction and complexity between various models. Some units have small holes which are easy to contaminate and block with dirty source water; some need frequent lubrication with grease, and various designs require different amounts of drive water to function. Typical ratios of drive water to chemical injected are (2 – 4) to 1.

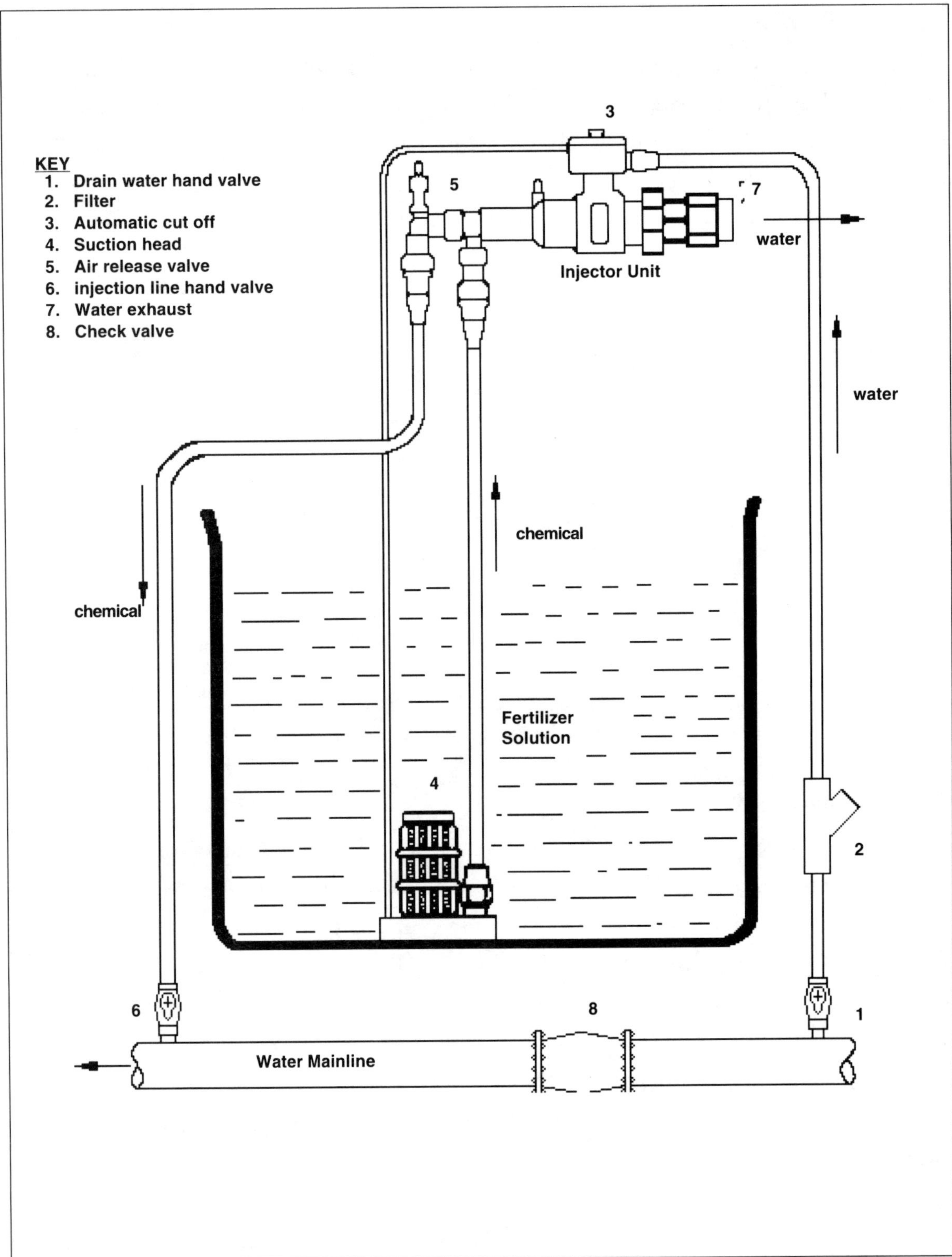

Figure 26. Water-powered pump using a linear hydraulic motor. (Courtesy Amiad)

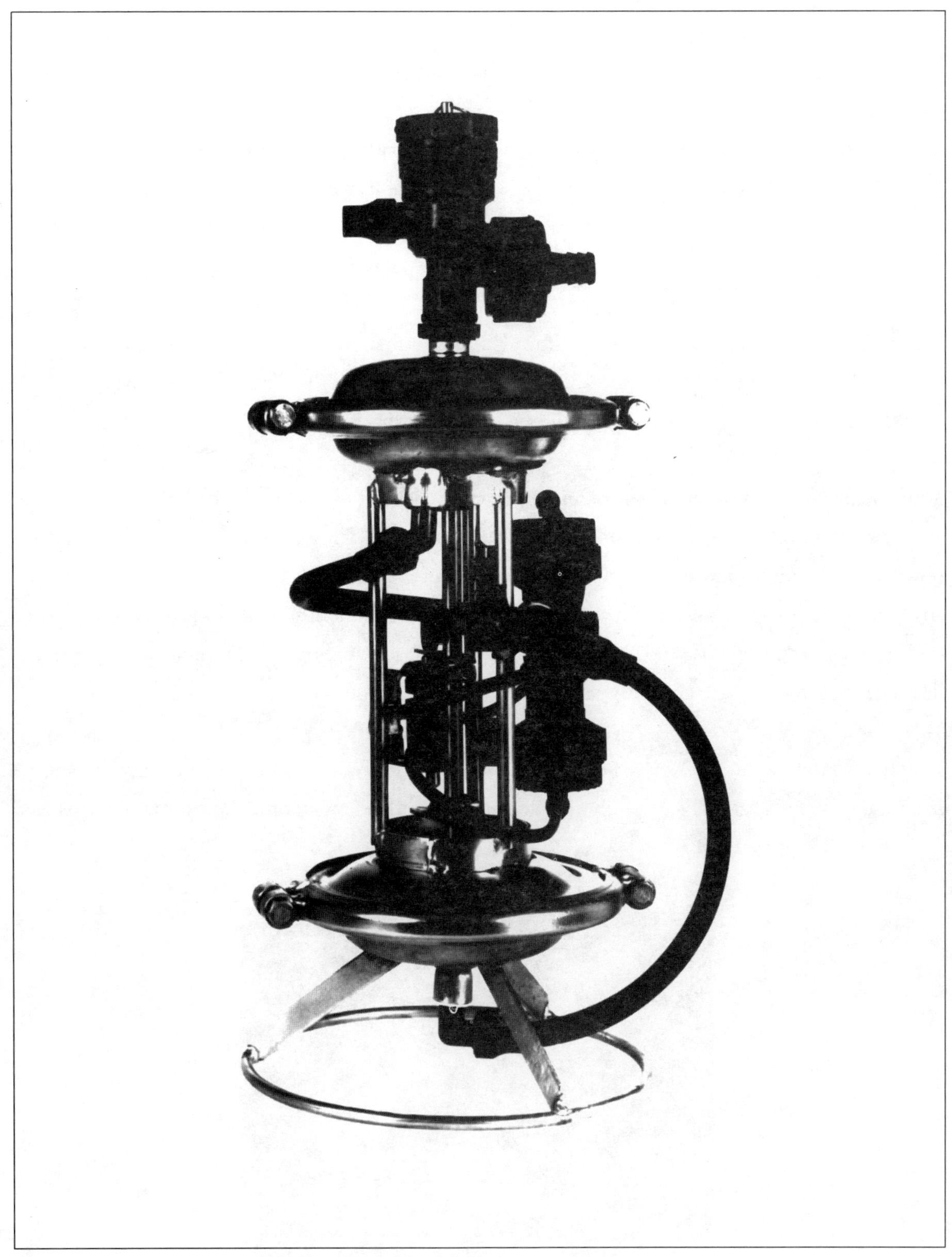

Figure 27. Water-powered pump using diaphragm action. (Courtesy Filtration Irrigation & Fertigation)

Diaphragm Pumps

Diaphragm pumps are very popular in the Midwest, but less so in California. Perhaps the reason is that many of the chemigation sites in the Midwest are located at center pivots (which have electricity available) receiving water from wells (rather than situations with booster pumps) and are used for pesticide injection as well as for fertigation. In California, most injection is done with fertilizers or fairly nontoxic chemicals, and some of the injection equipment is suitable for such applications but would not be suitable for more toxic chemicals due to the potential for human contact.

Diaphragm pumps are typically powered by electric motors, but may also be belt-driven or powered by small gasoline engines. They have traditionally had the distinct advantage over piston pumps in that they are easy to adjust while the pump is operating.

Many diaphragm pumps cannot maintain a constant discharge flow rate if the discharge pressure varies significantly. Therefore, they may not be the ideal selection if one wants to inject a constant flow rate of chemicals, and the irrigation mainline pressures change. Irrigation mainline pressures change as center pivots go up and down hills, and as fields of different sizes are irrigated with sprinkler or drip/micro irrigation systems. (Note that some brands/models of diaphragm pumps only change flow rates by a few percentage points with a large change in pressure.)

Diaphragm pump materials must be selected with care to be compatible with the chemicals to be injected. They must also be carefully rinsed after use, and generally must be overhauled (gaskets, o-rings, etc.) every season.

Figure 28. A typical diaphragm pump.

Care must be taken when purchasing inexpensive diaphragm pumps, as the seals and check valves are often rather fragile. Also, one should note that if a diaphragm pump is labeled as being able to deliver 40 GPH and 60 psi, that may mean that it can deliver 40 GPH at 0 psi, and 0 GPH at 60 psi. There are major differences in quality of pumps.

PISTON PUMPS

Piston pumps have the same characteristics as diaphragm pumps, in terms of power supply and maintenance. The big advantage of piston pumps is that as positive displacement pumps, the discharge flow rate will not change as the irrigation pipeline pressure varies. Their big disadvantage is that flow rates cannot be adjusted while the pump is operating. Therefore, one must measure the flow rate, shut the unit off, and adjust the piston stroke, measure the flow rate, and repeat the process until the desired discharge is obtained. Most units have flow rate indicators at the adjustment point. Some newer positive displacement units have unique designs which vary the piston stroke length automatically as adjustments are made, while the unit is running.

Figure 29. An electric powered piston pump. (Courtesy Edward Kessel, Santa Ynez, Ca.)

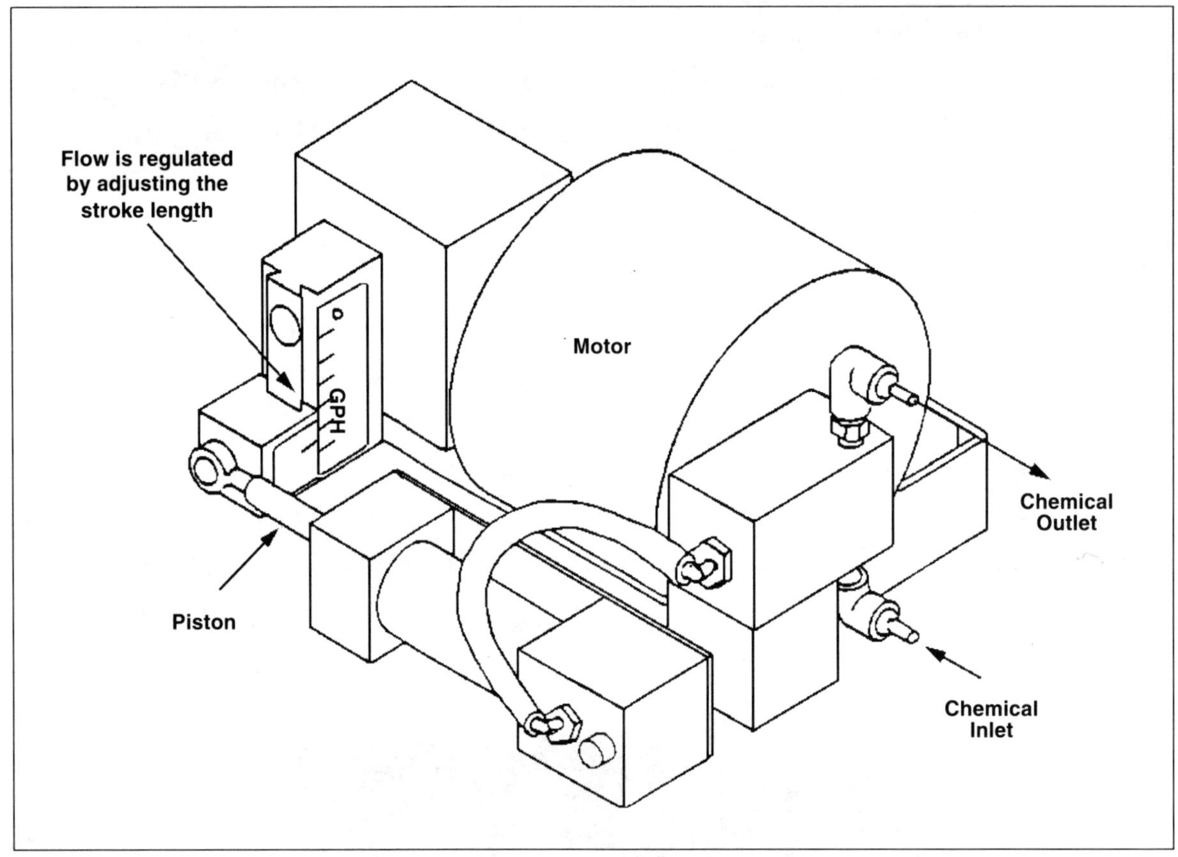

Figure 30. A traditional electric powered piston pump design.

GYPSUM INJECTORS

Recently there has been an interest in the injection of gypsum through irrigation systems. This has resulted in several injector designs. All have some type of agitator in a tank, combined with an injection pump. They are generally electric powered. Growers have found that in addition to injecting gypsum, they are useful for mixing and injecting a wide range of solid fertilizers. When purchasing a unit, one should consider:

- Injection rate
- Ease of loading
- Capacity
- Warranty on parts and labor
- Ability to maintain a fairly consistent injection rate of the chemical or slurry
- Availability of service

Figure 31. A typical commercial gypsum injector. (Courtesy Soil Solutions Corp.)

SO_2 GENERATORS

Sulfur dioxide generators of various designs have been available for several decades. Essentially, they are acid makers. They burn elemental sulfur to create a weak acid, sulfurous acid (H_2SO_3). They have been primarily sold to reduce the carbonate and bicarbonate concentrations in water, thereby enhancing water infiltration into the soil. Their primary problems are associated with dirty sulfur sources (in which case, a slag can be created in the bottom of the burner) and difficulties with SO_2 emissions during start up and shutdown. SO_2 is extremely toxic when breathed, so one should pay close attention to this aspect when selecting a machine.

The removal of carbonate and bicarbonate decreases the adjusted R_{Na} of the water. A pH of 6.5 removes most of the bicarbonates and all of the carbonates from the water, but it does not cause corrosion problems in the irrigation system. The reactions are as follows:

S	+	O_2	——>	SO_2		
Elemental Sulfur		Oxygen Gas	(burn)	Sulfur Dioxide Gas		
H_2O	+	SO_2	——>	H_2SO_3		
Water		Sulfur Dioxide Gas		Sulfurous Acid		
HCO_3^-	+	H^+	——>	H_2O	+	CO_2
Bicarbonate		Hydrogen (from acid)		Water		Carbon Dioxide Gas

Figure 32 illustrates the operation of one type of unit. It is a relatively efficient design, without any moving parts except where the water is pumped.

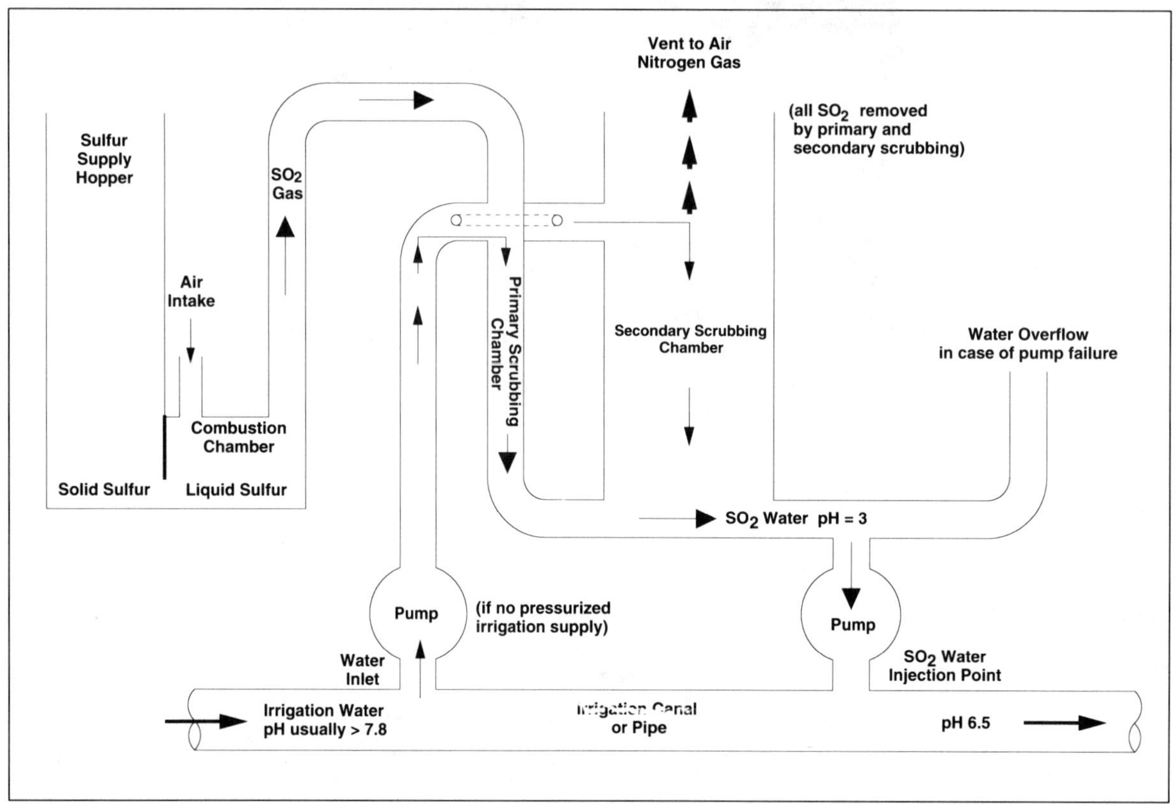

Figure 32. A typical sulfur dioxide gas generator.

INJECTOR CALIBRATION ACCURACY

There are several issues regarding injector accuracy. First, failure to select injectors according to chemical compatibility can have severe consequences on injector accuracy and the lifetime of the injector (Kranz and Eisenhauer, 1990).

In addition, calibrating injectors properly requires several considerations.

1. The flow adjustment valve (if used) should be small enough to provide good regulation over a range of flows.
2. Any regulation mechanism should be made of durable, noncorrosive material which will not deform with time.
3. If a flow meter is used, it should be properly installed.
4. The pump should be calibrated with the same liquid formulation that will be injected. Calibration with water may give incorrect results.

5. Some pumps and configurations change their discharge flow rate if the irrigation pipeline pressure changes. These injectors must be calibrated under regular operating pressures.
6. Viscosities of chemicals can change with temperature.

Kranz and Eisenhauer (1990) conclude:

> Differences exist in output among individual pumps and between pumps from the same manufacturer and between manufacturer. A common calibration curve cannot be used for all pumps of a given model/manufacturer, nor should the same calibration curve be used for all operating conditions. Each pump should be calibrated on site under expected inlet and outlet pressures.

See Chapter 4, "Injection Techniques for Various Irrigation Methods" for more details on changing injection rates and concentrations.

Table 5. Manufacturer's information on fertigation equipment.

Company	Products	Notes
Agri-Inject Inc. 5500 East Hwy 34 P.O. Box 437 Yuma, CO 80759 Phone: (800) 4-INJECT FAX: (303) 848-5338	Diaphragm pump Agitators Calibration tubes Injection check valves	Pumps are sold separately or with a tank/agitator unit. Injector capacities range from 0.25 – 227 GPH. Duplex models can inject two chemicals at the same time at different rates.
American Technical Services P.O. Box 1595 Clovis, NM 88101 Phone/FAX: (505) 762-1199	Injector check valve	Stainless steel construction; compatible with herbicide, insecticide, fungicide and fertilizers.
Amiad USA, Inc. P.O. Box A Reseda, CA 01337 Phone: (800) 969-4055 FAX: (818) 781-4055	Linear hydraulic injector Controller unit	Simplex or duplex models, injection rate capacity of 1.4 GPM.
Blue-White Industries 14931 Chestnut Street Westminster, CA 92683 Phone: (714) 893-8529 FAX: (714) 894-9492	Diaphragm injectors and peristaltic injectors, tanks, flow meters	Injection rates range from 0.3 – 32 GPH; some units have tanks attached to injector
CDS Injection Pumps 5621 Kimball Court Chino, CA 91710 Phone: (909) 597-1813 FAX: (909) 597-7785	Diaphragm and piston pumps, check valves, float valves	Simplex and duplex pumps with injection rates ranging from .66 – 100 GPH. Float valves have a 0 – 65 GPH gravity flow rate.

continued

Table 5. Continued.

Company	Products	Notes
Chlorinators Inc. 4125 Southwest Martin Hwy Suite 2 Palm City, FL 43990-5524 Phone: (407) 288-4854 FAX (407) 287-3238	Sulfur and chlorine gas injector units	Injection rates range from 4 – 100 pounds/24 hrs.
Cobb Sales Co. P.O. Box 337 Shallowater, TX 79363 Phone: (806) 832-4322	Strainers, drains, poppet valves	
CSI Chemigation Systems International P.O. Box 5000 Cedarburg, WI 53012-0500 Phone: (414) 375-8550 FAX: (414) 375-8559	Centrifugal pump injector with venturi. Chemical tanks, trailers and agitators	Injection rate capacities from 0– 400 GPH. Metering valve and visual flow meter aid calibration. Simplex and duplex models available.
Dosatron International Inc. 20090 Sunnydale Blvd. Clearwater, FL 34625 Phone: (813) 443-5404 FAX: (813) 447-0591	Piston driven hydraulic pump	Injection rates range from 2 – 100 GPM
Filtration Irrigation & Fertigation 18155 Elkwood St. Reseda, CA 91335 Phone: (818) 705-0215 FAX: (818) 343-0919	Hydraulic injector using a diaphragm action	Injection capacities range from 1 – 16 GPH. Working pressure is 20 – 100 psi.
Flojet Corp. 12 Morgan Irvine, CA 92718-2003 Phone: (714) 859-4945 FAX: (714) 850-1153	Diaphragm pumps (electric and N_2 gas powered)	Injection rate capacity is 5 GPM.
Injection Systems, Inc. 8761 Airport Road, Unit B Redding, CA 96002 Phone: (916) 221-0396 FAX: (916) 221-0397	Diaphragm pumps	Models available in simplex, duplex, or triple units.
Inject-o-Meter 820 Thorton Clovis, NM 88101 Phone: (505) 763-4461 FAX: (505) 762-2567	Diaphragm and piston pumps	Simplex and duplex models with injection rate capacities up to 163 GPH. Some units include a tank and agitator.

continued

Table 5. Continued.

Company	Products	Notes
Mazzei Injector Corporation 500 Rooster Drive Bakersfield, CA 93307-9555 Phone: (805) 363-6500 FAX: (805) 363-7500	Venturi injectors Metering valves Check valves Strainers Flow meters	Injector capacities range from 6 – 1130 GPH
Montague/Fisher Inc. 1215 E. Houston Ave. Visalia, CA 93291 Phone: (209) 732-4973 FAX: (209) 732-5364	Gypsum injectors	Units hold either one or two tons of gypsum. Injection rates are approximately 5.5 GPM.
Ozawa R & D, Inc. 3130 S.W. 6th Avenue Ontario, OR 979114 Phone: (503) 889-3644 FAX: (503) 889-3637	Piston injector pump	Injection rate capacity up to 45 GPH. Ability to supply four different chemicals which have independently operating flow rates. Ability to adjust flow rate during operation.
Neptune Chemical Pump Co. 336 Fitzwater St. Philadelphia, PA 19147 Phone: (215) 925-8942 FAX: (215) 592-7695	Diaphragm pump	Various models with injector capacities ranging from 1 – 130 GPH.
Netafim Irrigation Inc. 3025 E. Hamilton Fresno, CA 93721 Phone: (209) 442-3115 FAX (209) 442-3119	Venturi injectors	Injection rate capacities from 12 – 750 GPH.
Prominent Fluid Controls 490 Southgate Drive Guelph, Ontario N1G-4P5 Canada Phone: (519) 836-5692	Diaphragm pumps	Various models, maximum capacity at 3.6 GPH.
Soil Solutions Corp. 8000 W. Doe Ave. Ste. E Visalia, CA 93291 Phone: (209) 734-0275 FAX: (209) 651-4109	Gypsum injector	Tank sizes range from 150 – 600 gallons. Injection rate capacity is 9 GPM, depending on the model. Units include agitators, screens, tanks, and injectors.
Yardney 6666 Box Springs Blvd. Riverside, CA 92507-0736 Phone: (909) 656-6716 FAX: (909) 656-3867	Pressure differential tanks	Tank capacity ranges from 15 – 45 gallons. Larger sizes are available on special request.

CHAPTER 4. INJECTION TECHNIQUES FOR VARIOUS IRRIGATION METHODS

HIGHLIGHTS

- California has special conditions for chemigation compared to the Midwestern U.S.
- The irrigation method influences whether or not chemicals must be injected at a constant rate.
- The irrigation method influences the optimum duration of chemical injection.
- Chemical injection rates may inadvertently vary during an irrigation set for many reasons: improper mixing, system flow rate fluctuations, and system pressure fluctuations.
- Drip systems must account for the disposal of filter flush water.

CALIFORNIA CONSIDERATIONS

Most of the literature on chemigation originates east of the Rocky Mountains, and because some California irrigation systems are often distinct from Midwestern U.S. irrigation systems, differences in recommendations should be noted. First, many of the eastern studies are done on center pivot irrigation systems. Literature on "sprinkler" chemigation usually means "center pivot sprinkler." Second, herbicide and pesticide injection are more common through the center pivot systems than through most other types of irrigation systems.

In California, there are relatively few center pivots and therefore herbicide and pesticide injection, which requires foliar applications, is not as common as it is in some other regions. California irrigation systems may require people to be in the fields to move the sprinkler irrigation equipment, thereby creating a health hazard if non-fertilizer chemicals are injected. Second, the majority of California irrigation systems (drip, microspray, furrow, and border strip) are incapable of wetting the leaves, where many pesticides must be applied.

California does have some widespread chemigation situations which are more rare than in other areas. These are:

- Low infiltration, caused by water chemistry (either very pure water or with an imbalance of certain ions in soil or water). (See Chapter 13 for chemigation information.)
- Plugging in drip systems, either externally from root intrusion or internally from mineral accumulation or biological growth. (See Chapter 12 for more information.)

Moving vs. Stationary Irrigation Systems

Continuous Move Irrigation Systems – General

Examples of continuous move irrigation systems are:

- Center pivot sprinklers
- Linear move sprinklers
- Travelers (big sprinklers)
- Furrow
- Border strip

Moving irrigation methods have two special chemigation requirements:

1. Fertilizer injection *must* be done at a *constant rate* to ensure uniform chemical distribution.
2. The rate *must* be calibrated so that the injection continues for exactly *the duration of the irrigation event.*

The "constant rate" requirement for continuously moving sprinkler systems is fairly easy to understand. If the chemical injection rate changes while a linear move travels across a field, chemical application will vary across the field. Additionally, if the injection does not occur throughout the entire irrigation duration, a section of the field will not receive chemicals.

Moving Systems – Surface Irrigation

There are several key rules for effective fertigation with surface irrigation (e.g., furrow and border strip) systems:

1. Irrigate for a high irrigation distribution uniformity. This means the following will occur:
   a. Tailwater will run off, to be collected and reused.
   b. Water will advance to the end of furrows in about half (or less) of the total irrigation time on non-cracking soils.
   c. Different irrigation sets on the same field will be irrigated for the same duration, using the same flow rates.

2. Inject fertilizer into the incoming irrigation water at a constant rate.
3. Inject fertilizer continuously.

The tailwater will have the same fertilizer concentration (ppm) as the irrigation water at the head of the field. Therefore, blending source water with tailwater will not cause problems. This assumes that the flow rate coming onto the field remains constant, and it is not reduced when the tailwater system recycles water.

Some people recommend beginning the fertigation after water advances half way down a furrow. However, this is a very cumbersome procedure once the first set has been completed, because it generally does not account for how the tailwater is handled. Furthermore, if a good irrigation distribution uniformity is achieved, such complex actions are not warranted.

The following points may shed some light on the dynamics of surface irrigation, and how those dynamics influence the rules above, especially the recommendation of injecting chemicals at a constant rate throughout the irrigation set.

1. Time is necessary for water to advance down a border strip or furrow (Figure 33). This is somewhat analogous to a moving center pivot.

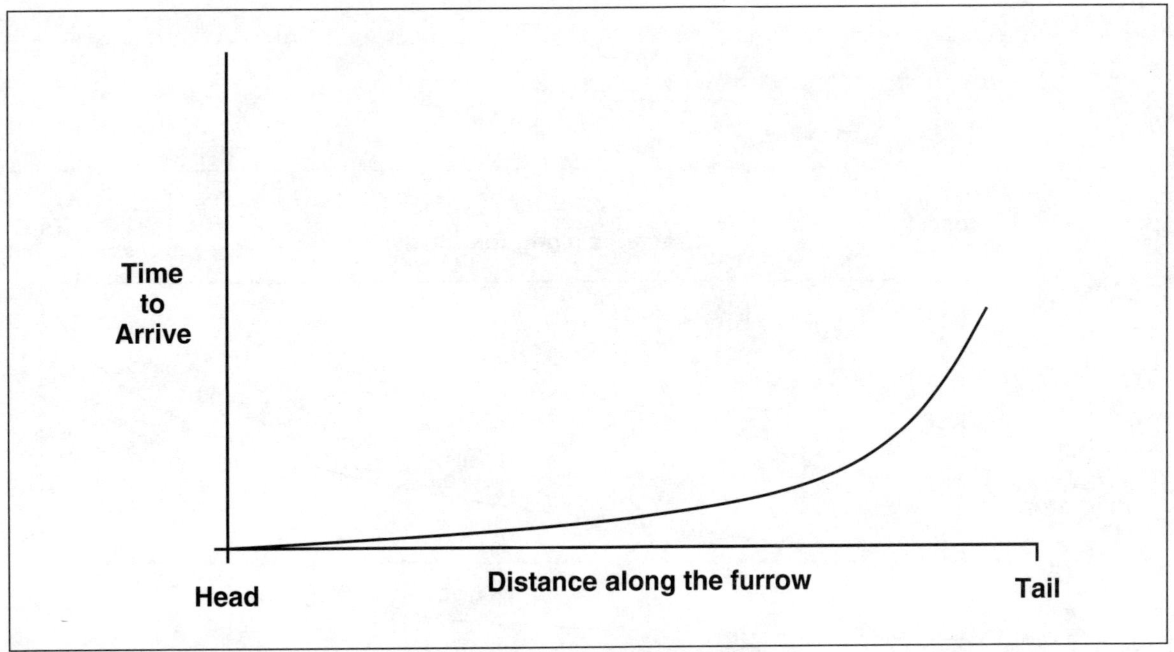

Figure 33. Distance vs. advance time along a hypothetical furrow.

2. The rate of water infiltration into the soil at a point varies with time. This is shown in Figure 34.

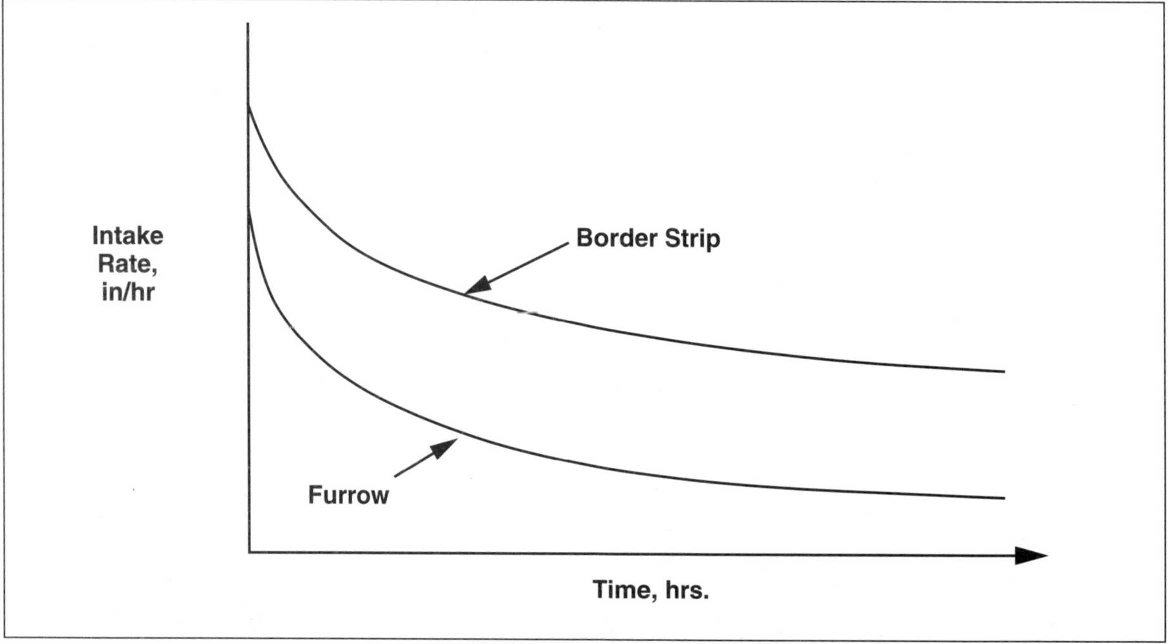

Figure 34. Infiltration rates decrease over time at a point along a furrow or border strip.

3. The concepts of unsteady (i.e., something which changes with time) advance and unsteady infiltration for furrows are combined in Figure 35.

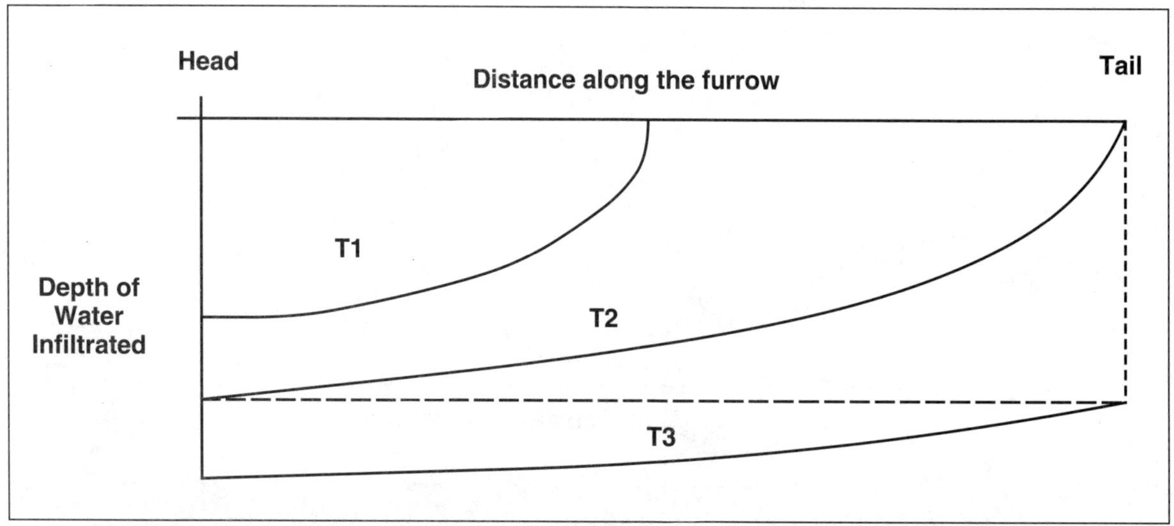

Figure 35. Depths of water infiltrated at various distances along a furrow at three times (T1, T2, and T3). T3 = 2 × T2. The dashed horizontal line indicates the depth of water needed if perfect irrigation scheduling occurred. Assumes a uniform soil.

Figure 35 shows that in a furrow with runoff (tailwater), more water always infiltrates at the head end than at the tail end. It is theoretically possible to reduce deep percolation of chemicals at the head end of furrows if one waits until the water has advanced about one-third of the distance down the furrow before beginning chemical injection. However, furrow and border strip management for a high Distribution Uniformity (DU) generally requires the existence of tailwater runoff, and the subsequent recycling of that tailwater. The tailwater will have the same concentration of chemical as the irrigation water itself. When one considers tailwater management schemes which are simple, one generally finds that it is impractical to withhold chemical injection during the first third of the advance phase of furrows, because tailwater is often used as part of the water source at that time. Instead, the key is to minimize the deep percolation of both water and fertilizer by having a good DU and the proper irrigation set duration.

Stationary Irrigation Systems

Stationary irrigation methods include:

- Hand move sprinklers (during a single set)
- Permanent undertree sprinklers (during a single set)
- Microirrigation (during a single set)

Fertigation with stationary irrigation methods is quite simple if the fertigation begins and ends within a single set. A *quantity* of fertilizer is applied to a set (i.e., an area). With stationary irrigation methods, one must consider two injection characteristics pertaining to uniformity:

1. *Spatial uniformity* (i.e., point-to-point) of fertilizer application within a set. If fertilizers are well mixed with the irrigation water, the fertilizer will be applied with the same uniformity as is the irrigation water.
2. *Time uniformity* of fertilizer application rate within a set. As long as all of the fertilizer is applied during that set, the fertilizer spatial variation does not depend on fertilizer injection *rate* variation.

The injection process is repeated for every set, until the whole field has been fertigated.

Fertigation can be accomplished continuously (as opposed to beginning and completing discrete fertigations for each irrigation set) on stationary irrigation systems having multiple sets, if *all* of the following conditions are met:

1. Fertilizer injection rate is constant.
2. Irrigation durations are the same for each set.

3. Each set has the same acreage.
4. Each set has the same total flow rate.

If these conditions are not met, and continuous fertigation occurs from set to set, different sets will receive different amounts of fertilizer, when measured as pounds per acre.

Variations in Chemical Injection Concentrations with Time

There are three general reasons why chemical concentrations in the irrigation water will vary with time:

1. Improper mixing/agitation of the chemical in the holding tank. This is especially important with solid fertilizers.
2. The injection hardware changes the injection rate with time.
   a. Differential pressure tanks start injection at a high rate, and end with a low rate of chemical injection (although the rate of fluid flow through the tanks remains constant).
   b. Diaphragm pumps may vary their discharge if the irrigation line pressure changes. The irrigation pump discharge pressure will change as center pivots move up- or downhill, or as new irrigation blocks are started with different distances or elevations from the irrigation pump.
   c. Venturi devices will have a change in suction pressure if the flow rate through them changes with time.
   d. Some gravity feed injection systems may vary their rate as the fertilizer tank empties.
   e. Suction screens on the injection devices may become dirty with time, and the chemical flow rate will be reduced.
3. The irrigation flow rate changes with time.
   a. Different irrigation sets may irrigate different acreage, with a different number of sprinklers or emitters per set.
   b. Elevation changes of a moving sprinkler irrigation system may result in different flow rates with time.
   c. Center pivot irrigation systems have different flow rates as end guns and corner units are activated and deactivated.
   d. Pipeline systems which are interconnected on a farm, and receive water from several sources and/or supply water to different fields, are notorious for wide pressure (and flow rate) fluctuations at outlets. This problem is especially severe for low pressure systems such as furrows and border strips which are supplied by such pipelines.

Acids and chlorine are typically injected into drip irrigation systems to prevent or clean out plugged lines. Chlorine is injected at a constant ppm rate, and acid is injected to maintain a constant pH. In both cases, the chemical must be injected at a constant proportional rate to achieve the desired result. Chlorine is generally injected through a gas chlorine injection system capable of constant rates.

Pesticides must be injected according to label requirements and generally require a specific concentration (i.e., constant chemical injection rate).

Consequences of not injecting these compounds at a constant rate include plant damage due to pH fluctuations or to toxic levels of chemical applications.

Centralized vs. Mobile Injection Units

Traditional fertigation discussions assume a stationary, centralized injection system. However, some growers use a mobile unit to fertigate fields which have many small blocks. This is particularly popular for vegetable fields in which blocks may be planted in intervals of several days or weeks, and for orchards with many different blocks of tree varieties.

If a drip system supplies a number of fields, it is often best to use a mobile unit, sometimes in conjunction with a central injection system. There are two types of "mobile" units:

1. An injector pump and chemical tank mounted on a trailer.
2. An injector pump and portable tank/stand (Figure 36).

Figure 36. An injector pump and portable tank and stand used to fertigate lettuce near Gilroy, Ca.

A central injection system can be used for water pH control and chlorine injection. Gas chlorinators, for instance, should be located behind a locked, chain link fence. The mobile unit would be used for fertigation. Mobile fertigation units allow one to customize fertigation according to the soil, planting date, and crop type within each block.

Another factor in selecting mobile units is the possibility of some fertilizers gravitating to the bottom of long irrigation pipelines, thus not distributing evenly at the various irrigation outlets. This does not appear to be a problem with smaller irrigation systems, but long mainlines may create such a problem. It is probable that poor mixing at the injector, plus high fertilizer dosages, accentuate the problem. Limited work at the Cal Poly ITRC has not been able to duplicate this problem, although it has been reported by a few growers. More research needs to be done to verify whether or not this is a significant problem.

If mobile units are used with drip systems, the injection points must be designed so that dirt does not enter into the drip system when the hose is connected. It may be necessary to have a second tank of rinse water on the trailer to be able to spray off fittings, thereby preventing this problem. Also, because the injection occurs downstream of the drip system filters, it is important to have a 120 mesh filter on the injection hose so that the drip system is not contaminated by dirt from the fertilizer tank.

Continuous vs. Non-Continuous Injection

There are two different practices used by growers regarding when to start and stop injecting fertilizers during an irrigation set in a stationary system. These are:

1. Only fertigate during the middle half of the irrigation set (quarter-half-quarter rule).
2. Fertigate continuously at a constant rate.

The quarter-half-quarter rule is justified as follows:

1. By waiting until the irrigation is one quarter completed, it is assured that the system is up to normal operation pressure and the water is being applied uniformly.
2. By not injecting fertilizers during the last quarter of an irrigation set, any remaining chemicals are flushed from the system.

The quarter-half-quarter rule is necessary if occasional, large dosages of chemicals or especially corrosive materials are applied. Obviously, such conditions could cause damage to fittings and emitters if there was prolonged contact with the chemical. In such a case, it is important to go one step further than the simple rule of "shutting off with one quarter time left". One must verify that the quarter time is sufficient to cleanse the system. The next section, dealing with chemical travel time computations, is helpful in this regard. For mobile injection units, this is particularly important because the dosages tend to be relatively high.

Growers who "spoon-feed" their crops by continuously injecting small dosages of fertilizers (a highly recommended practice) generally do not have to worry about flushing the lines because the chemical concentrations are so low they do not damage fittings and emitters.

For the non-continuous injection practices, it is important that the chemical be injected long enough to reach the last emission point (on sprinkler and drip systems). The irrigation duration must then be long enough to flush completely that last emission point. Knowledge of the chemical travel time in pipelines is important.

Chemical Travel Time in Pipelines

The travel time of a chemical, from the injection point to final application point, can be calculated by analyzing the travel time through each pipe segment, and then summing all of the segment travel times. For one pipe segment:

$$\text{Time} = \frac{\text{Distance}}{\text{Velocity}} = \text{Distance} \times \frac{\text{Inside Area of Pipe Segment}}{\text{Flow Rate through the Segment}}$$

Where: Time = seconds
Distance = pipe segment length, feet
Inside Area = square feet
Flow Rate = cubic feet per second

For an existing system, this computation is complicated due to the many different segments, flow rates, and lengths in a typical irrigation system. Furthermore, the exact pipe layout and dimensions may not be known on an existing system. Therefore, the next two steps are recommended instead.

For a new drip/microirrigation system design, the calculations are fairly simple because the pipe sizes and lengths are all known by the designer. Some drip hose hydraulic programs, such as the ITRC Drip Hose Hydraulics Program, compute travel times down a single hose. Typical hose travel times vary from 20 minutes to one hour. The total chemical travel time must also include travel time between the injection point and the beginning of the hose.

For an existing irrigation system, the travel time can be easily measured in the field. This is done by injecting a chemical and then going to the most distant irrigation outlet in the system. The water at that outlet should be continuously sampled, until the injected chemical concentration at that outlet begins to change. A simple salinity monitor (EC) is often used to detect the arrival of fertilizers (except urea, which has no charge).

Chapter 5. Irrigation Principles, Leaching, and Fertilizer Uniformity

Highlights

- Good water and fertilizer management requires that managers understand the concept of Distribution Uniformity of irrigation water.
- The amount of fertilizer which is leached (i.e., passes downward through the soil root zone) is affected by the amount of deep percolation water, and the concentration of leachable substance in the soil.
- Deep percolation is influenced by (i) the irrigation system Distribution Uniformity, (ii) irrigation scheduling, and (iii) rainfall.
- Concentration of leachable substances in the soil depends on the amount and form of the chemical applied. Some compounds do not leach because they adsorb to organic matter and to clay particles.
- Chemical leaching can be substantially reduced by minimizing the quantity of irrigation water deep percolation and by controlling the concentration of leachable material in the root zone.

Irrigation Distribution Uniformity and Scheduling

Many discussions regarding deep percolation losses of water and chemicals assume that all spots in the field are identical, and that water is applied uniformly across a field. This fundamental flaw in the understanding of fertigation management must be corrected.

All irrigation systems apply water non-uniformly. To manage irrigation water properly, especially as it relates to fertigation efficiency, one must understand the difference between non-uniformity and irrigation scheduling.

The preferred measure of uniformity is called "Low Quarter Distribution Uniformity (DU_{LQ})," and is defined as:

$$DU_{LQ} = \frac{\text{"Minimum" depth infiltrated in a field}}{\text{Average depth infiltrated}} \times 100$$

Where: "Minimum" depth infiltrated = average depth infiltrated in the region receiving the lowest 25% of the water.

The techniques used to measure DU in the field are not covered in this book. However, the Cal Poly ITRC provides software and short courses on this specific topic (Burt et al., 1995). For this book, it is sufficient to present some of the basics.

Figures 37 through 40 depict the concepts of DU and proper irrigation scheduling. They represent depths of water infiltrated at various parts of a field. These figures do not include information on evaporation, uncollected runoff, or total water applied; they merely depict the *infiltrated water.*

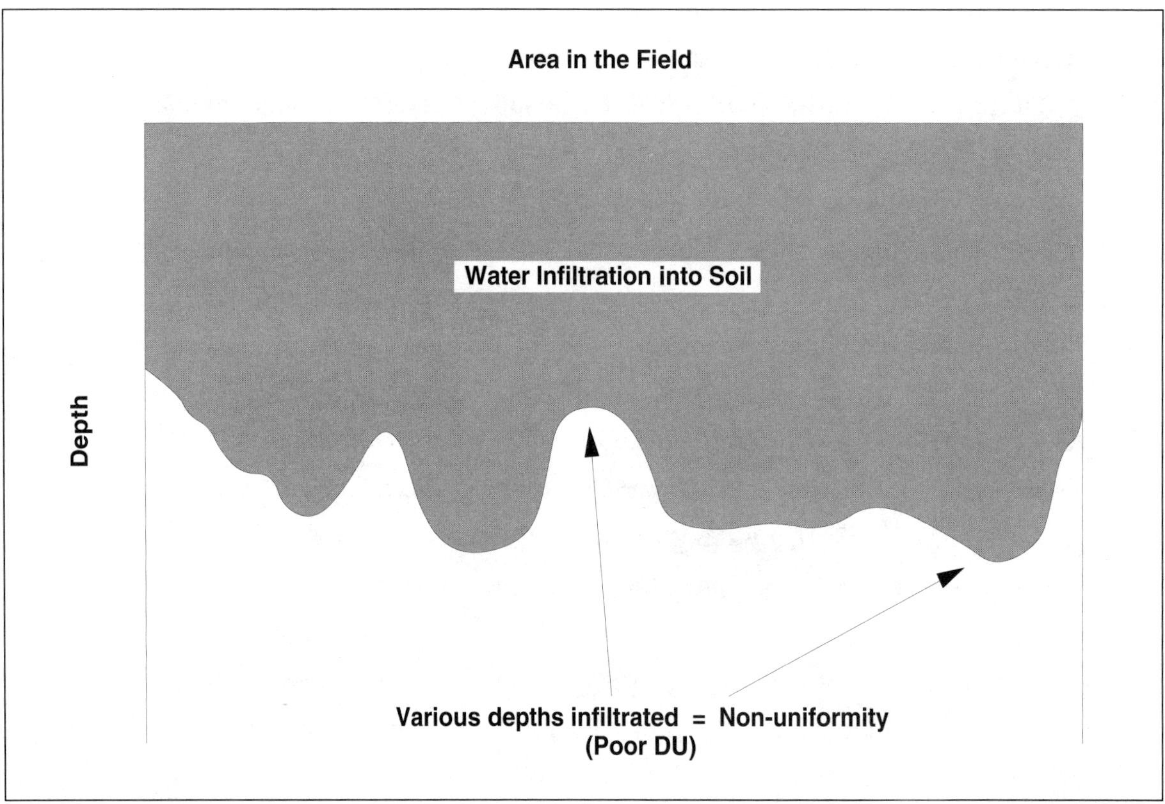

Figure 37. The concept of non-uniformity: regardless of the irrigation method, there are differences in the amount of water infiltrated throughout a field.

The causes of non-uniformity vary depending on the irrigation system type, design, and management. For example, some sprinklers and emitters can be partially plugged from dirt, lime, algae, or insects. Such plugging affects the uniformity of water applied and consequently the uniformity of water infiltrated. Sprinklers and emitters also have pressure differences among them, thus they apply different amounts of water on different parts of the field. For surface irrigation methods, non-uniformity is caused by factors such as water sitting at the head end of a furrow longer than at the tail end, and by soil differences throughout a field.

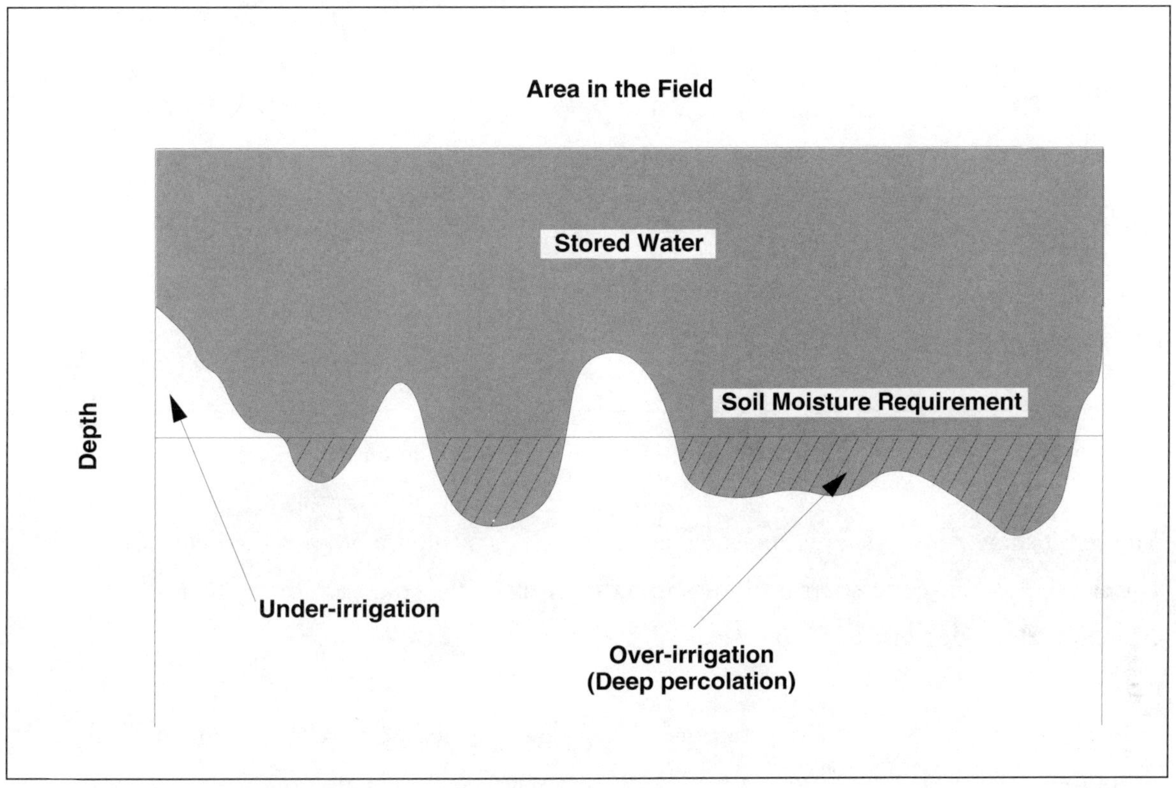

Figure 38. The concept of adequate irrigation (i.e., some of the field receives adequate water and some is under-irrigated). The deep percolation is "lost" to that field, along with some nutrients and chemicals which have leached down with the water.

Figure 38 is the same as Figure 37, except that the "soil moisture requirement" is incorporated into the drawing. This represents the amount of water which the soil could hold prior to the irrigation. Obviously, if the depth of water infiltrated is greater than this requirement, then there is deep percolation. Similarly, if the depth of water infiltrated is less than the soil can hold, or less than what the plants need, there is under-irrigation.

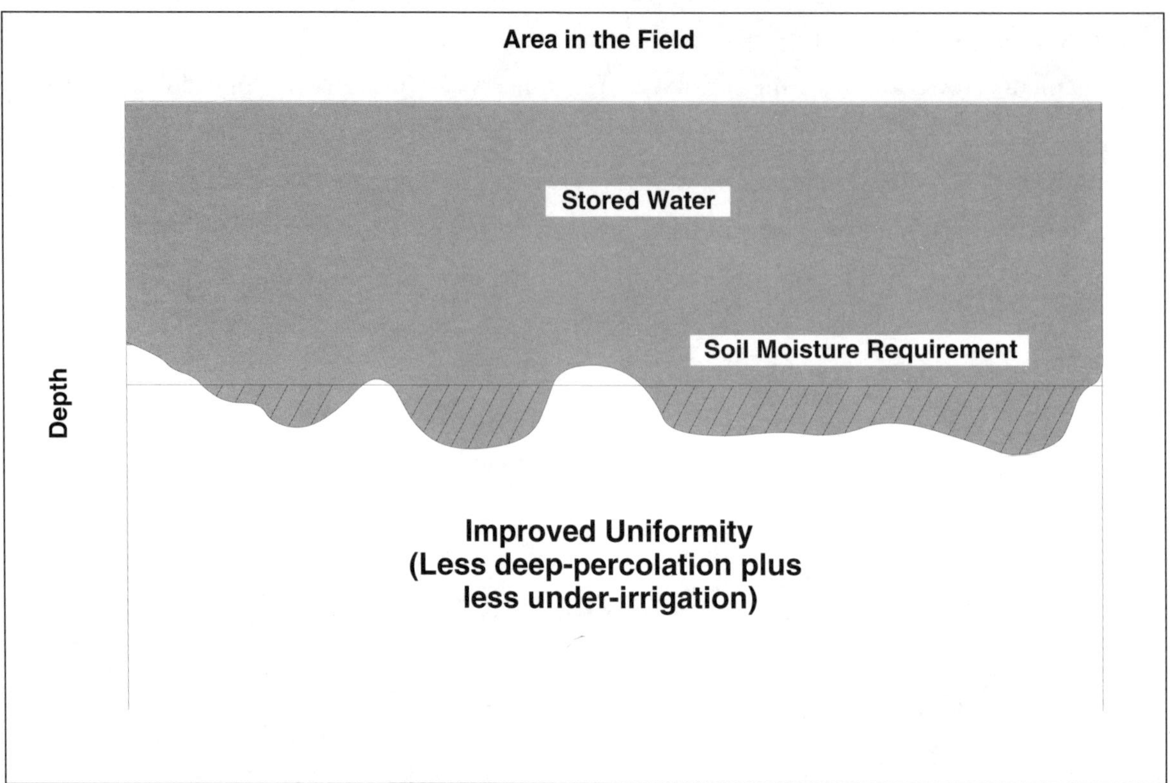

Figure 39. The concept of improved uniformity.

Figure 39 shows the same amount of water infiltrated as in Figure 38, but the field now has a better DU. Because the water infiltrates more uniformly, the result is a decrease in under-irrigation and a decrease in deep percolation.

A *very good* DU is 85 – 95%, depending on the irrigation method and circumstances (soil, topography, etc.). If there is *perfect* timing of the irrigations (there are no dry spots after irrigation, but also no over-irrigation at the spot which receives the least), an 85% DU suggests that 15% of the infiltrated water will deep percolate. A more typical field DU for a well managed field is 75 – 80%, resulting in 20 – 25% deep percolation if the irrigation duration is managed "perfectly" so that there are no dry spots and no over-irrigation at the spot in the field which receives the least amount of water.

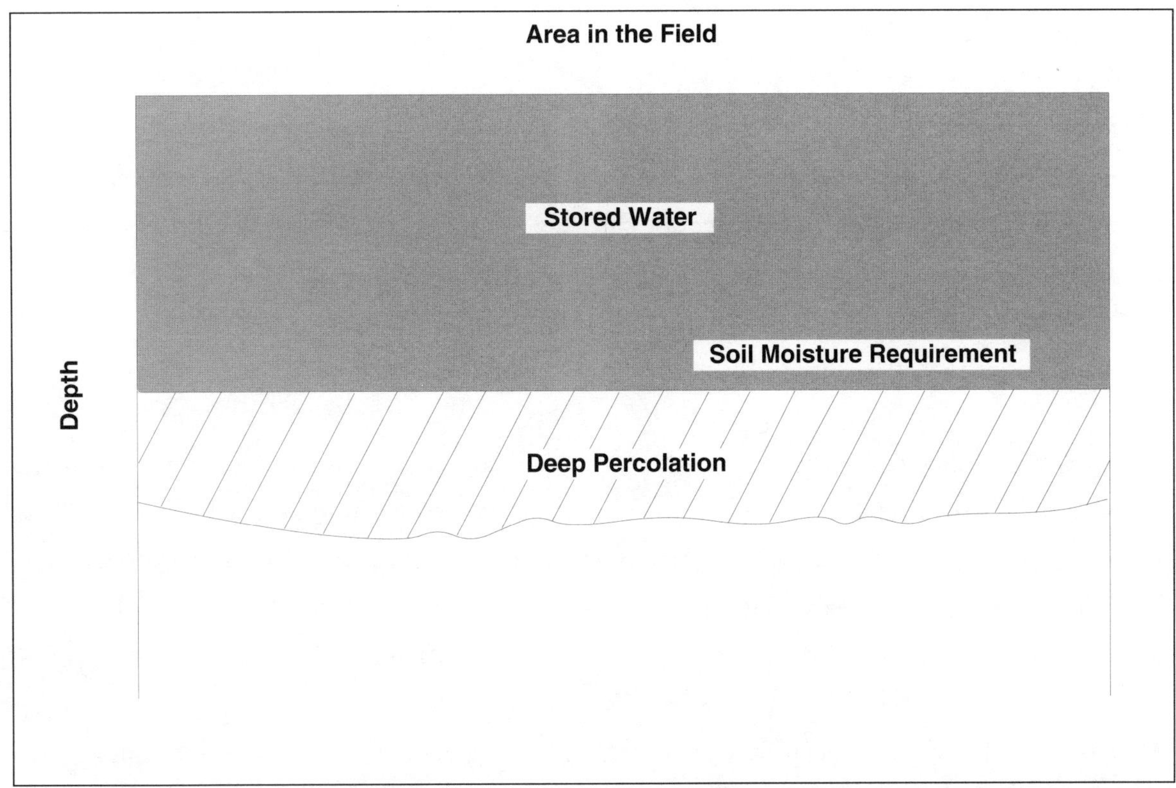

Figure 40. The concept of poor timing.

Figure 40 depicts the importance of the two major items affecting irrigation efficiency:

1. Distribution Uniformity
2. Irrigation scheduling

In this scenario there is a good DU (the water infiltrates uniformly), but the water was not shut off at the correct time. This results in large amounts of deep percolation.

In the case of Figure 40, the water infiltrates uniformly; but a large amount of water deep percolates due to poor timing (poor irrigation scheduling). Fertilizers will not leach if water does not deep percolate, and *the distribution of fertilizer cannot be better than the distribution of water.* Thus, a major aspect of controlling leaching is to decrease deep percolation by improving the system DU and irrigation scheduling (i.e., improving the system's irrigation efficiency).

Chemical Leaching

Nitrogen

Nitrogen in the nitrate (NO_3^-) form is the nutrient most noted for leaching. Since NO_3^- is an anion (i.e., it has a negative charge), it is not held onto the negatively charged cation exchange sites which exist on clay and organic matter particles. Consequently, NO_3^- will flow through the soil with irrigation water. If any water deep percolates, then the nitrate in that water will also pass through the root zone. If one continually fertigates with nitrate and 15% of the water deep percolates, approximately 15% of the applied nitrate will also deep percolate.

Ammonium (NH_4^+) cations are adsorbed onto the soil cation exchange sites within a few inches of the soil surface. Therefore, there is relatively little leaching of ammonium ions, as compared to nitrate ions. After ammonium converts to nitrate through nitrification, the nitrate will be redistributed throughout the root zone during subsequent irrigations. If there is deep percolation during subsequent irrigations, then a portion of the nitrate will deep percolate with the water.

Urea ($CO(NH_2)_2$) has no electrical charge, and therefore penetrates into the soil until it is transformed by the urease enzyme. This process can occur within a few inches of the soil surface, and there is rarely deep percolation of urea on most soils. The urea is transformed first to ammonium, and later to nitrate. Once it has converted to nitrate, the nitrogen can be moved freely by the irrigation water.

For sprinkler and surface (flood) irrigation methods, the best form of fertilizer may be a combination of nitrate and urea. The nitrate will be immediately available to the plants, and will penetrate with the water. The urea will eventually be converted to NH_4^+ and NO_3^-. Urea ammonium nitrate (UAN-32) is a popular liquid mix which combines all three forms of nitrogen.

Urea has the advantage over ammonium fertilizers in that the deeper penetration of urea will reduce the amount of NH_3 (ammonia gas) which is lost by volatilization into the air. Ammonium fertigation, especially when applied to alkaline soils, can result in appreciable NH_3 volatilization loss when the top of the soil becomes dry.

Other Nutrients

Both phosphorus and potassium are relatively immobile in the soil if they are applied as phosphate ($H_2PO_4^-$) or Potassium (K^+). Therefore, P and K leaching is minimal unless there are (i) very high applications, (ii) very sandy soils, or (iii) applications of other forms of these nutrients, such as polyphosphates. These and other nutrients will be discussed in later sections regarding their mobility and leaching potential.

Pesticides

Pesticide contamination of groundwater is generally caused by pesticide movement with irrigation and/or rainwater through the soil profile. Many pesticides are either absorbed into or adsorbed onto organic matter near the soil surface. Soils with high sand and low organic matter contents (e.g., low cation exchange capacity) tend to have more pesticide leaching than do other soils. The fact that pesticide leaching is so closely related to organic matter content brings up an interesting question regarding pesticide application through buried drip irrigation systems. With those systems, the pesticides are applied at depths below the large organic matter concentrations in the soil, and therefore there may be a greater leaching potential than with above-ground drip systems.

For the vast majority of pesticide applications, though, the keys to reducing pesticide leaching include improving the irrigation management as follows:

1. Schedule irrigations efficiently so that very little water goes below the root zone.
2. Make certain that the irrigation system has a good DU.

Preferential Flow

The previous discussions of chemical leaching assumed that water moves through a matrix of soil particles and small pores. This type of water movement is referred to as laminar flow. The flow through the tortuous soil matrix allows ample opportunity time for adsorption of fertilizer ions onto the cation exchange sites. However, recent studies have focused on the concept of preferential flow, in which water and solutes flow through cracks and continuous large pores (e.g., old root paths).

Preferential flow is very significant in areas with cracking clay soils, such as the Imperial Valley of California. In those soils, surface irrigation is accomplished by first filling up the soil cracks with water; the water then infiltrates *laterally* into the soil. In a cracking clay soil, perhaps 3 – 10% of the surface irrigation volume may flow vertically directly past the root zone via the cracks, with little or no soil particle contact. This 3 – 10% of the irrigation volume may represent more than 50% of the leached water on a heavy cracking soil. This type of leaching cannot be controlled by conventional irrigation management practices.

Decreasing the leaching caused by preferential flow must focus on (i) minimizing the amount of cracks which reach the soil surface, (ii) on closing the cracks with a minimum of water application, and (iii) on decreasing the concentrations of fertilizer applications, thus decreasing the total chemical amount which will leach. The cracks can be closed with a minimum of water application if one switches from border strip irrigation to sprinkler or drip irrigation. There may be an economic downside with drip irrigation due to the low intake rates or low crop values in these cases.

The amount of preferential flow in a non-cracking clay soil will be less than that of a cracking soil, but it will still exist. Various researchers have quantified preferential flow, including Everts and Kanwar (1990) who measured 9 – 20% of the nitrate leaching as preferential flow on a loam soil. A simple example of preferential flow leaching occurs in typical soil science laboratory demonstrations, when NH_4 is found in the leachate from a soil column treated with ammonium fertilizer. This finding contradicts what people assume about ammonium—that it does not leach. The laboratory demonstration shows that leaching does not occur as a result of laminar flow (on which most leaching discussions have been based), but rather it is a result of preferential flow.

Broadcasting vs. Injecting Nitrogen

Leaching

In general, there is considerably less leaching of nitrates with fertigation than with broadcast fertilization. The amount of irrigation deep percolation can be the same in either case. The cause in the leaching lies in the different fertilizer management strategies.

Broadcast applications are generally made in heavy doses, assuming that subsequent irrigations will leach a portion of the fertilizer. The heavy fertilizer applications are made so that sufficiently high levels of nutrients persist throughout the growing season. Consequently, the higher doses provide higher nutrient concentrations in the soil. Since the nutrient supply exceeds the crop's demand, the nutrient will be present in excess of the crops' needs. If the fertilizer is in a leachable form (i.e., NO_3-) and water deep percolates, the fertilizer will leach because it has not been used by the plant.

With fertigation, the fertilizer can be "spoon-fed" with every irrigation. There is no need to over-apply in order to "save some fertilizer for later". Consequently, there is never a high concentration of the nutrient in the soil. The fertilizer supply meets the demand; and, the potential for leaching is significantly reduced because the soil does not contain extra nutrients which can leach.

Soil Nitrogen Uniformity

The uniformity of soil nitrogen (N) across a field can be better with fertigation applications than with broadcast applications. The uniformity of fertilizer *application* can equal the irrigation water DU.

There is a second factor, in addition to application uniformity, which affects the uniformity of chemicals in the soil across a field. This second factor deals with the uniformity of fertilizer *retention* across a field.

With conventional, broadcast applications of fertilizer, one may start with a fairly uniform concentration of fertilizer at various points across a field. However, with time the uniformity will decrease because some spots in the field receive more water than other spots. Therefore, there will be more leaching of nitrates at these spots than at others.

With fertigation, areas which receive more water (thus having higher leaching) will also receive more N. Therefore, the soil nitrogen levels across the field remain fairly uniform over time if the fertilizer is spoon-fed. The qualifying conditions for this to be true include the following:

1. There can be no under-irrigation of any spot on the field.
2. The nitrogen must be applied in the *nitrate* form.

CHAPTER 6. NITROGEN TRANSFORMATIONS AND PROCESSES

HIGHLIGHTS

- Nitrogen fertilizers are the most commonly used nutrient substances in fertigation.
- Effective use of these nitrogen fertilizers requires an understanding of how these materials interact with the irrigation water and how they are transformed and move within the soil.
- Nitrogen fertilizers may be divided into four major forms: ammonia (NH_3), ammonium (NH_4^+), nitrate (NO_3^-) and organic (e.g., urea).

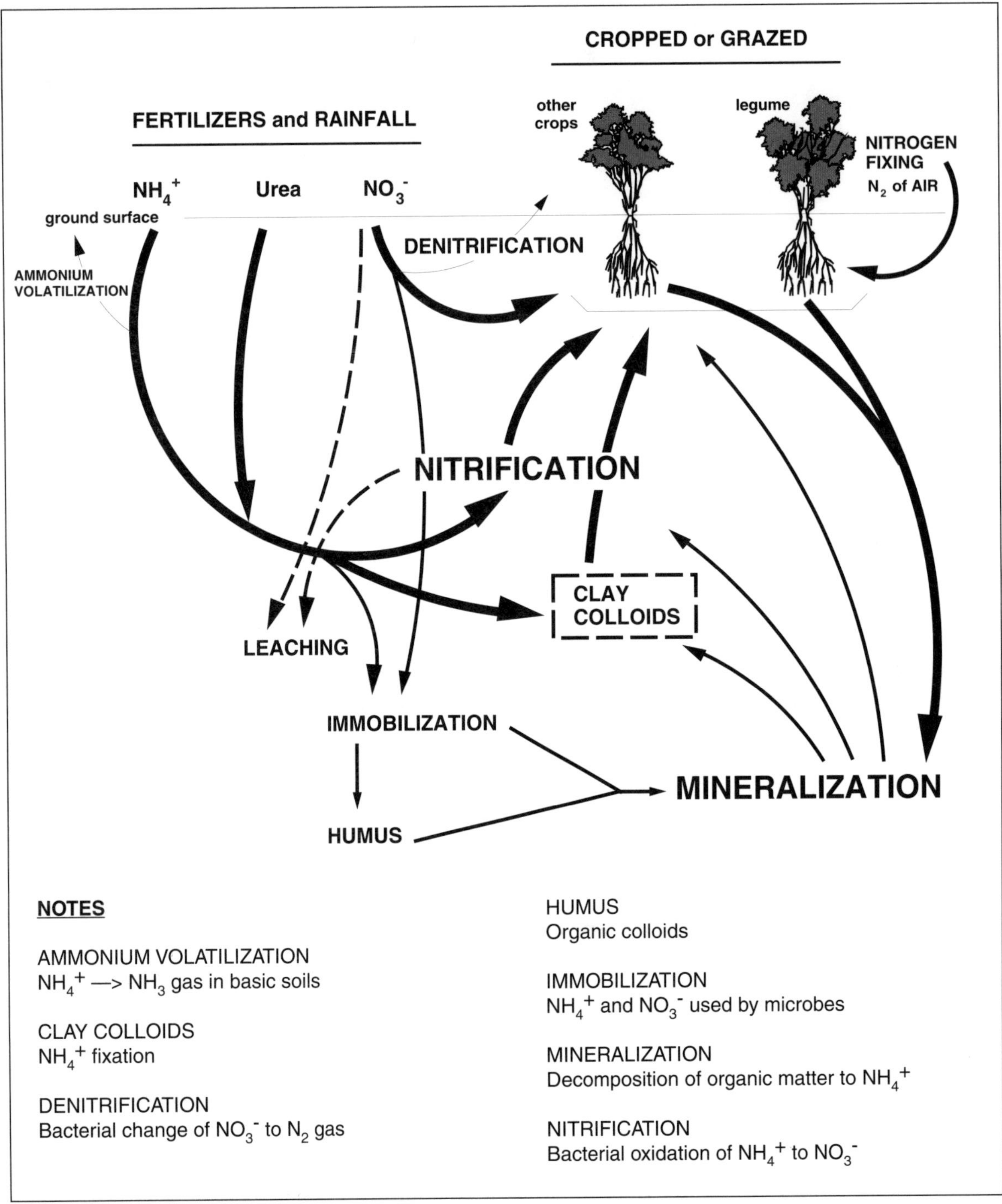

Figure 41. The nitrogen cycle.

NITROGEN CYCLE

Since nitrogen is the most commonly applied fertilizer in agriculture, understanding the processes surrounding nitrogen is essential for sound nitrogen management. Nitrogen is very dynamic in the plant and soil systems—both chemical and biological processes constantly transform nitrogen from one form to another. Figure 41 depicts the nitrogen cycle.

NITROGEN TRANSFORMATIONS

NITROGEN FIXATION

Nitrogen fixation is the transformation of dinitrogen gas (N_2) from the atmosphere to ammonium that is converted into organic nitrogen in plant tissue. Leguminous plants (e.g., alfalfa, peas, beans, clover, and vetch) can fix up to 300 pounds of nitrogen per acre per year depending on the plant type and environmental conditions. If these plant materials are incorporated into the soil, the organic nitrogen held in these plants will eventually become available for subsequent crops through mineralization.

N_2	————————>	Organic N
N gas from the atmosphere	Fixation by Rhizobium bacteria in the nodules of legumes	N held in the tissue of plants

MINERALIZATION

Mineralization represents the conversion of organic nitrogen (i.e., that held in plant tissue in the form of proteins or amino acids) to mineral nitrogen (i.e., ammonium) by soil microbes.

Organic N	————————>	Ammonium
N held in the tissue of plants	Mineralization (decomposition by microbes)	Mineral N (NH_4^+)

NITRIFICATION

Nitrification represents the biological transformation from ammonium to nitrate. This is a two step process which forms nitrite in the intermediate step. The nitrification process is outlined below:

					Nitrosomonas			
NH_4^+	+	O_2	+	HOH	————>	NO_2^-	+	2 H^+
Ammonium		Oxygen		Water		Nitrite		2 Hydrogen ions

			Nitrobacter	
NO_2^-	+	O_2	————>	NO_3^-
Nitrite		Oxygen		Nitrate

IMMOBILIZATION

Immobilization represents the conversion of inorganic nitrogen (NH_4^+ or NO_3^-) to organic nitrogen (proteins or amino acids) through plant or microbial uptake.

Mineral N	————————>	Organic N
NH_4^+ or NO_3^-	Immobilization (plant or microbial uptake)	Proteins and amino acids in living tissue

Denitrification

Denitrification represents the conversion of nitrate or ammonium to gaseous nitrogen forms (mostly N_2 and a little N_2O) under anaerobic conditions.

Volatilization

Volatilization represents the chemical transformation of ammonium to ammonia gas (NH_3) in high pH soil/water conditions.

Ammonium	——————>	Ammonia gas
NH_4^+	Volatilization	NH_3

While mineralization, nitrification, and nitrogen fixation are considered gains to the system, denitrification and volatilization are considered losses. Immobilization is considered only a temporary loss of nitrogen to the system because the organic nitrogen is only temporarily unavailable for plant uptake. The following sections will focus on the *forms* of nitrogen which are relevant to fertigation. It will also describe the impact of these transformations in an agronomic context.

Ammonium

General

Ammonium (NH_4^+) is a positively charged cation which is held electrically on the soil cation exchange sites. Therefore, there is very little leaching of ammonium. This ion stays on the cation exchange sites until another positively charged nutrient (i.e., calcium, potassium, or magnesium) replaces the ammonium on the same site. At that time, the ammonium ion will move into the soil solution where it is immediately available for plant uptake.

In California soils, most of the applied ammonium fertilizer is expected to be transformed into nitrate (through nitrification) within three weeks after application to the soil during the summer. This quick process in the summer can be attributed to favorable conditions for the nitrifying bacteria. Nevertheless, the ammonium ion concentration in the soil can be high for several weeks after certain types of fertigation. The following list describes several ammonium sources which temporarily increase ammonium concentrations in the soil.

1. Obviously, all ammonium containing fertilizers will release ammonium (NH_4^+) ions into the soil. Typical ammonium fertilizers include the following: ammonium nitrate (AN), ammonium sulfate (AS), monoammonium phosphate (MAP), diammonium phosphate (DAP), and other mixed fertilizer blends made from ammonia and other acids.

2. Anhydrous ammonia, aqua ammonia, and free ammonia fertilizers are initially converted into ammonium (NH_4^+).
3. All organic nitrogen compounds (such as fishmeal or manure) are initially converted to ammonium (NH_4^+) ions through mineralization.
4. The urease enzyme converts urea nitrogen into ammonium (NH_4^+).

Accumulation of ammonium ions can cause infiltration problems if the irrigation water is extremely pure (i.e., snow melt, $EC_W < 0.2$ dS/m). In these cases, the soil will be susceptible to swelling, clogging, and sealing.

Soil Acidification with NH_4^+ and NH_3 Fertilizers

The first nitrification step involving the *Nitrosomonas* bacteria produces two hydrogen (H^+) ions. These hydrogen ions lower the soil pH.

NOTE	Whenever any organic nitrogen, urea, ammonia or ammonium fertilizer is applied to the soil, nitrification produces acidity and lowers the soil pH.

The more non-nitrate based nitrogen fertilizers are applied, the more the pH is lowered. When the soil becomes too acidic (about pH 4.5), these nitrifying bacteria stop functioning. This can happen directly under a drip emitter which has received many applications of organic nitrogen, urea, ammonia or ammonium fertilizers over the years. Under these conditions, ammonium can build up under the drip emitters because nitrification is inhibited by lowered pH.

One impact of localized acidity is that specific ion toxicities and deficiencies appear and the health of most agronomic crops suffers. Another impact is the detrimental effect on soil microorganisms.

Correcting Acidity

Limestone needs to be applied to correct this localized acidity problem. See Chapter 14, "Soil pH Modification."

Nitrification Inhibitors

There are several products sold to inhibit or slow the conversion of urea to ammonium and the conversion of ammonium to nitrate. Growers use these products for several reasons: the acidification process associated with nitrification is retarded; nitrogen remains in the ammonium form longer, thus reducing nitrate leaching, and releasing inorganic nitrogen slowly throughout the season.

Such inhibitors are of primary interest to growers who apply large dosages of fertilizer by conventional (i.e., non-fertigation) means. Nitrification inhibitors are registered products, and include N-SERVE™, Dicyandiamide (with trade names such as Super N, Super N pluS, or Super U), and NBPT (which retards the hydrolysis of urea to ammonia) (Overeem, 1992).

NITRATE

GENERAL

The nitrate ion (NO_3^-) is a negatively charged anion which is *not* held on the cation exchange sites. Due to nitrification, much of the nitrogen occurring naturally in the soil system is in the nitrate form. Nitrate in the soil solution is considered immediately available for plant uptake. Plant uptake of nitrate is discussed in detail in Chapter 7, "Nitrogen Uptake."

LEACHING

Nitrate is mobile and moves freely with the water. If nitrate concentrations in the soil solution are in excess of the plant's need, and irrigation water deep percolates beyond the potential root zone, then some of the nitrate will leach below the root zone where it is lost to the system. Environmental hazards associated with nitrate leaching (i.e., nitrate contaminated ground water) have increased the focus of improving fertilizer and irrigation efficiencies in agriculture.

AMMONIA

GENERAL

Ammonia (NH_3) is a gas and is not immediately available for plant uptake. For this form of nitrogen fertilizer to be available to plants, it must first undergo a chemical process to become ammonium. The reaction is illustrated as follows:

NH_3	+	H_2O	<——>	$NH_3 \bullet H_2O$	<——>	NH_4^+	+	OH^-
Anhydrous Ammonia		Irrigation Water		Aqua Ammonia		Ammonium		Hydroxide

Notice that with ammonia applications, a single hydroxide ion (OH^-) is produced to raise the soil or water pH. However, this is eventually counterbalanced by the formation of two hydrogen (H^+) ions during nitrification. The final effect is a reduction of soil pH.

VOLATILIZATION

Under certain conditions, applications of ammonium (NH_4^+), aqua ammonia (NH_4OH) and anhydrous ammonia (NH_3) fertilizers can result in ammonia losses through the process of volatilization. Such a reaction represents a loss of nitrogen to the system. This important reaction is illustrated below:

NH_4^+	+	OH^-	——>	NH_3	+	HOH
Ammonium		Hydroxide (High pH)		Ammonia Gas		Water

The extent of volatilization is directly related to pH. If NH_3 is applied to basic (high pH) water, very little will convert to ammonium; instead the NH_3 will volatilize. If NH_4^+ is applied to basic water, it will convert to NH_3 and volatilize. For a given amount of nitrogen in the NH_3 or NH_4^+ state, there will be a higher concentration of free ammonia, and therefore a higher potential for volatilization, in high pH water than in low pH water. If the pH of water or soil is 9.2 (with the fertilizer in it), then 50% of the nitrogen from ammonia application is in the form of NH_3 gas (Table 6).

Table 6. The effect of pH on ammonia volatilization.

Soil/Water pH	Potential N Volatilization (%)
7.2	1
8.2	10
9.2	50
10.2	90
11.2	99

The strong pH dependence indicates why growers will lose a significant portion of the ammonia nitrogen fertilizer if it is injected into irrigation waters with a pH of 8 or higher. In addition, the reaction of free ammonia with the water drives the already high pH even further upward, thereby compounding the volatilization problem. Injecting aqua ammonia into normal irrigation water will raise the pH beyond 9.4 where losses of ammonia can be from 30 to 50% of the applied nitrogen. Applying as little as 10 pounds of ammonia per acre-inch of irrigation water can result in 30% loss of ammonia by volatilization.

This pH relationship holds for all ammonia fertilizers and any ammonium fertilizers added to high pH irrigation water. It makes no difference whether the ammonium came from either free ammonia, aqua ammonia, or ammonium fertilizers. In addition, volatilization losses will be more rapid from nitrogen treated water when the air is windy and dry because evaporation losses are greater under these conditions.

Once the fertilizer or fertilizer-treated water has moved into the soil, the soil itself may have a high hydroxide (OH^-) concentration and thus a high pH. Soils which are high in lime often contain a pH of 8.0 to 8.3. Thus, as the soil dries out after an irrigation, the soil pH itself can force the reaction to the right and cause ammonia gas to escape from the soil surface as it dries out. Warmer soils will have more ammonia loss than will cool soils.

AVOIDING VOLATILIZATION LOSSES

It is best to fertigate with ammonia or ammonium fertilizers on cool, humid days with little wind. The best way to handle the application of free ammonia fertilizers is to acidify the irrigation water *prior* to injecting the fertilizer. This will reduce the pH of the water and keep more of the nitrogen in the ammonium (NH_4^+) form. Acidification of water is discussed in Chapter 12, "Drip System Maintenance."

Ammonia fertilizers should not be injected through sprinkler systems. Not only will the ammonia be lost into the air more rapidly from each droplet of water, but the ammonia gas can severely damage the leaves. The ammonia molecule has no electrical charge and can freely enter plant cells. Once it is inside the cell it forms OH^- ions and causes the cell pH to rise. This causes a major disruption of the plant physiology and can kill the plants. In fact, ammonia is sometimes intentionally sprayed on the leaves of cotton plants to promote defoliation prior to picking in the Southeastern U.S. Within one day of spraying most of the leaves have died and dropped from the plants.

Little or no ammonia or anhydrous ammonia is fertigated into drip systems in California due to dangers in handling, precipitation, and volatilization problems. These fertilizers are mainly used with furrow irrigation.

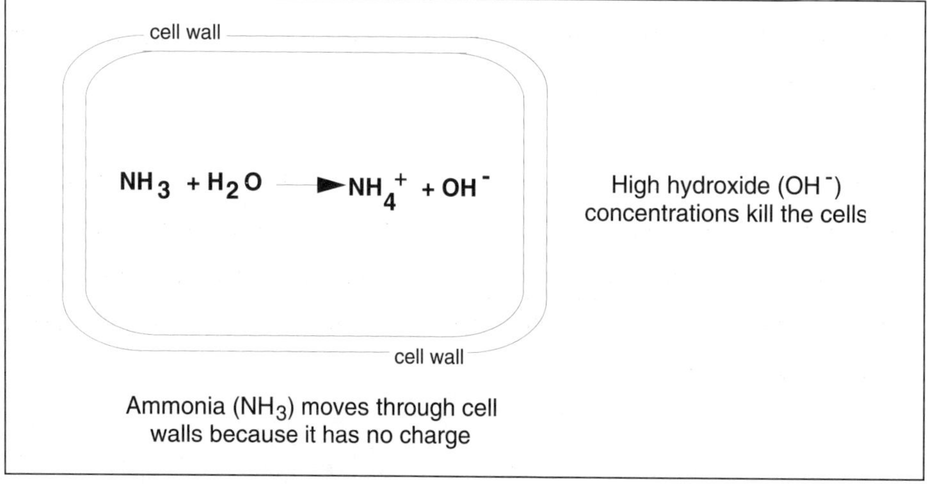

Figure 42. Plant leaves can be damaged if exposed to ammonia.

ORGANIC NITROGEN FERTILIZERS

Organic nitrogen compounds can be from various sources (e.g., crop residues, manure, fishmeal). These nitrogen compounds are not immediately available for plant uptake; they must mineralize (be decomposed by microbes) before they are available for plants. Most organic fertilizers are NOT immediately soluble in water and are not suitable for use in fertigation. The non-synthetic organic fertilizers which are water soluble (i.e., fishmeal) are approximately 10 times more expensive than comparable inorganic fertilizers (with the exception of urea). However, all organic nitrogen fertilizers when applied to the soil undergo microbial degradation which releases ammonium (NH_4^+) ions into the soil. This is true regardless of the form of the organic material which was initially applied to the soil.

UREA

GENERAL

Urea fertilizer ($CO(NH_2)_2$) is a synthetic organic compound which contains 46% nitrogen. Urea is very soluble in water and is well suited for fertigation.

Because urea has no electrical charge and is highly soluble in water, it will move with the water into the soil. Urea will continue to move with the water until it is transformed by microorganisms. The depth it travels into the soil depends on the microbial activity of the soil and the soil texture, but it generally ranges from 1 to 5 inches.

UREA HYDROLYSIS

Urea must undergo a biological transformation in order for the nitrogen to become available to plants. Microorganisms and plants produce the urease enzyme which transforms urea into ammonium and carbonate as illustrated below:

$$\underset{\text{Urea}}{H_2N\text{-}\overset{\overset{\displaystyle O}{\|}}{C}\text{-}NH_2} + \underset{\text{2 Waters}}{2(HOH)} \xrightarrow{\text{UREASE}} \underset{\text{2 Ammonium Ions}}{2\,NH_4^+} + \underset{\text{Carbonate}}{CO_3^{2-}}$$

Urease enzymes can transform all urea into ammonium carbonate within a few hours after application in most soils.

When the urea moves downward and encounters the urease enzyme, it becomes transformed into the ammonium (NH_4^+) ion which becomes held by the clay and organic matter particles in the soil. Thus, once the urea contacts the urease enzyme, it stops moving with the irrigation water and remains in place.

CHAPTER 7. NITROGEN UPTAKE

HIGHLIGHTS

- The form of nitrogen nutrition (NH_4^+ vs. NO_3^-) will affect the amount of other nutrients used by the plant due to cation-anion balance.
- The form of nitrogen nutrition (NH_4^+ vs. NO_3^-) will affect the soil pH because plants exude acids or bases from their roots to balance the charges.
- Nitrate nutrition generally results in quicker visual responses because nitrates are immediately translocated to the leaves once they enter the plant.
- Ammonium, on the other hand, will remain in the roots to fulfill plant requirements if the roots need nitrogen. If the roots do not need nitrogen, it will be transported to the leaves also.
- A balance between ammonium and nitrate nutrition is generally recommended for optimum plant growth.

CATION-ANION BALANCE

The concept of cation-anion balance in plants is a useful mechanism to discuss nitrogen uptake by plants. This concept simply means that the total number of nutrient cations (positively charged ions) in a plant must equal the total number of nutrient anions (negatively charged ions) in a plant. If this were not true, the plant would become electrically charged.

Kirkby (1968) studied the effect of applying either ammonium, nitrate, or urea upon nitrogen uptake by the plant. Because urea has no charge, some urea can directly enter the plant as urea and become converted to ammonium carbonate by the urease enzyme within the plant root cells. However, almost all urea will be converted into ammonium by urease in the soil.

The ammonium ion (NH_4^+) will be in competition with other cations for uptake. Plants may take up primarily ammonium if a nitrification inhibitor is applied, or if the soil is so strongly acidic that nitrification is inhibited. The nitrate ion (NO_3^-) will be in competition with other anions for uptake.

Table 7 illustrates the concept of cation-anion balance: the form of nitrogen nutrition influences other nutrient uptake because the plant must maintain a charge balance. The specific numbers are only used for illustration of this concept.

Table 7. Plant composition as influenced by the type of nitrogen nutrition in order to maintain charge balance.

	Plant Composition	
Source of Nutrition	Cations	Anions
NH_4^+	8 NH_4^+	
	4 K^+	9 $H_2PO_4^-$
	1 Ca^{2+}	3 SO_4^{2-}
	1 Mg^{2+}	1 Cl^-
Sum of Charges	16 +	16 -
NO_3^-		8 NO_3^-
	8 K^+	5 $H_2PO_4^-$
	2 Ca^{2+}	1 SO_4^{2-}
	2 Mg^{2+}	1 Cl^-
Sum of Charges	16 +	16 -

In the above example, a plant receiving nearly all of its nitrogen in the form of ammonium will result in a low to deficient level of potassium (K^+), calcium (Ca^{2+}), and/or magnesium (Mg^{2+}). The plant will compensate the positive charge by absorbing more phosphate ($H_2PO_4^-$) and sulfate (SO_4^{2-}) than normal. In addition, a plant which takes up primarily ammonium nitrogen will produce a more acidic root environment because the roots release H^+ ions to balance the charge from taking up NH_4^+ ions.

In contrast, the high nitrate uptake competes with and decreases the total uptake of phosphate ($H_2PO_4^-$) and sulfate (SO_4^{2-}). The uptake of potassium, calcium and magnesium is enhanced when nitrate is taken up by plant roots.

Nitrogen Source and Effect on Soil pH

A plant which takes up primarily nitrate-nitrogen will produce a more basic root environment because the roots exude bicarbonate HCO_3^- with NO_3^- uptake (Barber, 1974). A high level of nitrate nutrition also promotes the production of a higher level of organic acids in the plant (Hageman, 1984).

NITROGEN MOVEMENT IN THE PLANT

If the ammonium is needed in the root system, it will be used in the roots before any ammonium will be translocated to the stems and leaves. Nitrate is taken up by roots, and is immediately moved upward in the plant to the stems and leaves. In order for the nitrogen to be used by the plant for growth, the plant must convert the nitrate to ammonium, which it does with nitrate reductase enzymes. That conversion is done in the stems and leaves.

Young plants (less than 3 weeks of age) have not developed the nitrate reductase enzyme yet, so there is a preference for ammonium uptake.

Once plants get older, there is generally a quicker visual response to nitrate applications because of the immediate movement of nitrate into the leaves. Also, the nitrate moves more freely in the soil solution than ammonium.

AMMONIUM VS. NITRATE NUTRITION

As explained earlier, ammonium uptake can interfere with potassium uptake and potassium movement within the plant. The result is decreased sugar production and reduced photosynthesis. In addition, the plant will have an increased demand for potassium (Dickinson, 1995). At the end of the season, ammonium may interfere with calcium uptake, which may detrimentally affect the fruit quality. These problems do not occur with nitrate nutrition.

With excessive nitrate uptake, sugar production is increased which then increases the formation of amino acids. In excess, this process is not considered good for healthy crop growth. Excess nitrate nutrition will also interfere with anion uptake.

Consequently, a balance of nitrate and ammonium nutrition is recommended for optimal plant growth. In general, ammonium should not exceed 50% of the total nitrogen supply, and nitrate should not exceed 60% of the total nitrogen supply (Dickinson, 1995).

CHAPTER 8. OTHER NUTRIENT PROCESSES

HIGHLIGHTS

- There are three mechanisms for nutrient uptake: diffusion, root interception, and mass flow.
- Fertigation with microirrigation systems offers new potentials with regards to improving phosphorus and potassium movement in the soil.
- Soil pH is a major factor in the solubility, and therefore the availability, of many nutrients.
- Sulfur does not generally have to be added to California soils because much well water and soils contain sulfur. If the land has been cropped for a long time and sulfur has been "mined," amending with sulfur may be necessary.
- The *balance* of nutrients in the soil solution is critical because an excess of one nutrient may impede uptake of another; some nutrients compete with each other for space on the roots' carrier sites.

MECHANISMS FOR NUTRIENT UPTAKE

Barber and Olson (1968) identified three primary mechanisms for nutrient uptake by plant roots: diffusion, root interception, and mass flow. Diffusion indicates that nutrient ions move from high to low concentration (i.e., from the solution into the root). Root interception suggests that the roots actually come into contact with the ions. Mass flow indicates that ions move in the soil solution to the plant roots in response to the transpiration loss of water from the leaves.

The following nutrients are taken up by plant roots primarily through mass flow: boron, calcium, copper, magnesium, manganese, molybdenum, nitrogen, and sulfur. Diffusion of a nutrient from the soil to the root appears to be most important for phosphorus and potassium, and to a lesser extent for iron and zinc. The uptake of calcium is enhanced by roots intercepting the calcium on the clay particles as the roots grow through the soil (Barber and Olson, 1968).

NUTRIENT INTERACTIONS

Fertilizer applications should not be viewed as isolated, independent actions. Rather, fertilizers alter the chemistry of the soil, and consequently the plant, thereby influencing crop nutrition. Therein lies a fundamental advantage of spoon feeding nutrients through fertigation. Small, incremental fertilizer applications are less likely to alter the nutrient equilibrium in the plants and soil. Growers must remember that an application of one nutrient will affect the uptake of another (Table 8).

Table 8. Documented nutrient interactions. (Adapted from Tisdale et al., 1985)

Column A Uptake of this nutrient	Column B Decreases uptake of these nutrients	Column C Increases uptake of these nutrients
NH_4^+	Mg, Ca, K, Mo	Mn, P, S, Cl
NO_3^-	Fe, Zn	Ca, Mg, K, Mo
P	Cu, Zn	Mo
K	Ca, Mg	Mn (on acid soils)
Ca		Mn (on acid soils)
Mg	Ca, K	Mo
	S	Mo
Fe	Cu, Zn	
Zn	Cu	
Cu	Zn, Mo	
Mn	Zn, Ca, Mo	

There are several possible explanations for the above mentioned interactions. Soil pH may be altered by the form of nitrogen applied, thus affecting the solubility and availability of other nutrients. Some nutrients are similar in size and charge and compete for carrier sites on the root surfaces. Sometimes a plant's need for charge balance will promote uptake of a cation with an anion.

The discussion regarding roots excreting substances in response to either cation or anion uptake to maintain charge balance (discussed in the previous chapter) also holds true for nutrients other than nitrogen. When a plant root takes up a cation, the root exudes hydrogen (H^+) and when it takes up an anion, the root exudes bicarbonate (HCO_3^-).

PHOSPHORUS

GENERAL

Phosphorus (P) provides challenges in nutrient management because it generally reacts with other nutrients to become insoluble and thus unavailable for plant uptake. At pH < 6.0, P reacts with iron (Fe^{3+}), aluminum (Al^{3+}) and manganese (Mn^{2+}) to become unavailable. When pH > 6.5, phosphorus precipitates with calcium (Ca) and magnesium (Mg). The *ideal* pH for optimal phosphorus solubility is 6.5, although realistically, soil pH between 6.0 and 7.0 is acceptable (Tisdale et al., 1985).

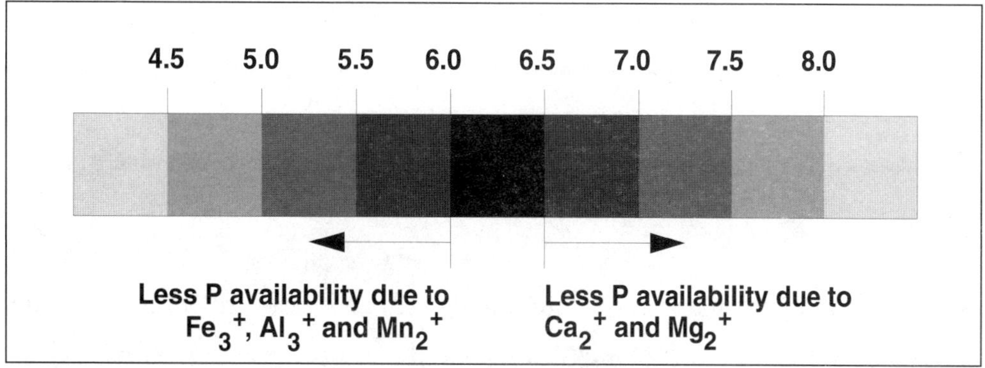

Figure 43. The relationship between soil pH and phosphorus availability.

Phosphorus precipitation in the soil suggests that when phosphorus fertilizers are applied they can become transformed into a less soluble form in the soil in a few minutes, hours, or days, depending on the form, concentration, and placement. Phosphorus generally accumulates in the top eight inches of soil for two reasons: phosphorus fertilizers are generally surface applied, and plants use phosphorus from within the soil profile. This phosphorus accumulates in plant tissues, and when the plant dies, the phosphorus is released through decomposition of organic matter.

PHOSPHATE MOVEMENT

The consequence of the above mentioned reactions is that the phosphorus tends to remain at the surface of the soil, or near a buried emitter. Phosphate applied through sprinklers penetrates 2 to 6 inches into the soil (Lauer, 1988). California's climate dries the soil and prevents appreciable root growth near the soil surface for most crops. Thus, the plants often may not be able to utilize much of the phosphate fertilizer that is applied through a normal fertigation application.

Phosphorus movement is related to soil texture. Phosphorus will move farther in a sandy soil than in a clay soil. This is due to the lack of clay minerals and other substances which would otherwise react with the phosphate to make it insoluble. Researchers found that with an application of 18 pounds P/acre, the phosphorus moved eight inches deep into a loamy sand under drip irrigation. However, for the same conditions with a clay loam the phosphorus moved only one inch below the emitters (Goldberg et al., 1971).

Phosphorus fertigation with drip irrigation systems provides interesting potential for increasing phosphorus penetration into the soil. If very concentrated applications of orthophosphate ($H_2PO_4^-$) fertilizers are applied, it is possible to move phosphates deeper into the soil profile. In another experiment, researchers applied 35 pounds of orthophosphate per acre and studied phosphorus movement below drip irrigation emitters (Rauschkolb et al., 1976). In a Panoche clay loam the phosphorus moved downward 12 inches vertically and 10 inches horizontally. Even though the application rate per *acre* was not particularly large, the rate per *wetted area* was large. This relationship allowed the soil to become saturated with phosphorus. The soil became completely saturated at the surface with phosphate so more phosphorus was able to move further downward in the soil.

An interesting possibility for phosphorus movement occurs with the use of long chain linear polyphosphates in fertigation. Very limited studies at the ITRC have shown that such compounds move into the soil just as nitrate does, which means that it can be distributed throughout the complete root zone. The long chain linear polyphosphates are eventually transformed into orthophosphate through a breakdown by polyphosphatase enzymes which naturally occur in the soil. Polyphosphates of smaller size (e.g., triammonium pyrophosphate $[(NH_4)_3HP_2O_7 \cdot H_2O]$) also may have more mobility than the traditional orthophosphate ($H_2PO_4^-$) forms (Tisdale et al., 1985). The authors of this book anticipate that more field trials will be done in the future with fertigation of various forms of polyphosphates, to determine if there are significant benefits over orthophosphates.

It should be noted that neutral ammonium polyphosphate has been reported to penetrate to only 60 to 70% of the depth of urea phosphate or white phosphoric acid (both of which are strongly acidic).

Phosphorus Application

For sprinklers and surface irrigation methods, it is generally best to apply most phosphate fertilizers as preplant, or during planting, by conventional fertilizer application. With drip irrigation on *permanent* crops, the phosphorus is best supplied through the irrigation system due to root damage associated with tractor application. With drip on *row crops*, most growers apply some or all of the phosphorus as a preplant for two reasons:

1. Preplant phosphorus fertilizers are less expensive.
2. Phosphorus is especially important at seed emergence and transplanting. If rain occurs near that time, the last thing a grower wants to do is irrigate just to apply P. Soil phosphorus also becomes less available for uptake at cooler temperatures. This is another reason to avoid phosphorus fertigation at early growth stages; the irrigation may be unnecessary and cool the soil down. Therefore, there is more of a tendency to find P fertigation in arid climates and sandy soils than in more humid, cool climates.

Some consultants recommend applying phosphorus as a preplant fertilizer, but they should also be prepared to supplement phosphorus through fertigation later in the season. This is because the preplant phosphorus will most likely become insoluble in the soil (and unavailable to the plants) before the end of the growing season. In addition, for long-term crops such as strawberries, the phosphorus is mined from the soil. Since the plants are on plastic, the organic phosphorus (in the plant tissue) will never be recycled into the soil. Additional phosphorus applications are necessary.

Phosphorus Uptake

Phosphorus uptake by plants is very fast for very young plants and tends to decline rapidly as the plant matures (Jungk and Barber, 1975). Chapter 11, "Plant and Soil Testing" details phosphorus utilization by various crops (Table 22).

Soil phosphorus is most available to the plant when the soil is at or near the field capacity moisture condition. When the soil reaches three atmospheres of soil moisture tension (a common condition for the top 6" of soil just prior to irrigation) the uptake efficiency of the phosphorus decreases by 50% (Watanabe et al., 1960).

Potassium

General

Potassium (K^+) occurs in the soil in four forms: exchangeable, soluble, fixed, or mineral. The sum of these forms of K^+ in the soil refers to "Total Potassium." Exchangeable K^+ refers to the potassium ions held on the soil cation exchange sites. Since K^+ is a positively charged cation, it is attracted to the negatively charged soil particles. It stays on the exchange site until another positively charged ion with a stronger attraction (such as NH_4^+) can replace it from the exchange site; K^+ may also leave the cation exchange sites through diffusion when soil solution (soluble) K^+ decreases through plant uptake, but only if some other cation replaces the K^+ on the cation exchange sites.

Soluble K^+ refers to the potassium ions which are in the soil solution (i.e., the water in the soil) and are immediately available for plant uptake. Fixed potassium refers to the K^+ ions which are trapped in between clay layers and are not available for plant uptake. Potassium fixation primarily depends on soil's mineralogy. Soils weathered from mica minerals common in the Sierras and the east side of the Central Valley fix K^+. Mineral K^+ refers to the portion which is tied up in the soil particle and only released by weathering processes. Mineral K+ is not immediately available for uptake; rather, it must be released via weathering before it is available. This process takes years.

These forms of K^+ are present in an equilibrium within the soil complex (Figure 44). When the plant depletes the K^+ from the soil solution, the equilibrium shifts and exchangeable K^+ goes into solution to restore the equilibrium. Similarly, if large potassium fertilizer applications are made, soluble K^+ will become exchangeable K^+ in order to restore the equilibrium. In many soils, soil can not supply sufficient K^+ fast enough during periods of heavy plant uptake. There can be advantages to fertigating potassium on the K^+ release rate limiting soils even though soil levels appear high.

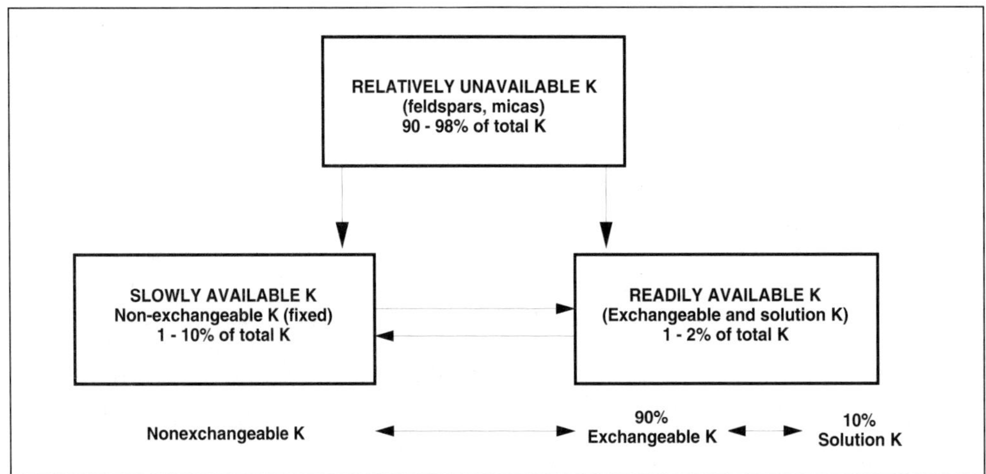

Figure 44. The potassium environment (relationships between mineral, fixed, exchangeable, and solution potassium). Double arrows indicate dynamic equilibrium. (Follett et al., 1981)

POTASSIUM MOVEMENT

Potassium fertilizers will remain primarily at the top of the soil because of adsorption onto the cation exchange sites of the soils.

Researchers found that fertigated potassium at a high concentration (e.g., 190 ppm K) moved two feet downward and spread three feet laterally in a Sycamore clay loam under drip emitters near Colusa (Uriu et al., 1977). On a Wyman clay loam near Gridley, the potassium moved one foot downward and two feet laterally from drip emitters. This deep movement of potassium is only possible with fertigating through drip irrigation. Fertigating potassium with normal sprinkler irrigation usually results in the potassium moving only two to four inches because it is spread over more surface area.

One way to get potassium fertilizers to move deeper into the soil is to saturate all of the cation exchange sites with potassium. The danger with this strategy is that the potassium equilibrium will be upset with super saturation. The K injection rate must be monitored carefully, or the potassium equilibrium may be upset, resulting in an undesirable soil fertility reaction. Some people recommend no more than 40 pounds of K/acre per injection.

POTASSIUM UPTAKE

The plant's need for potassium nearly equals that for nitrogen. Thus, similar concentrations of potassium and nitrogen are found in plant tissue, and similar amounts of these materials need to be supplied to the plant for adequate growth. Fortunately, many California soils have an abundance of potassium. However, this potassium may not be released to the plant root at the rate that a rapidly growing plant takes it up from the soil.

Plant requirements for potassium often reach a peak uptake rate of 5 to 8 pounds of K per acre per day, depending on the crop. Chapter 11, "Soil and Plant Testing," details potassium utilization by various crops (Table 30).

SECONDARY NUTRIENTS

Secondary (Ca, Mg, S) and micronutrient (Cu, Fe, Mn, Zn, Mo, B, Cl) availability are highly pH dependent. Metal micronutrient (Cu, Fe, Mn, and Zn) solubilities decrease at high soil pH; thus, plant availability of these nutrients is low at a high pH. Conversely, sulfur, molybdenum, boron, calcium, and magnesium tend to be more abundant and thus more available at high soil pH. Chlorine is equally available at all pH levels (Figure 45).

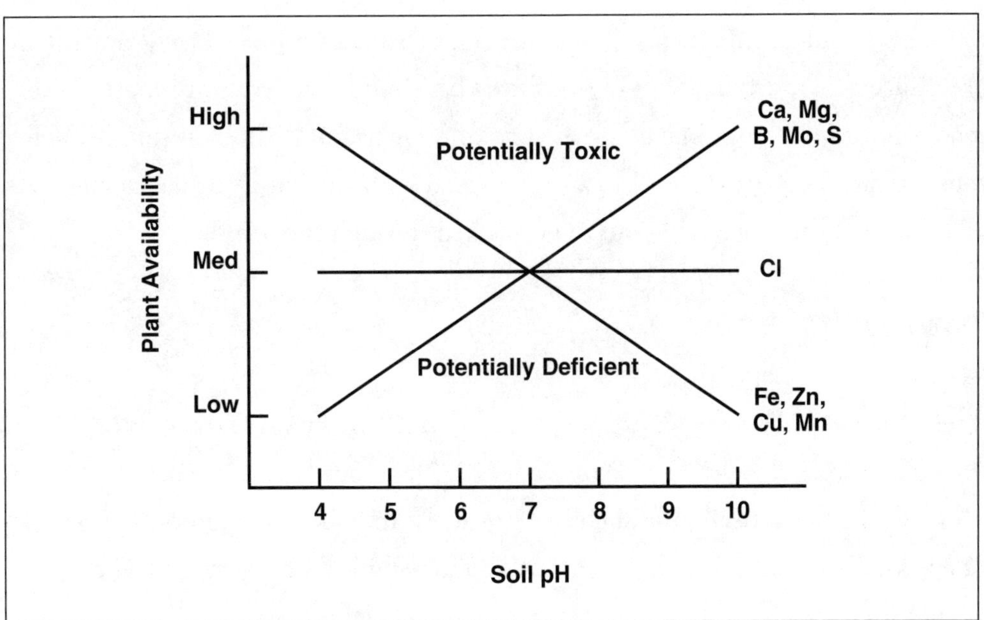

Figure 45. Relationship between soil pH and nutrient availability.

CALCIUM

Calcium is required in high concentrations in the soil to provide for healthy plant roots and strong cell walls (e.g., for firm tomatoes or peppers). Calcium and magnesium often compete with each other due to their similar size and charge; excess magnesium will depress calcium uptake and visa versa. Soil imbalances in ammonium and potassium can also interfere with calcium uptake. Low calcium levels are often found on acidic and/or sandy soils.

MAGNESIUM

Magnesium in a plant-available form is also a divalent cation (Mg^{2+}) and is held on the cation exchange sites. Excessive magnesium in the soil can interfere with both calcium and potassium uptake.

SULFUR

Sulfur is utilized by plants as the sulfate (SO_4^{2-}) ion. Because of its two negative charges, it is highly mobile and moves freely with the water. Elemental sulfur (S) must be oxidized by bacteria (*Thiobacillus*) to become available for plant uptake. This process provides acid as a by-product—often growers will add elemental sulfur to lower the soil pH of high lime soils.

Most well waters contain sufficient sulfur so the soil often does not need to be amended with sulfur in fertilizers. Most valley soils in California have moderate levels of sulfate. If the soils have been cropped for a long time or are in areas of higher rainfall, the soil sulfate levels may be low because sulfate leaches easily. Also, areas irrigated with snow melt water may be lower in sulfate than normal because it will tend to be leached through the soil.

MICRONUTRIENTS

GENERAL

A good reference regarding the nature and use of all micronutrients for crop production is the book *Micronutrients in Agriculture* by Mortvedt (1972). A newly revised edition of this book by Mortvedt et al., 1991 has been published by the American Society of Agronomy (677 South Segoe Road, Madison, WI 53711). It is a very useful technical reference in regards to chemistry and plant physiology.

A micronutrient deficiency should always be confirmed by a soil and plant tissue analysis. Several symptoms can easily be masked by other nutrients or environmental conditions. In addition, it is wise to have a water analysis completed (especially for boron) before making any decision regarding the application of any micronutrient to the soil. Chapter 11, "Plant and Soil Testing" addresses details of soil, plant, and water testing.

CHLORIDE

Chloride (Cl^-) is never applied to alleviate any deficiency *symptom.* It is common in most soils and is too high in some desert soils and in some irrigation waters. Some liquid fertilizers contain Cl^- and growers may inadvertently create chloride toxicities if they do not pay attention to the total composition of their fertilizers. It is equally plant-available at all soil pH values. As an ion, it remains in solution with nearly all compounds. Chloride enhances disease resistance in some corps. Avoid applying chloride to chloride sensitive crops like avocados, berries, and chipping potatoes.

BORON

Boron (B) is slowly leached from most soils. Consequently, it will most likely become deficient on highly leached, sandy soils with a low pH. Boron is found in toxic levels in some desert soils which have formed in association with evaporated sea water. Parts of the west side of the San Joaquin Valley contain high boron soils. The boron in some of these areas is too high to be easily removed by leaching with extra irrigation water. The range between boron deficiency and toxicity is relatively narrow and varies significantly between crops. A deficient level of soil boron for one crop like sugar beets could be a toxic level for another crop. Consequently, it is very important not to over apply any boron material. If in doubt, apply only a portion of the needed boron to determine if the plant actually responds to the application.

Boron materials are relatively few. Boric acid is a common eye wash. It is a very weak acid and has a low solubility which limits its use for some applications. All other common boron materials contain sodium. The amount of sodium being applied with boron compounds is so low that growers should not worry about adding this extra sodium to the soil.

Borax, which was mined from Death Valley, is the most common form of boron. Sodium borate materials include Solubor® and other compounds with different amounts of crystallized water. All of the boron materials should be equally effective for crop production.

MOLYBDENUM

Molybdenum (Mo) becomes more insoluble in leached, acidic soils; thus, Mo is not available for the crop in low pH soils. Most growers correct molybdenum deficiencies by applying lime to the soil, thus increasing the soil pH and thereby increasing molybdenum solubility. A soil test for the lime requirement is the first thing which should be checked before applying molybdenum.

An excessive level of molybdenum is sometimes found on very alkaline soils, especially on those used for grazing livestock or alfalfa production. The high pH increases the amount of soluble molybdenum; this Mo enters the plant and is consumed by the livestock. As a result the animals suffer from a Mo toxicity called molybdenosis (a Mo-induced copper deficiency).

Molybdenum is an integral part of the enzyme systems which perform nitrogen transformations (i.e., nitrogen fixation). Consequently, a molybdenum deficiency can easily be confused with a nitrogen deficiency.

Molybdenum materials are usually quite expensive per pound. They are generally limited to either ammonium molybdate or sodium molybdate. Both are moderately soluble and easily injected into irrigation water.

METAL MICRONUTRIENTS

The metal micronutrients of copper, iron, manganese and zinc all behave similarly in the soil; and, iron, manganese and zinc function similarly in the plant. Thus, some general considerations apply to all four of these nutrients.

A deficiency of these metal ions is most likely to occur in soils with a high pH. This is particularly true for high lime soils. Usually, the pH buffer (greater than 1% free lime) of these soils is so high that it is not economically practical to apply acids to neutralize the lime.

Iron, manganese and zinc function in the plant in various ways, but all are involved with chlorophyll synthesis. A deficiency of one or more of these ions will result in the formation of interveinal chlorosis in the plant. Interveinal chlorosis is not definitive for these metals because a magnesium deficiency also gives the same symptom.

A metal micronutrient toxicity can occur in very acidic soils (soil $pH < 4.5$); such acidic soil conditions dissolve all the metal ions. The continued use of acid-forming or acid-containing fertilizers over a long period of time may cause the soil pH to drop far below what had initially been a normal soil pH.

COPPER

Copper (Cu) functions differently in plants than iron, manganese and zinc. Copper deficiencies are particularly difficult to diagnose if only plant symptoms are used. Cu can be very strongly bound by natural chelates in manures and other organic wastes or by organic soils. Consequently, high organic matter conditions may cause a deficiency of copper.

IRON

Iron (Fe) becomes very insoluble as iron oxide (rust) in soils. This form of iron is unavailable to plants. Most soluble iron salts will immediately form iron oxide when applied to the soil. Thus, iron sulfate will have limited value in the soil. Correcting an iron deficiency will require the use of chelates under most circumstances, particularly with crops grown on high lime soils.

MANGANESE

Manganese (Mn) is often supplied to plants when the soil becomes wetted by irrigation or rain water. Wet soil lacks oxygen and this condition causes manganese oxides to dissolve slightly. The small amount of manganese which does dissolve is often available for plant growth for several days after an irrigation or rain. Thus, manganese deficiencies are less common and tend to correct themselves in the soil unless the parent material of the soil is inherently low in manganese.

Manganese is the metal ion least strongly bound by synthetic chelates. Growers are often disappointed after applying a manganese chelate and not finding an improvement in crop growth. This is explained by the fact that although the manganese was applied to the soil, the chelate dropped the manganese ion and picked up an ion which binds more strongly to the chelate. The other metal ion would be brought to the root and taken up by the plant instead of manganese. When the plant takes up another metal ion it can actually accentuate a manganese deficiency through a metal *imbalance* in the plant. The best way to correct a manganese deficiency is probably to apply manganese oxides, hydroxides, carbonates, or sulfates broadcasted and mixed thoroughly with the soil.

ZINC

Zinc (Zn) is commonly deficient in California soils. Apparently, the rocks and minerals of California are inherently low in zinc; consequently, the soils are low in zinc. This condition often requires a Zn application for adequate crop growth. High lime soils are particularly prone to zinc deficiencies because Zn will complex with carbonates to become insoluble. Organic soils or soils treated with a high loading rate of manure or other organic waste may also result in a zinc deficiency because the natural chelates in the organic material bind strongly to the zinc prohibiting release of the Zn ion for the plant roots.

In California, the correction of a zinc deficiency should be based on incorporating an appropriate amount of zinc material into the soil prior to planting and applying supplements of zinc chelates. Rehm and Hergert (1978) in Nebraska found that fertigation with various zinc materials resulted in the zinc remaining primarily at the soil surface when zinc oxide, zinc sulfate or zinc ammonia complex were used. When zinc was fertigated as the chelated ZnEDTA, the zinc moved downward in the soil into the root zone. This application was more effective for crop uptake than was a surface application of other zinc materials.

Metal Chelates

Chelated micronutrient metals are the most convenient means of applying the metal ions and having them move into the plant root zone. Chelates are organic compounds which wrap around a metal ion and neutralize it electrically. Since the metal chelate is neutral in charge, it can move down into the soil and is unaffected by the soil cation exchange capacity.

Chelates of metal ions work in the soil by moving to plant roots along with the water used during transpiration. At the root surface the root produces a local concentration of acidic hydrogen (H^+) ions which attach to the chelate in exchange for releasing the metal nutrient ion. The root can then take up the metal ion.

The important point with this process is that the chelate without the metal ion is still present. This chelate ion will move away from the root until it encounters another metal ion. The chelate has a very strong affinity for metal ions and will pick up the metal ion, convert it into a metal chelate, and transport the chelate to the root. In this way, the chelates make better use of the micronutrient metal ions already in the soil.

Chelates are very expensive. They sell for about ten times the cost of most normal micronutrient materials. However, as noted in the previous paragraph, chelates make better use of the soil's natural level of the micronutrient.

Chelates have a general preference for binding most tightly to 3^+ charged ions like iron (Fe^{3+}) or aluminum (Al^{3+}). Next, they prefer 2^+ ions. Many of the toxic heavy metals (mercury, nickel, and cadmium) have a 2^+ charge and are bound tightly. The metal nutrients usually are chelated most strongly as copper (Cu^{2+}), followed by zinc (Zn^{2+}), then iron (Fe^{2+}), and finally by manganese (Mn^{2+}), which is the least strongly chelated of the nutrient metal ions. In addition, calcium (Ca^{2+}), magnesium (Mg^{2+}), and hydrogen (H^+) ions can compete with metal ions for a position on the chelate.

Chelates will generally have reduced effectiveness in slightly saline soils. High level of calcium (Ca^{2+}) and magnesium (Mg^{2+}) will compete with the metal ions for binding onto the chelate. Similarly, chelates generally are not very effective in acidic soils; high concentrations of acidic hydrogen (H^+) ions will compete with the metal ions for binding onto the chelate. Chelates become insoluble in very acidic conditions.

Each chelate varies with pH in terms of its ability to chelate metal ions. It is difficult to generalize for all synthetic chelates and metal ions. However, the EDDHA chelate usually binds iron at all pH values. The DTPA chelate binds copper, zinc and iron at pH values commonly found in high lime soils. This will probably be the chelate of choice under these conditions. The EDTA chelate has been used for many years. EDTA is less effective in binding metal ions at higher pH values (above pH 7.5), consequently, EDTA is less effective in high lime soils. Some of the other synthetic chelates are not very effective below pH 6.5.

Natural chelate materials are often made from wood or food processing wastes (e.g., lignosulfates). These natural chelates are often cheaper than the synthetic chelates. However, the natural chelates generally have a lower binding capability for micronutrient metal ions than do the synthetic chelates. Thus, the natural chelates will work but not to the same extent as do the synthetic chelates.

All chelates are organic and are decomposed by soil microorganisms over time. Consequently, the use of chelates requires that more chelate be applied for the next crop.

CHAPTER 9.
SOLUBILITY AND COMPATIBILITY

HIGHLIGHTS

- Fertilizer solubility and purity must be considered when using solid fertilizers to prepare fertilizer solutions.
- Growers must consider compatibility with other fertilizers, the irrigation water, and the irrigation hardware when combining fertilizer materials.
- Several factors may inhibit proper mixing of a solution, improper mixing may lead to inconsistent, incorrect nutrient concentrations and/or irrigation system plugging.
- Certain solid fertilizers will cool the water or fertilizer solution when added to the solution, thus retarding or delaying their solubility.
- Special consideration must be given to the product used and irrigation water quality when injecting phosphorus materials.

FERTILIZER SOLUBILITY

GENERAL

The solubility of various fertilizers must be considered when they are used for injection into irrigation systems. Table 9 provides useful information for some of the most commonly injected fertilizers and agricultural minerals. The data given in the extreme right column of Table 9 shows the amounts of various fertilizer materials that can be dissolved when any given material is dissolved by itself. Combining some materials with others may, indeed, decrease the solubility of the materials involved. There will also be instances when the combining materials have no effect on one another.

DRY FERTILIZER CONDITIONERS

Most dry solid fertilizers are manufactured by coating them with a special conditioner to keep the moisture from being absorbed by the fertilizer pellets. Three common conditioners applied to commercial fertilizers include: attapulgite clay, diatomaceous earth, and hydrated silica. To avoid having these materials create plugging problems, it is best to prepare a small amount of the mix to observe if the coating agent settles to the bottom of the container, if it forms a scum at the surface or if it remains in suspension. The conditioner should rapidly settle to the bottom of the container. If it does, the entire batch can be made. After the conditioner settles, the clear transparent liquid can be drained, pumped, or siphoned from the top portion without disturbing the bottom sediment which contains the conditioner. This procedure will prevent injecting materials that can cause plugging. Some fertilizer dealers carry chemicals that will cause scum to form on the surface, facilitating the removal of the conditioner by skimming.

It should be kept in mind that when using dry fertilizers as ingredient sources, only 100% water soluble dry fertilizer materials should be used to prepare fertilizer solutions. Several dry fertilizer products used for making fertilizer solutions are marketed with or without a protective conditioner. Whenever possible, the "SOLUTION GRADE" form of these products should be purchased to avoid having to deal with the conditioners and the potential plugging problems they can cause. Moreover, if a 100% water soluble dry fertilizer is not available in the "SOLUTION GRADE" form (e.g., ammonium nitrate 33.5% to 34% nitrogen content), it is best to substitute the liquid form of that fertilizer, if possible (e.g., AN-20 in place of dry ammonium nitrate).

COOLING EFFECT WITH MIXING

Most dry solid nitrogen fertilizers will absorb heat from the water when they are mixed. This will result in a very cold liquid which may cause the water in the atmosphere to condense on the outside of the mixing tank. This cold temperature should not pose a problem for injection into irrigation water. The cold fertilizer solution will be mixed with a considerably larger volume of water that will result in fertilizer-treated irrigation water of about the same temperature as the water pumped from the well, irrigation canal, or reservoir.

The real problem from the "cooling effect" is that the total solubility of the fertilizer is usually the least when the solution is very cold. Consequently, it may be difficult to dissolve as much fertilizer as expected. If this is the case, it will be necessary to warm the solution by allowing the mixture to stand for several hours or by continuously recirculating the mix until all the fertilizer is dissolved. This will allow the desired amount of fertilizer to be mixed into the solution. If the ambient temperature is low enough to prevent the solution from warming, it would be best to prepare a fertilizer solution of lower concentration. For example, in warm climates it is possible to make a 2-0-8 solution using potassium nitrate. In cool climates, a 2-0-8 is difficult to prepare, so the analysis of the solution can be reduced to a 1-0-4. This presents no problem because the solution can be applied at twice the intended rate of the 2-0-8, thereby achieving the same level of applied nutrients as originally intended.

Table 9. Solubility information on various fertilizer compounds commonly used to prepare fertilizer solutions and/or for application through irrigation water. (California Fertilizer Association, 1980)

	Grade	Form	Temp °F	Solubility gm/100 ml	Solubility pounds/gal
Nitrogen Fertilizers					
Ammonium Nitrate	34-0-0	NH_4NO_3	32	18.3	9.87
Ammonium Polysulfide	20-0-0	NH_4S_x		high	high
Ammonium Sulfate	21-0-0	$(NH_4)SO_4$	32	70.6	5.89
Ammonium Thiosulfate	12-0-0	$(NH_4)_2S_2O_3$		v. high	v. high
Anhydrous Ammonia	82-0-0	NH_3	59	38.0	3.17
Aqua Ammonia	20-0-0	NH_3 H_2O/NH_4OH		high	high
Calcium Nitrate	15.5-0-0	$Ca(NO_3)_2$	62	121.2	10.11
Urea	46-0-0	$CO(NH_2)_2$		100.0	8.34
Urea Sulfuric Acid	28-0-0	$CO(NH_2)_2 \cdot H_2SO_4$		high	high
Urea Ammonium Nitrate	32-0-0	$CO(NH_2)_2 \cdot NH_4NO_3$		high	high
Phosphate Fertilizers					
Ammonium Phosphate	8-24-0	$NH_4H_2PO_4$		moderate	moderate
Ammonium Polyphosphate	10-34-0	$(NH_4)_5P_3O_{10}$ & others		high	high
Ammonium Polyphosphate	11-37-0	$(NH_4)_7P_5O_{16}$ & others		high	high
Phosphoric Acid, green	0-52-0	H_3PO_4		45.7	high
Phosphoric Acid, white	0-54-0	H_3PO_4		45.7	high
Potash Fertilizer					
Potassium Chloride	0-0-60	KCl	68	34.7	2.89
Potassium Nitrate	13-0-44	KNO_3	32	13.3	1.10
Potassium Sulfate	0-0-50	K_2SO_4	77	12	1.00
Potassium Thiosulfate	0-025-17S	$K_2S_2O_3$		150	12.5
Monobasic Potassium Phosphate	0-52-34	KH_2PO_4		33	2.75
Micronutrients					
Borax	11% B	$Na_2B_4O_7 \cdot 10H_2O$	32	2.10	0.17
Boric Acid	17.5% B	H_3BO_3	86	6.35	0.53
Solubor	20% B	$Na_2B_8O_{13} \cdot 4H_2O$	86	22	1.84
Copper Sulfate (acidified)	25% Cu	$CuSO_4 \cdot 5H_2O$	32	31.6	2.63
Cupric Chloride (acidified)		$CuCl_2$	32	71	5.93
Gypsum	23% Ca	$CaSO_4 \cdot 2H_2O$		0.241	0.02
Iron Sulfate (acidified)	20% Fe	$FeSO_4 \cdot 7H_2O$		15.65	1.31
Magnesium Sulfate	9.67% Mg	$MgSO_4 \cdot 7H_2O$	68	71	5.93
Manganese Sulfate (acidified)	27% Mn	$MnSO_4 \cdot 4H_2O$	32	105.3	8.79
Ammonium Molybdate	54% Mo	$(NH_4)_6Mo_7O_{24} \cdot 4H_2O$		43	3.59
Sodium Molybdate	39% Mo	Na_2MoO_4	44.3		3.70
Zinc Sulfate	36% Zn	$ZnSO_4 \cdot 7H_2O$	68	96.5	8.05
Zinc Chelate	5% – 14% Zn	DTPA & EDTA		v. sol.	v. sol.
Manganese Chelate	5% – 12% Mn	DTPA & EDTA		v. sol.	v. sol.
Iron Chelate	4% – 14% Fe	DTPA, HOEDTA & EDDHA		v. sol.	v. sol.
Copper Chelate	5% – 14% Cu	DTPA & EDTA		v. sol.	v. sol.
Zinc Lignosulfonate	6% Zn	Lignosulfonate		v. sol.	v. sol.
Manganese Lignosulfonate	5% – 14% Mn	Lignosulfonate		v. sol.	v. sol.
Iron Lignosulfonate	6% Fe	Lignosulfonate		v. sol.	v. sol.
Copper Lignosulfonate	6% Cu	Lignosulfonate		v. sol.	v. sol.
Lime Sulfur		$CaS_5+Ca_2SO_3 \cdot 5H_2O$		high	high
Sulfuric Acid	95%	H_2SO_4		v. high	v. high

Dry fertilizer products that significantly lower the temperature of the water they are being dissolved in are urea, ammonium nitrate, calcium nitrate, and potassium nitrate.

FERTILIZER COMPATIBILITY

GENERAL

It should be noted that compatibility problems may occur only when liquid fertilizers are combined, whether or not the liquid fertilizers were originally in liquid form, or were prepared from soluble dry materials. When preparing fertilizer solutions from various ingredient sources, one must consider the following:

1. The safety involved in making the solutions.
2. The effects of the liquid fertilizer solutions upon one another when added to the same tank.
3. The reactions of the liquid fertilizers within the irrigation system.
4. The type of irrigation system and its susceptibility to plugging or other problems.

Figure 46 shows compatibilities of some fertilizer products. If there is any question about the safety of making a mix or the compatibilities of the ingredients, it would be prudent to consult fertilizer dealers or local Farm Advisors. A good rule of the thumb is: *WHEN IN DOUBT, LEAVE IT OUT!*

THE JAR TEST

A "jar test" involves putting some of the fertilizer solution into a jar of irrigation water, and watching for any precipitate or milkiness which may occur within one to two hours. If cloudiness does occur, there is a chance that injection of that chemical will cause line or emitter plugging. If different fertilizer solutions are to be injected simultaneously into the irrigation system, mix the two in a jar. When testing the compatibility or acceptability of any fertilizer solution or combination of solutions, use the approximate dilution rate that would be expected when the products are injected into the irrigation system. For example, assume a fertilizer solution is to be injected at the rate of 30 gallons per hour into an irrigation system delivering 1200 gallons per minute to the field. The dilution rate would be 1 to 2400. One (1) milliliter of the fertilizer solution added to 2400 milliliters of water would be the same as the intended field injection rate. Be sure to wear protective clothing and safety glasses when performing any jar test.

NOTE	It is best to make a small quantity of the solution to test fertilizer compatibilities *before m* preparing large batches of a fertilizer solution. Before injecting the fertilizer solution into the irrigation system one should ***always* perform a jar test.**

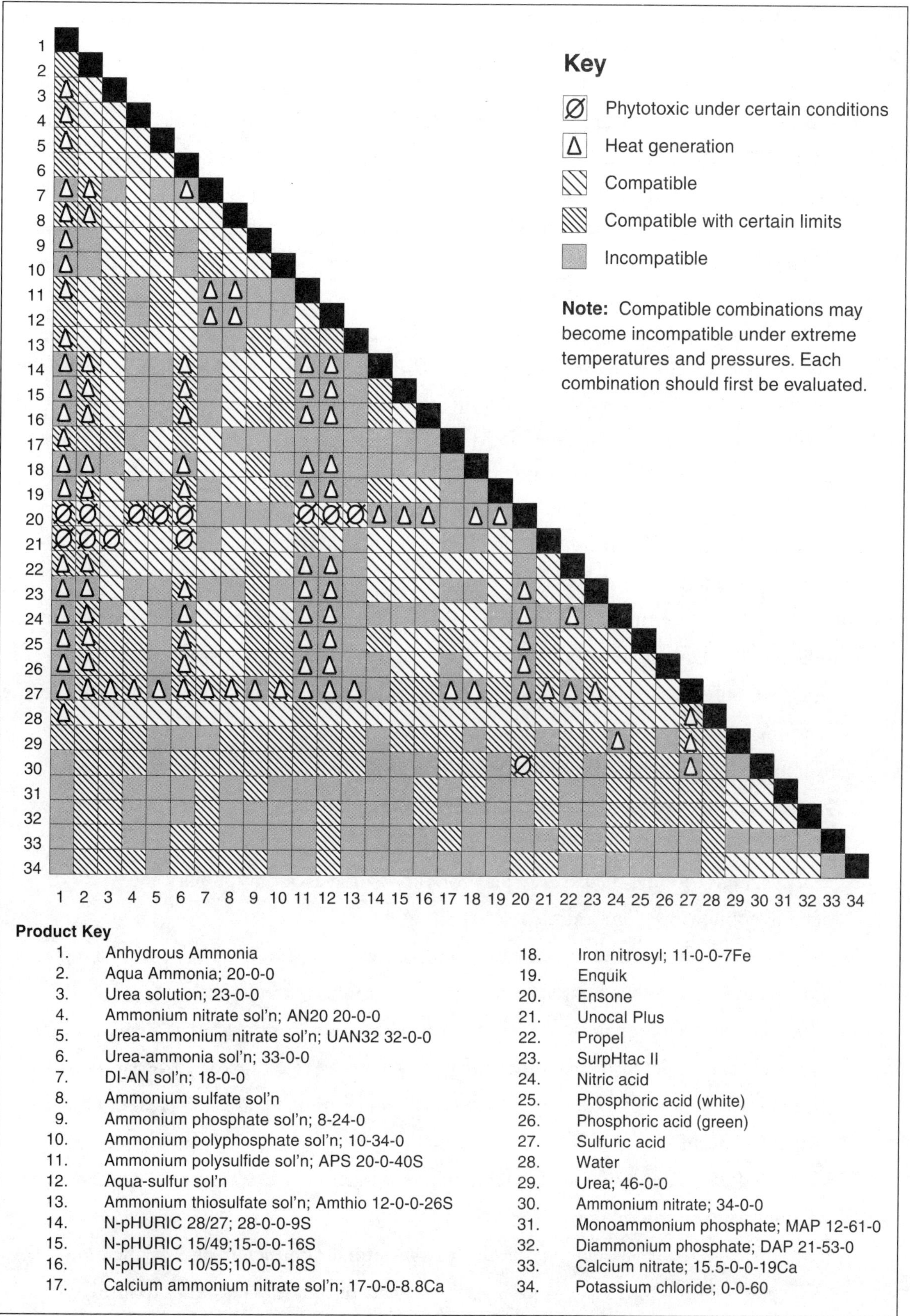

Figure 46. Compatibility chart for common fertilizers and agricultural minerals. (Courtesy of Unocal)

BASIC MIXING RULES

Some of the basic mixing rules of compatibility include the following:

1. ***Always*** fill the mixing container with 50 – 75% of the required water to be used in the mix.
2. ***Always*** add the liquid fertilizer materials to the water in the mixing container before adding dry, soluble fertilizers. The additional fluid will provide some heat in case the dry fertilizers have the characteristic of making solutions cold.
3. ***Always*** add the dry ingredients slowly with circulation or agitation to prevent the formation of large, insoluble or slowly soluble lumps.
4. ***Always*** **put acid into water, not water into acid.**
5. When chlorinating water with chlorine gas, ***always*** add chlorine to water, and not vice versa.
6. ***Never*** **mix an acid or acidified fertilizer with chlorine**, whether the chlorine is in the gas form or liquid form such as sodium hypochlorite. A toxic chlorine gas will form. *Never* store acids and chlorine together in the same room.
7. *DO NOT* attempt to mix either anhydrous ammonia or aqua ammonia directly with any kind of acid. The reaction is violent and immediate.
8. *DO NOT* attempt to mix concentrated fertilizer solutions directly with other concentrated fertilizer solutions.
9. *DO NOT* mix a compound containing sulfate with another compound containing calcium. The result will be a mixture of insoluble gypsum. For example, injecting both calcium nitrate and ammonium sulfate fertilizers into the same irrigation water will cause the formation of calcium sulfate (gypsum)—calcium sulfate has a very low solubility. Although the calcium nitrate is very soluble and the ammonium sulfate has good solubility, they create problems when mixed together in the same container or when poured together from separate mixing tanks. Gypsum crystals will form and can clog drip emitters or filters.
10. ***Always*** check with the chemical supplier for information about insolubility and incompatibility (See Figure 46).
11. Be extremely cautious about mixing urea sulfuric fertilizers (e.g., N-pHURIC®) with most other compounds. Urea sulfuric is incompatible with many compounds.
12. Since fertilizer solutions are applied in very small dosages, and if injected at separate locations in the irrigation line, many incompatibility problems tend to disappear. The jar test is essential when it comes to deciding if solutions can be simultaneously injected into the irrigation system.
13. *DO NOT* mix phosphorus containing fertilizers with another fertilizer containing calcium without first performing the jar test.
14. Extremely hard water (containing relatively large amounts of calcium and magnesium) will combine with phosphate, neutral polyphosphate or sulfate compounds to form insoluble substances.

NITROGEN

GENERAL

Urea, ammonium nitrate, calcium nitrate, potassium nitrate, ammonium sulfate, ammonium phosphate 21-53-0, and ammonium phosphate 12-61-0, are very soluble in water. These nitrogen containing fertilizer materials are readily available on the market and are used extensively in the preparation of single nutrient or multinutrient fertilizer solutions. The right hand column in Table 9 shows the amounts of these fertilizer materials that can be dissolved in one gallon of water.

As a general rule, nothing should be mixed with *concentrated* urea sulfuric acid, especially where the pH is less than 4.5 (Purdy, 1994). There is more of a risk of producing reactions that result in the formation of insoluble, gummy substances than causing hazardous reactions. There are two exceptions regarding the compatibility of urea sulfuric acid with other fertilizer materials: Urea sulfuric acid can be combined in certain proportions with either phosphoric acid, or ammonium ortho/poly phosphate 10-34-0. These two materials can be mixed together in certain proportions without adverse affects occurring. It is more common to combine urea sulfuric acid in association with other compatible fertilizers in aqueous mixes. Apparently the dilution with water provides a greater latitude in compatibilities than when concentrated urea sulfuric acid and other concentrated fertilizer products are directly mixed together. In practice, some growers will inject low rates of urea sulfuric acid while simultaneously injecting other fertilizer materials in the irrigation water, but the injection port of urea sulfuric acid must be well located away from the other fertilizer injection port.

Urea sulfuric acid can be mixed with phosphoric acid, sulfate sources of micronutrients, muriate of potash, and other chloride-containing materials. The proper order of mixing must be followed though when combining fertilizer materials with urea sulfuric acid. The next chapter addresses specific fertilizers and describes in detail the compatibility rules and mixing procedures for urea sulfuric acid.

LIMING EFFECT WITH AMMONIA FERTIGATION

When anhydrous ammonia or aqua ammonia fertilizers are injected into irrigation water, the ammonium is quite water soluble. However, the pH of the water is raised, and some of the ammonia will escape and volatilize into the atmosphere. In addition, the high pH of the irrigation water will cause the calcium and magnesium ions in the water to precipitate as calcium and magnesium hydroxides and carbonates (e.g., $Ca(OH)_2$, $CaCO_3$).

This precipitate is lime in a very finely divided form. The harder the water is, the more lime precipitates. This precipitate will clog drip lines and filters, as well as valves, gates, and sprinklers. In addition, this fine precipitate will tend to clog soil pores over time, reducing the water intake rate into the soil.

Hard water sources used for irrigation are desirable for crop production. They usually have a good balance between the concentrations of calcium, magnesium, and sodium ions. The formation of lime precipitates usually indicates a reduction in the quality of the irrigation water. The soluble calcium and magnesium are taken out of solution. With the loss of soluble calcium and magnesium, sodium salts become the dominant species. Sodium is associated with water intake problems, and it is possible that fields flood or furrows irrigated with water containing ammonia will show water standing between the borders or in the furrows longer than usual. This effect is slowly reversed with subsequent applications of fresh irrigation water.

This precipitation effect is commonly overcome by using phosphorus containing water conditioners, such as rose-stone (Calgon®) and other products marketed for this purpose. These products are added to the water upstream to the point of ammonia fertilizer injection. Refer to Chapter 13, "Infiltration Problems" for more information.

Phosphorus Solubility

All commercial fertilizers contain the guaranteed percentage of P_2O_5 on the label as water soluble and citrate soluble phosphate. This indicates that under laboratory conditions, a certain amount of phosphate will dissolve in water and in ammonium citrate (a solution similar to orange juice with household ammonia added). Citrate soluble phosphate is considered to be available for crop uptake during the growing season even though it is not immediately water soluble.

Most dry phosphorus fertilizers marketed for general farm application *cannot* be injected into irrigation water because they are too insoluble. All farm-grade ammonium phosphate fertilizer and all superphosphate fertilizers are too low in solubility to be of value in fertigation.

Monoammonium (MAP) phosphate 12-61-0, diammonium phosphate (DAP) 21-53-0, monobasic potassium phosphate 0-52-34, phosphoric acid, urea phosphate, liquid ammonium polyphosphate 10-34-0, and long chain linear polyphosphates represent several water soluble or water miscible phosphate fertilizers. Nevertheless, they can still have precipitation problems when injected at high application rates into hard irrigation water.

Many farmers who inject phosphorus through drip and microirrigation systems should use acidic forms of phosphorus fertilizers rather than neutral forms. Some growers use phosphoric acid either alone, or in combination with other fertilizer materials. Phosphoric acid not only provides phosphorus and lowers the pH of the water; it also helps to keep the emitters clean of bacterial growth. White phosphoric acid has fewer impurities compared to green phosphoric acid. White phosphoric acid should be the preferred form when injecting in drip irrigation systems, but its availability on the market and considerably higher cost compared to green acid limit its use to very specific situations. Some growers have reported emitter plugging problems with the use of green acid, but plugging is more associated with injection rates too low to offset the buffering effects of the calcium and magnesium content of the water. Phosphoric acid injection will be effective only as long as the pH of the fertigated water remains very low. As the pH rises (due to dilution with the irrigation water) the phosphate precipitates with the calcium and magnesium. Some growers supplement phosphoric acid injections with urea sulfuric acid to insure that the irrigation water pH will remain low (3.0 or lower).

Caution should be exercised with either of these practices because low pH values (below 5.5) can increase corrosion of metal hardware, increase toxicity of certain micronutrients, or damage plant roots.

Urea phosphate materials are quite excellent fertilizers, although their availability on the market is not particularly extensive. They have a low enough pH to prevent calcium and magnesium precipitation. In addition, they are excellent N source. They can also retard P precipitation in highly calcareous soils which improves phosphorus availability in the soil (Purdy, 1994).

As previously mentioned, phosphorus reacts with many other combinations to create insoluble precipitates. These reactions must be considered when combining phosphorus fertilizers with other fertilizers. The following are general rules regarding phosphorus compatibility:

1. Phosphorus and calcium, when in solution together, form di- and tricalcium phosphate which are insoluble phosphates. These phosphate forms can clog irrigation lines. Calcium can come either from the irrigation water or other fertilizers.
2. Phosphorus, ammonium-nitrogen, and magnesium, when in solution together, sometimes can form magnesium phosphates or magnesium ammonium phosphates, which are insoluble and clog lines. Magnesium can come either from the irrigation water or other fertilizers.
3. Phosphorus and iron, when in solution together, can form iron phosphates which are insoluble. However, most waters contain only low levels of iron, and this is rarely a problem. Controlling (lowering) the irrigation water pH or using long chain linear meta- or polyphosphates can avoid the problems with iron phosphate precipitation.

4. For ammonium ortho/polyphosphate materials (10-34-0 and 11-37-0) and monoammonium and diammonium phosphates (12-61-0 and 21-53-0), the quality of the irrigation water must be known before attempting to inject these fertilizers. These phosphorus materials react in hard water to form precipitates which cause plugging. Consequently, the water hardness must be ascertained before injecting these materials. Combined Ca and Mg should remain below 50 ppm and bicarbonate should remain less than 150 ppm. The concentration of Ca and Mg can be as high as 75 ppm if the bicarbonate concentration is less than 100 ppm with a neutral P source (Purdy, 1994).

POTASSIUM

All potassium fertilizers are water soluble. Potassium sulfate (K_2SO_4) can be run through various gypsum injecting machines by itself, but it might form a mealy substance if it is run together with the gypsum in high calcium waters.

Potassium thiosulfate (KTS) is compatible with urea and ammonium polyphosphate solutions in any ratio. Potassium thiosulfate should not be mixed with acids or acidified fertilizers. When blended with UAN solution, a jar test is recommended before mixing large quantities. Under certain mixing proportions, particularly when an insufficient amount of water is used in the mix, potassium in KTS can combine with nitrates in the mix to form potassium nitrate crystals. If this happens, adding more water and/or heating the solution should bring the crystals back into solution.

CALCIUM

Fertilizers containing calcium should be flushed from all tanks, pumps, filters, and tubing prior to injecting any phosphorus, urea-ammonium nitrate, or urea sulfuric fertilizer. The irrigation lines must be flushed to remove all incompatible fertilizer products before a calcium containing fertilizer solution is injected. Calcium should not be injected with any sulfate form of fertilizer. It combines to create insoluble gypsum.

MICRONUTRIENTS

Several metal micronutrient forms are relatively insoluble, and therefore not used for fertigation purposes. These include the carbonate, oxide, or hydroxide forms of zinc, manganese, copper, and iron. These relatively inexpensive materials can be broadcast and incorporated into the soil. However, they constitute a long term source of micronutrients and will supply only a low level of nutrients for many years.

The sulfate form of copper, iron, manganese and zinc is the most common and usually the least expensive source of micronutrients. These metal sulfates are water soluble and are easily injected into irrigation water. However, using these materials for fertigation is not very successful in alleviating a micronutrient deficiency when applied through sprinklers, furrows, and border strips. The reason is that the metal ion has a strong electrical charge (2^+) and becomes attracted to the cation exchange sites of clay and organic matter particles where it tends to sit near the soil surface. Consequently, the micronutrient usually does not reach the major plant root zone. With buried drip, sulfates can be effective.

The soil pH controls the availability of sulfate forms of manganese, iron, copper, and zinc. Although manganese sulfate, iron sulfate, and copper sulfate are sufficiently soluble for preparing solutions for injection into drip systems, if the soil pH is high (i.e., pH 7.4 or higher) manganese, iron, and copper are changed into unavailable forms and little or no benefit will be obtained from their use. If the metal sulfate solutions are acidified, however, the availability of the micronutrient can be prolonged in the soil. Zinc is less susceptible to the effects of soil pH, and therefore zinc sulfate can be a very effective source of this micronutrient.
Some of the metal micronutrients may be available as the chloride or nitrate forms. These forms are very soluble and expensive. They can be easily used for fertigation, but like the sulfate forms, the metal ion will accumulate at the soil surface rather than moving into the soil. Chloride and nitrate metal compound forms are much more corrosive and special handling procedures must be used with these materials when they are in concentrated solutions.

Micronutrients can be applied as foliar applications to supplement micronutrients injected into the irrigation system. The mixes currently injected are chelate type (EDTA, DTPA, EDDHA (iron, only) or lignosulfonate complexes of the micronutrient metals) or inorganic metal sulfate solutions. Chelated metals of the EDTA, DTPA, and EDDHA types are not compatible with acids or acidified fertilizer mixes. Lignosulfonates have limited compatibility with acids or acidified fertilizers. Inorganic metal sulfates are usually not compatible with neutral fertilizer mixes, but are compatible with acids and acidified fertilizers up to about 1.5% metal content of the mix on a weight-to-weight basis. Again, a small amount of micronutrient/macronutrient mix should be made to ascertain compatibility. Once compatibility is assured, the micronutrients can be applied according to the needs of the crop. Growers contacted by the ITRC have mentioned no problems with properly formulated mixes.

There has been a problem with mixing certain zinc chelates with CAN-17 and storing this mix overnight. Variation in the formulation process of making EDTA zinc chelates appears to be associated with the incompatibility. The problem is that there is a formation of insoluble compounds that settles on the bottom and sides of the tank. Because all brands of EDTA zinc chelate are not incompatible with CAN-17, the jar test is the appropriate method of determining compatibility. However, separate and simultaneous injection of CAN-17 and zinc chelates does not result in precipitation problems. This is due to the extensive dilution that occurs. The problem only arises when the incompatible brands of EDTA zinc chelate are stored together with CAN-17. If growers want to inject CAN-17 and zinc chelate as a mix, they should do so immediately after receiving the product (Dickinson, 1995). When growers order an EDTA/CAN-17 mix from their fertilizer dealer, they should require the CAN-17 and zinc chelate mix be prepared within only a few hours before delivery and then use the mix promptly.

Corrosion

Fertilizers and other injected chemicals can be corrosive to irrigation equipment. Some classic work was done by Martin (1953) with six metals immersed in eight different fertilizer solutions for four days. Table 10 shows the results of his work.

Table 10. Relative Corrosion of Various Metals (Martin, 1953).

Kind of Metal	Calcium Nitrate	Sodium Nitrate	AN-20	Ammonium Sulfate	Urea	Phos. Acid	Di-Ammon. Phos.	17-17-0 mix*
Galvanized Iron	2	1	4	3	1	4	1	2
Sheet Aluminum	No	2	1	1	No	22	2	1
Stainless Steel	No	No	No	No	No	1	No	No
Phospho-Bronze	1	No	3	3	No	2	4	4
Yellow Brass	1	No	3	2	No	2	4	4
pH of Fertilizer Solution	5.6	8.6	5.9	5.0	7.6	.4	8.0	7.3

No, none; 1, slight; 2, moderate; 3, considerable; 4, severe.
Solutions made by dissolving 100 pounds of material in 100 gallons of water.
* Mix of ammonium sulfate, diammonium phosphate, and potassium sulfate

The table above indicates that ammonium nitrate, phosphoric acid, and ammonium sulfate are all quite corrosive. Sprinkler heads and bearings made of brass and bronze are corroded by phosphate, particularly in the presence of ammonium (Viets et al., 1967).

Copper is very corrosive to aluminum in even small dosages, and copper compounds (such as copper sulfate) should be avoided with all aluminum equipment. Care must be taken to not inadvertently use copper sulfate-treated reservoir or canal water.

There are several grades of stainless steel, the most common of which is No. 304. No. 304 stainless steel is generally an unsuitable material for fertilizer tanks because of corrosion. No. 316 is required for fertilizer tanks, and should be considered for many filter designs.

CHAPTER 10.
SPECIFIC FERTILIZERS AND AGRICULTURAL MINERALS

HIGHLIGHTS

- Clear, liquid fertilizers are acceptable for fertigation.
- Many manufacturers and/or fertilizer dealers are "color coding" their products to identify them, but the industry has not standardized the colors yet.
- Suspension fertilizers are not appropriate for application through closed irrigation systems, and rarely are used for application through flood or furrow irrigation systems.

Most liquid fertilizers which are used for fertigation are clear liquids. This allows them to be applied with any irrigation program which the grower may choose to use.

Some fertilizer manufacturers recently have begun to color code all of their liquid products. There is not yet a uniform, accepted color standard in the industry. The color usually consists of an organic dye which is added in trace amounts. The colors indicate the kind of fertilizer, i.e., UN-32, AN-20, CAN-17, urea sulfuric acid, etc. The color of a fluid fertilizer has no relationship to its compatibility with other colored fluid fertilizers. It is essential to be thoroughly familiar with the chemical characteristics and specifications of any given fertilizer in order to assess its compatibility with other fertilizers. Sometimes the color is used to distinguish among the three different grades of a fertilizer material, e.g., urea sulfuric acid grades 28/27, 10/55, or 15/49.

Suspension fertilizers are sometimes called "liquid fertilizers," and while they are fluid in nature, they are not true fertilizer solutions. They are actually suspensions consisting of solid fertilizer materials suspended in a solution of other fertilizer nutrients. Suspension fertilizers were devised by the Tennessee Valley Authority's National Fertilizer Development Center a number of years ago to provide growers with a means of applying high analysis fertilizers in a fluid form.

Suspension fertilizers are formulated by adding large amounts of dry fertilizers to a limited volume of water, to the extent that the solubility of those fertilizers are exceeded. The excess, non-solubilized fertilizer is then suspended in the solution. Attapulgite clay is often added as the gelation agent which keeps the fine crystal solid particles suspended. Suspension fertilizers are not suitable for application through closed irrigation systems such as drip or sprinkler systems and rarely, if at all, are they applied via flood or furrow irrigation. Suspension fertilizers are intended to be applied with ground applicators fitted with high pressure pumps, booms, and flood-type nozzles.

Trends in Fertilizer Use

The type of fertilizer applied in the irrigation water varies depending on region and crop. For the most part, the most popular fertilizers for fertigation are CAN-17, AN-20, UN-32, urea sulfuric acid, and specialty formulations. ITRC conducted a limited survey of major fertilizer dealers in different areas of California regarding the extent of fertilizer sales for *fertigation*. Table 11 summarizes the findings.

Table 11. A summary of the fertilizers sold specifically for fertigation in California according to a 1994 ITRC survey conducted of major fertilizer dealers.

Fertilizer	Grade	Cost Range	Unit	Number of Dealers Responding Relative amount sold for fertigation			
		($)		V. Much	Avg.	Little	None
Ammonium Bisulfate*	8-0-0-11	170	ton			2	5
Ammonium Molybdate							7
Ammonium Nitrate	34-0-0	200	ton		1		6
Ammonium Nitrate Solution	20-0-0	160 – 210	ton	2	1	1	3
Ammonium Phosphate	8-24-0					1	6
Ammonium Polyphosphate	10-34-0	250 – 400	ton	1	4	2	
Ammonium Polyphosphate	11-37-0						7
Ammonium Polyphosphate	7-21-7					7	
Ammonium Polysulfide	20-0-0-45					1	6
Ammonium Sulfate	21-0-0	170	ton		1	15	
Ammonium Thiosulfate	12-0-0-26	168 – 220	ton	1	2	2	2
Borax							7
Boric Acid							7
Calcium Ammonium Nitrate	17-0-0	173 – 300	ton	5	1		1
Calcium Nitrate	15.5-0-0	200 – 351	ton	2		2	3
Calcium Polysulfide		270	ton		1		6
Concentrated Super Phosphate	0-45-0						7
Copper Sulfate		1.50	lb			1	6
Cupric Chloride							7
Diammonium Phosphate	18-46-0						7
Gypsum	0-0-0	121	ton		1	1	5
Iron Sulfate	0-0-0	251	ton			1	6
Lime Sulfur	0-0-0	230	ton		1	1	5
Long Polyphosphate Chain							7

continued

Table 11. Continued.

Fertilizer (see note)	Grade	Cost Range ($)	Unit	Number of Dealers Responding Relative amount sold for fertigation			
				V. Much	Avg	Little	None
Magnesium Sulfate	0-0-0	0.47	lb			1	6
Manganese Sulfate	0-0-0						7
Monoammonium Phosphate	11-48-0						7
Monoammonium Phosphate	11-55-0						7
Phosphoric Acid (white)	0-54-0	1,000	ton			1	6
Phosphoric Acid (green)		600	ton			1	6
Potassium Chloride	0-0-60	100 – 137	ton		1	2	4
Potassium Nitrate	13-0-44	570 – 607	ton			2	5
Potassium Phosphate	0-10-6-14.5						7
Potassium Sulfate	0-0-50	354	ton			1	6
Potassium Thiosulfate (KTS)	0-0-25-17	491	ton			1	6
Sodium Molybdate	0-0-0						7
Solubor	0-0-0	1.40	lb			2	5
Sulfur Dioxide Gas	0-0-0						7
Sulfuric Acid	0-0-0	110	ton			1	6
Superphosphoric Acid	0-68-0						7
Urea	46-0-0	220 – 300	ton		2	1	4
Urea Ammonium Nitrate	32-0-0	200 – 270	ton	5	2		0
Urea Phosphate	17-44-0						7
Urea Phosphate	10-34-0						7
Urea Solution	20-0-0	190	ton			1	6
Urea Sulfuric Acid	10-0-0-18	300	ton	1			6
Urea Sulfuric Acid	15-0-0-49	183 – 240	ton	1	2	1	3
Urea Sulfuric Acid	28-0-0-9						7
Zinc Sulfate	0-0-0	224 – 315	ton			2	5
OTHERS:							
CN-9	9-0-0-11	150	ton	1			
Special Blends	18-18-18	1.00	lb	1			
	5-0-12	150	ton	1			
	8-8-8	200	ton	1			
	10-0-10	170	ton	1			
	12-0-9	180	ton	1			
Zn Chelate (9%)		11.00	gal		1		
Fe Chelate (4.5%)		15	gal		1		
Mn Chelate (6%)		16	gal		1		

* Two dealers responded that they sold a "little" ammonium bisulfate for fertigation; five dealers responded that they sold "none" for fertigation.

Complete fertilizers (i.e., 8-8-8, 4-10-10, 5-15-5) are popular for applications on vegetable crops, fruit, tree and vine grown in California. With the added concern about crop quality, growers are also using more complicated formulations which often include micronutrients. For example, some lettuce growers in the Salinas Valley are applying zinc, manganese, iron, and copper on lettuce to improve yields, increase head uniformity, and increase head size. Some fertilizer dealers blend their own specialty mixes, thus providing special blends at competitive prices. Other dealers are selling "Concentrated Water Soluble Fertilizers" which vary in grade

(15-10-30 or 6-30-30). These fertilizers are completely water soluble and are similar to Miracle Grow™. Because they are completely water soluble, they are quite expensive. Their use is limited to a variety of specialty crops such as produced by nurseries. Table 12 describes liquid fertilizer usage in California for 1993.

Table 12. Liquid fertilizer and agricultural minerals used in California including the tons of each nutrient applied. (California Department of Food and Agriculture, 1993)

Fertilizer	Fertilizer Used (tons)	Grade	N	P_2O_5 (tons)	K_2O
N-P-K Grades	682,627	11-11-6	76,447	73,877	39,511
Anhydrous ammonia	161,593	82-0-0	132,506	0	0
Aqua ammonia	203,280	20-0-0	41,621	0	0
Ammonium nitrate solution	94,849	20-0-0	18,970	0	0
Ammonium polysulfide	8,052	20-0-0	1,610	0	0
Ammonium thiosulfate	11,014	12-0-0	1,322	0	0
Calcium ammonium nitrate	147,125	17-0-0	25,011	0	0
Nitrogen solutions < 28% N	7,717	17-0-0	1,324	0	0
Nitrogen solutions 28% N	9,442	28-0-0	2,644	0	0
Nitrogen solutions 30% N	2,270	30-0-0	681	0	0
Nitrogen solutions 32% N	460,824	32-0-0	147,464	0	0
Urea solution	30,479	20-0-0	6,097	0	0
Phosphoric acid	14,317	0-53.4-0	0	7,666	0
Ammonium polyphosphate liquid	90,314	10-34-0	9,002	30,584	0
Superphosphoric acid	8,030	0-71.5-0	0	5,743	0
Lime sulfur solution	210	0-0-0	0	0	0
Sulfuric acid	42,743	0-0-0	0	0	0
Zinc sulfate solution	688	0-0-0	0	0	0
Total	4,126,845	8.7-2.3-1.8	588,965	191,087	161,142

Note: Some liquid fertilizers may be applied directly to the soil rather than used for injection into irrigation systems. Agricultural minerals are liquids which contain nutrients other than nitrogen, phosphate or potash have a grade of 0-0-0.

Several fertilizers which are used for fertigation are briefly described and their properties identified. Specific precautions for use are indicated.

ANHYDROUS AMMONIA (82-0-0)

NH_3 (82-0-0) is a liquefied gas which must be handled with special equipment to maintain the high pressure required to keep it in liquid form. It is a colorless gas with a very pungent smelling odor. Your nose can detect as little as 50 ppm in the air. It is stored in pressurized steel tanks near its boiling point (-28°F) and at a pressure of less than 1 psi. The pH will often be well above 11. Ammonia injected into irrigation water will increase the pH of the water significantly. Anhydrous ammonia is no longer sold for fertigation in many areas of California, due to the potential of safety hazards. It should never be used in drip or sprinkler systems.

CAUTION	Although there is a very narrow window of mixtures which can be detonated with a spark, it is possible to have an explosion of ammonia when it is mixed with air. However, it is generally considered nonflammable. Inhaling ammonia can be deadly.

Aqua Ammonia (20-0-0)

$NH_3 \cdot H_2O$ is a liquid fertilizer (20-0-0) sold in some areas of California, but safety hazards in handling this product limit its use. It is a low pressure ammonia solution. That is, it is a solution containing both gaseous ammonia and ammonium hydroxide (NH_4OH). The lid must always remain securely over the tank because the ammonia gas can quickly volatilize. The tank should be of the type to hold pressurized liquids and should be fitted with a pressure relief valve to avoid over-pressurization of the tank. It should never be used in drip or sprinkler systems.

Ammonium Nitrate Solution or AN-20 (20-0-0)

$NH_4NO_3 \bullet H_2O$ is ammonium nitrate fertilizer dissolved in water. It has a density of 10.5 pounds per gallon. This product is commonly used for fertigation in California. *Under no conditions should concentrated AN-20 be mixed with concentrated urea sulfuric acid, concentrated sulfuric acid, concentrated hydrochloric acid, or concentrated phosphoric acid.*

Urea-Ammonium Nitrate Solution or UN-32 (UAN-32) (32-0-0)

$(NH_2)_2CO \bullet NH_4NO_3$: Urea-ammonium nitrate solution is manufactured by combining urea (46% N) and ammonium nitrate (35% N) on an equal nitrogen content basis. No other combination of dry nitrogen fertilizer products can produce solutions as high in nitrogen content as that obtained with the combination of urea and ammonium nitrate. UN-32 contains the highest concentration of nitrogen of all the nitrogen solution products. Urea-ammonium nitrate solutions are marketed as 32% nitrogen solutions in warmer agricultural climates; and as 28% nitrogen solutions in cooler agricultural areas. Urea-ammonium nitrate solutions should not be combined with CAN-17 or solutions prepared from calcium nitrate. A thick, milky-white insoluble precipitate forms, presenting a serious potential plugging problem. This product is commonly used for fertigation in California.

Calcium Ammonium Nitrate or CAN-17 (17-0-0-8.8Ca)

$Ca(NO_3)_2 \bullet NH_4NO_3$: This popular specialty nitrogen fertilizer is high in nitrate-nitrogen, low in ammonium-nitrogen, and supplies calcium. It is used on crops that have a high calcium requirement. Moreover, certain crops such as pepper, strawberry, tomato, and eggplant appear to produce higher yields of higher quality fruit when fertilized with predominantly nitrate-nitrogen fertilizer programs. CAN-17 fulfills this requirement. CAN-17, however, should not be applied as a major source of calcium since the crop will be fertilized with excessive amounts of nitrogen. CAN-17 can be combined with ammonium nitrate, magnesium nitrate, potassium nitrate, and muriate of potash (0-0-60), and should not be combined with any products containing sulfates or thiosulfates. While certain CAN-17 mixes can be made with food grade phosphoric acid, it is best to avoid attempts to make these mixes since application methods could possibly result in drip irrigation plugging.

AMMONIUM PHOSPHATE (8-24-0)

$NH_4H_2PO_4$ is a clear liquid which can be injected into irrigation water, but only at very low rates. It has a density of 10.5 pounds per gallon. Ammonium phosphate 8-24-0 has limited availability on the market since it is made from either anhydrous or aqua ammonia and food grade phosphoric acid. Food grade phosphoric acid is not widely available in the agricultural market because of cost, and demand by the food processing industry.

If this material is injected at high rates of application, or if it is injected into hard water, it will form an insoluble calcium and/or magnesium phosphate material which may plug the drip emitters or clog the irrigation lines. Acid pretreatment of the irrigation water with urea sulfuric acid could resolve this problem, but the additional cost of using urea sulfuric acid would preclude this type of fertilization program from being cost effective.

AMMONIUM POLYPHOSPHATE (9-30-0, 10-34-0 AND 11-37-0)

This material consists of a variety of various mixtures of polymeric forms of phosphates (i.e., diammonium pyrophosphate $(NH_4)_2H_2P_2O_7$) and ortho ammonium phosphates (i.e., $NH_4H_2PO_4$).

Table 13. Densities of various ammonium polyphosphate fertilizers.

Compound	Density pounds/gal
9-30-0	11.3
10-34-0	11.4
11-37-0	11.7

In hard water (moderately high in calcium and/or magnesium) this fertilizer will form monocalcium diammonium pyrophosphate $Ca(NH_4)_2P_2O_7 \cdot H_2O$ which is stable in neutral or basic ($pH > 7$) solutions. This allows it to be used in chemigation or fertigation with most systems as long as the injection rate does not exceed too high of a value. However, if the water has a high buffering capacity (high carbonate/bicarbonate content associated with a high pH, i.e., > 8.0) along with a high calcium content, highly insoluble calcium pyrophosphates will form and result in severe plugging problems in low volume drip, trickle, or micro-jet irrigation systems.

Ammonium polyphosphate fluid fertilizers are widely used to make a variety of neutral and acidified fertilizer solutions. It should be noted that micronutrients can be added to only the acidified kinds of fertilizer solutions prepared with ammonium polyphosphates.

AMMONIUM POLYSULFIDE (20-0-0-45)

This is a mixture of polymer forms of sulfide stabilized and maintained in solution with ammonia. It is usually represented as $(NH_4)_2S_x$, where the x indicates some multiple sulfur atoms which form the sulfur polymer. The grade is (20-0-0-45) where the 45 represents the percentage of sulfur in the material. It exists as a reddish colored liquid with a strong odor of ammonia and hydrogen sulfide (rotten egg) gas. Although this solution has a very high pH, the formation of microscopic size particles of sulfur when the solution undergoes dilution gives this product a very unusual characteristic: the solution frequently is used as a soil acidifying agent. It has a density of approximately 9.2 pounds per gallon. Ammonium polysulfide can be mixed with aqua ammonia (which is basic) in all proportions. For optimum results it is best to inject ammonium polysulfide by itself.

Ammonium polysulfide is unstable when mixed into water or neutral solutions. When contact is made with any kind of acid, dangerous hydrogen sulfide gas is generated. It is generally considered to be incompatible with any phosphate fertilizers. One should ***NEVER*** enter a tank where ammonium polysulfide has been stored or transported to the field for use. Ammonium polysulfide should be stored in a cool place and kept tightly closed (but with a pressure relief valve on the tank) to prevent ammonia volatilization losses and to contain the hydrogen sulfide gas.

AMMONIUM THIOSULFATE (12-0-0-26)

$(NH_4)_2S_2O_3$ is used as both a fertilizer and as an acidulating agent. When applied to the soil, *Thiobacillus* bacteria oxidize the free sulfur to sulfuric acid. The acid then dissolves lime in the soil and forms gypsum. The gypsum helps to maintain a good, well granulated, aerated, and porous soil structure. Ammonium thiosulfate is ideal for treatment of calcareous (high lime) soils. It is compatible in any proportions with *neutral* or *alkaline* phosphate liquid fertilizers, aqua ammonia, and other nitrogen fertilizers. Ammonium thiosulfate can be applied in liquid mixes or by itself.

Ammonium thiosulfate should not be mixed with acidic *compounds* because it will decompose into elemental sulfur and ammonium sulfate at pH < 6. Application to neutral and acidic *soils* (without free lime) may result in a pronounced drop in soil pH over several weeks. The extent of the pH drop in these types of soils depends upon the total amount of this fertilizer applied, the cation exchange capacity of the soil, and the buffering capacity of the soil. The higher the clay content and the higher the lime content of the soil (i.e., the larger the buffering capacity), the slower the pH will drop with the same fertilizer application.

CALCIUM POLYSULFIDE OR LIME SULFUR

This is a mixture of polymer forms of sulfide stabilized and reacted with calcium hydroxide with the general composition of CaS_x, where the x represents multiple sulfur atoms in the polymer. It is not considered as a fertilizer since it has no nitrogen, phosphorus or potassium, and therefore has a grade of 0-0-0. Calcium polysulfide contains 6% calcium and 22% sulfur. It has a density of 10.6 pounds per gallon. It has the odor of hydrogen sulfide (rotten egg) gas. The calcium and sulfur content, however, make it an excellent source for these nutrients. It is a strong soil acidulating agent.

CaS_x is often used to increase the water permeability with furrow and flood irrigation on calcareous (high lime) soils that may also have sodium related water intake problems. The furrows may have a white appearance after the material is applied. While the soil in the furrows is moist the *Thiobacillus* bacteria will oxidize the sulfur to sulfuric acid. This acid dissolves the lime to form gypsum directly in the soil. The gypsum which forms helps to keep the soil open and well aerated.

Calcium polysulfide application to neutral and acidic soils may result in a pronounced drop in soil pH over several weeks. The extent of the pH drop in these types of soils depends upon the total amount of the material applied, the cation exchange capacity , and the buffering capacity of the soil. The higher the clay content and the greater the buffering capacity of the soil, the slower the pH will drop with the same fertilizer application. Keep the container tightly closed to prevent the loss of the hydrogen sulfide gas. ***DO NOT ENTER STORAGE TANKS OR MATERIAL TRANSPORT TANKS, UNDER ANY CIRCUMSTANCES.***

METAL CHELATES

Chelated micronutrient metals are the most convenient and sometime the only means of applying the metal ions and have them move into the plant root zone. Chelates are organic compounds which wrap around a metal ion and neutralize it electrically. Since the metal chelate is neutral in charge, it can move down into the soil and is unaffected by the soil cation exchange capacity, other elements that would ordinarily precipitate the inorganic forms of the metal ions, or the soil pH that would affect the availability of the inorganic micronutrient metals.

Chelates of metal ions work in the soil by moving to plant roots along with the irrigation water. The exact mechanism of chelated micronutrient uptake by plants is yet to be defined. Some researchers have found that some plants are capable of detaching the micronutrient metal from its chelating agent, and absorb the metal directly. Other researchers have found that some plants absorb the chelated metal intact with its chelating agent. The definition of uptake mechanism(s) is beyond the scope of this publication. Suffice it to say that chelated micronutrients are more likely to resolve an immediate micronutrient deficiency than the inorganic counterpart. However, the correction of the deficiency with a single application of chelated micronutrients may not be of sufficient duration to satisfy the needs of the crop through to harvest.

The chemistry of chelation is complicated, as is the chemistry of equilibria of chelates in the soil. It has been found that the most economic and effective forms of chelates are ethylenediaminetetraacetic acid (EDTA), diethyleneaminepentaacetic acid (DTPA), and ethylenediamine di-(O-hydroxyphenyl-acetatic acid) (EDDHA). Zinc, manganese, iron and copper are most effective in the DTPA form, but still have a good level of effectiveness in the EDTA forms on most soils in California. However, on highly calcareous soils, iron in the EDDHA form is most effective.

Rarely, if ever, are chelating agents applied to soils without the association with a micronutrient metal. There are situations where heavy metals already present in the soil can displace the micronutrient metal in the chelate. When this occurs, the micronutrient metal is dropped from the chelate, and the other heavy metal (i.e., cadmium) can take its place. This effectively increases the probability that the unwanted heavy metal can be absorbed by the plant. This point is raised at this time because it is necessary to know the concentration of heavy metals in soils that have been treated with municipal sewage sludges. It must be remembered that chelating agents have a very strong affinity for metal ions, can pick up the metal ion, and can convert it into a metal chelate. The metal will then be available for plant uptake.

Chelates are expensive. They sell for about ten times the cost of most normal micronutrient materials. However, as noted above chelates are excellent sources of micronutrients, particularly when a micronutrient deficiency has been identified after a crop is established, but methods for correcting the problem with less expensive, long term products are not possible. Chelates have an immediate efficacy factor of 3× to 5× that of the inorganic source of micronutrients. A distinct disadvantage of chelated micronutrients is that they cannot be used to increase the over all micronutrient levels in soils for long term availability. Their low rate of use, disappearance during the crop season, and cost preclude their use in this management practice.

Chelates have a general preference for binding most tightly to 3+ charged ions like iron (Fe^{3+}). Next, they prefer 2+ ions. Many of the toxic heavy metals (mercury, nickel, and cadmium) also have a 2+ charge and are bound tightly. The metal nutrients usually are chelated most strongly as copper (Cu^{2+}), followed by zinc (Zn^{2+}), then iron (Fe^{2+}), and finally by manganese (Mn^{2+}), which is the least strongly chelated of the nutrient metal ions. In addition, calcium (Ca^{2+}), magnesium (Mg^{2+}), and hydrogen (H^+) ions can compete with metal ions for a position on the chelate.

Chelates are effective in alkaline soils. The lower the alkalinity, the more effective the chelate. When the soil alkalinity (pH) is high enough to cause one to question whether or not a chelate should be applied because of the effect of the alkalinity on the chelate, per se, the crop being considered for treatment is usually in poor condition just as a result of the high alkalinity alone. There are times when a good crop can be produced in somewhat alkaline soils (pH 7.5), but manganese, iron, and copper deficiencies are encountered. DTPA or EDTA applications of manganese and copper, and EDDHA applications of iron often can correct the deficiency shortly after application, but the deficiencies will reoccur if the chelated micronutrient applications are discontinued.

Chelated micronutrients should not be the choice as the source of micronutrients on acidic, micronutrient deficient soils. The inorganic sulfate micronutrients are usually the most effective sources. It is rare, however, to encounter zinc, manganese, iron and copper deficiencies on acidic soils. Manganese and iron, in particular, and sometimes copper can be present at levels toxic to crop plants. Correcting the soil pH with limestone or dolomite usually eliminates the toxicity problem and brings the available manganese, iron and copper to normal levels. Zinc availability is less affected by soil pH. If zinc is determined to be deficient, this problem, too, can be economically corrected with applications of zinc sulfate.

Natural chelate materials are often made from wood or food processing wastes. These natural chelates are often cheaper than the synthetic chelates. However, the natural chelates generally have a lower binding capability for micronutrient metal ions than do the synthetic chelates. Thus, the natural chelates will work, but not to the same extent as do the synthetic chelates. All chelates are organic and are decomposed by soil microorganisms over time. Consequently very little, if any, chelated micronutrients remain available for the following crop.

DTPA, EDTA, and EDDHA chelates are compatible with neutral solutions, but there may be a problem when EDTA chelates (especially zinc) are stored with CAN-17. Apparently differences in EDTA formulation by different formulators can result in the formation of an insoluble precipitate on the bottom and sides of the storage tank. It usually takes about 12 hours for the precipitate to form, if it is going to form. There is not a problem injecting these chemicals together, but they should not be stored together unless it is predetermined that no problem will occur (via the jar test). (Dickinson, 1995).

Phosphoric Acid (0-54-0 "White" & 0-52-0 "Green" Acids)

H_3PO_4 contains between 51 and 54 per cent P_2O_5, and the most common concentrated acid has a density of approximately 14.1 pounds per gallon. White phosphoric acid is also referred to as food grade acid. Green phosphoric acid refers to the acid obtained from the "wet" process whereby phosphorus is extracted from the ore with phosphoric acid. Wet process green acid contains impurities which impart the green color. Both acids are syrupy liquids which are normally stored in stainless steel (No. 316) tanks. They release a small amount of heat when added to the irrigation water, but that heat is soon dissipated. White phosphoric acid is the preferred grade of acid for formulating mixes intended to be applied through drip irrigation systems. The cost and availability of this high grade acid limit its use. The green acid can be successfully used for certain formulations of nitrogen, phosphorus, and potassium mixes.

NOTE — Phosphoric acid should never be mixed with any calcium fertilizer. It will form insoluble calcium phosphate which can plug irrigation lines.

Potassium Chloride

Potassium chloride (KCl) is generally the least expensive source of potassium and is the most popular K fertilizer applied through fertigation. It may not be desirable for use on crops such as avocados, berries, peppers, and other crops that are sensitive to high levels of chloride.

Potassium Nitrate

Potassium nitrate is expensive but the consumer benefits from both the nitrogen and the potassium in the product. It is probably the second most popular potassium fertilizer because farmers prefer to have a nitrate-nitrogen source of nitrogen and not have to concern themselves about applying chlorides. It is less soluble than potassium chloride, but more soluble than potassium sulfate.

POTASSIUM PHOSPHATE (0-52-34)

These solutions consist mainly of KH_2PO_4 and have a moderately low pH. The density of the solution is 10.5 pounds per gallon. Potassium phosphate is imported; therefore its use is limited to very special purposes, i.e., greenhouse crops, and bedding plant and ornamental nurseries. Potassium phosphate can be used in neutral and acid fertilizer mixes.

POTASSIUM SULFATE

The advantage of K_2SO_4 is that it is also an excellent source of sulfur. It is fairly popular for fertigation. It is less soluble than potassium chloride and potassium nitrate.

POTASSIUM THIOSULFATE (0-0-25-17 and 0-0-22-23)

$K_2S_2O_3$ (KTS) is marketed in two grades and is a neutral to basic, chloride free, clear liquid solution. Depending on the formulation, each gallon of KTS contains 3 pounds of K_2O and 2.1 pounds of sulfur. This product can be blended with other fertilizers, but KTS blends should not be acidified below pH 6.0. The proper mixing sequence for KTS is water, pesticide, KTS and/or other fertilizer. Always perform a jar test before injecting blends.

Growers with calcareous soils have found that potassium thiosulfate has given good responses from crop growth. Note that potassium thiosulfate provides not only potassium, but the thiosulfate is oxidized by *Thiobacillus* bacteria to produce sulfuric acid. This acid reacts with calcium carbonate in the soil which releases additional calcium for the plant. The sulfuric acid is also an available source of sulfate from which plants obtain sulfur. Thus, potassium thiosulfate use on calcareous soils not only supplies potassium and sulfur, but aids in increasing the availability of calcium to the crop plants.

SULFURIC ACID

H_2SO_4 is not a fertilizer and thus has a grade of 0-0-0. It has a density of approximately 15.3 pounds per gallon when concentrated. Sulfuric acid is a clear liquid when pure, however much of the agricultural material may have a brown to black color. It has no odor and pours as an oily liquid. It is injected into high bicarbonate water to control the pH by reducing it to about pH 6.5 to 7.0. It is sometimes injected directly into calcareous soils (high lime) where the reaction produces gypsum. When it is mixed with irrigation water it gives off an appreciable amount of heat which is absorbed with the water and its temperature rises. The container should be kept tightly closed and water should be prevented from entering the container. It will immediately react with any organic material quickly. Contact with plant material may cause it to become black and have a charred appearance.

Do not attempt to acidify irrigation water and inject calcium ammonium nitrate (or any other calcium fertilizer) in the same irrigation run. Calcium sulfate (gypsum) will form and create a creamy suspension very much like cottage cheese which can easily plug the lines.

Sulfuric acid is an extremely hazardous chemical. O.S.H.A. requirements for safe handling preclude fertilizer dealers from storing sulfuric acid on the premises; therefore it is difficult to find a source of sulfuric acid. At one time "spent" sulfuric acid was available as a by-product from industrial sources, but the State of California has decreed that this material can no longer be used for agricultural purposes.

CAUTION Sulfuric acid is extremely corrosive and must be handled with proper equipment and clothing. Never combine urea and sulfuric acid in the field.

Urea Solid (46-0-0) and Urea Solution (23-0-0)

Urea is sold as 46-0-0 dry fertilizer or as a liquid 23-0-0 urea solution. Generally, growers purchase urea solution for fertigation, although they can make their own solution by dissolving solid urea in water. Urea makes solutions extremely cold when it is being dissolved. The high solubility of urea and non-ionized form allows growers to apply dilute solutions of urea directly onto the leaves of plants as a means of supplying foliar nitrogen. Solutions of 0.5 to 1.0% urea are usually safe to apply directly on plant leaves. Urea is noncorrosive.

Commercial urea contains about 2.25% biuret. Biuret is a urea polymer which forms only during the manufacturing process. It can inhibit plant growth or damage plants if urea with 2.25% biuret content is used for foliar applications. Thus, if growers use urea for foliar fertilization as a liquid spray, it is best to purchase the special recrystallized urea which is guaranteed to contain a low content of biuret, usually about 1.25% biuret. A source of urea in the liquid form containing only 0.005% biuret has recently appeared on the market. It is a very safe and effective source of urea intended for foliar applications. Most urea fertilizer manufactured today has a low content of biuret, and the biuret content in the product is stated on the bag. Growers wishing to apply urea as a foliar spray should look for the words, "Low Biuret" on the fertilizer bag.

CAUTION Urea should never be mixed with sulfuric acid.

Urea Phosphate (17-44-0)

$CO(NH_2)_2 \cdot H_3PO_4$ consists of crystals of urea mixed with phosphoric acid. In its diluted liquid form, urea phosphate can be fertigated for nutritional and system cleaning purposes. It is easy to handle and requires no special precautions for application. It is completely soluble in water. When it dissolves in water it forms both urea and phosphoric acid (Mikkelsen, 1989). The acidic reaction of urea phosphate prevents ammonia losses by volatilization from the soil surface when these fertilizers are used.

Urea phosphate has been effective when applied with drip irrigation for tomatoes on high lime soils (Mikkelsen and Jarrell, 1987) and with cabbage and squash (Rubeiz, 1984). The use of urea phosphate fertilizer for drip irrigation has been examined by Mikkelsen (1989).

Urea Sulfuric Acid

General

Urea sulfuric acid ($CO(NH_2)_2 \cdot H_2SO_4$) is an acidic fertilizer which combines urea and sulfuric acid. By combining the two materials into one product, many disadvantages of using these materials individually are eliminated. The sulfuric acid decreases the potential ammonia volatilization losses from the soil surface and ammonia damage in the root zone that can occur with the use of urea alone. Urea sulfuric acid is safer to use than sulfuric acid alone.

Urea sulfuric acid is well suited for fertigation and can be used for other purposes.

1. To add nitrogen to the soil. This application can be used as a general fertilizer strategy.
2. To acidify the irrigation water, thus reducing the amount of carbonates and bicarbonates in the water. This application can be used as a maintenance strategy to keep drip lines and emitters clear of calcium carbonate deposits.
3. To clean irrigation lines. Once lines have been plugged, urea sulfuric acid can be injected to clean the irrigation lines as a reclamation strategy.
4. To acidify the soil, thus increasing the solubility, availability and mobility of several nutrients.
5. To increase the availability of calcium in soils with free lime.
6. To prevent soil crusting thus improving germination and soil structure.

This product is commonly sold under various names, such as N-pHURIC® by Unocal. The nitrogen and sulfuric acid contents of these products vary depending on their specific formulation (Table 14).

Table 14. Summary and constituents of N-pHURIC® products. (Unocal, 1993)

Product	Density pounds/gal	Nitrogen %	Sulfuric Acid %	Sulfur %
N-pHURIC 10/55	12.8	10	55	18
N-pHURIC 15/49	12.7	15	49	16
N-pHURIC 28/27	11.8	28	27	9

SAFETY CONSIDERATIONS

It should be noted that the manufacturers of these products have developed special processes to combine these materials.

CAUTION Under no circumstances should urea and sulfuric acid be mixed in the field. A compound toxic to plants will form.

As with many other chemicals, protective clothing and eyewear should be worn when handling these products. Skin contact should be avoided, and contaminated clothing and shoes should be washed before using again. A safety shower and eyewash should be available near the transfer site. Although urea sulfuric acid is not flammable, care should be taken to avoid welding or burning fires near the holding tank. If the tank heats up, it should be cooled with water—there is no heat of reaction if water is added to urea sulfuric acid. Spills must be treated as acid spills: they should be isolated, neutralized (with baking soda, soda ash, limestone, hydrated lime or other suitable base materials), and diluted.

Tanks and injection equipment should be constructed of polyethylene, polypropylene, or No. 316 stainless steel. Safety and identification signs should be placed at the valves and connections of storage, transport, and field equipment. Equipment and tanks should be rinsed, neutralized, and cleaned after each use and before using any other chemical. Due to its corrosive nature, urea sulfuric acid can not be used in irrigation systems consisting of nylon or Delrin®. More detailed compatibility lists may be obtained from Unocal. In addition, concentrated injection should occur downstream of the filter system so that the filter is not damaged.

UREA SULFURIC ACID AS A FERTILIZER

Urea sulfuric acid can be mixed with ammonium phosphate, phosphoric acid, and micronutrient sulfates to create special blends to meet a particular grower's needs.

CAUTION Under no circumstances should sulfuric acid be mixed with urea sulfuric acid.

Urea sulfuric acid should not be mixed with any solution containing nitrate due to compatibility problems. In addition, mixing urea sulfuric acid with chloride (i.e., muriate of potash) creates highly corrosive mixtures and should be mixed in approved plastic tanks.

It should be noted that a grower should consult an expert before trying to formulate their own blend. It is critical to follow the proper mixing sequence guidelines (Table 15). Not following the sequence rules may result in either dangerous reactions or insoluble products.

Table 15. Examples of safe mixtures of N-pHURIC®. (Unocal, 1993)

N-P-K-pHURIC®s	Percent by Weight	Ingredients	Sequence of Addition
3-6-6-5.2%S	50.7	Water	1
	20.0	N-pHURIC 15/49	2
	11.6	Sulfate of Potash	3
	17.7	10-34-0	4
4-0-10-4.3%S	56.9	Water	1
	26.7	N-pHURIC 15/49	2
	16.5	Muriate of Potash	3
4-4-10-4.3%S	49.9	Water	1
	7.7	Phosphoric Acid 52%	2
	26.7	N-pHURIC 15/49	3
	16.4	Muriate of Potash	4
5-1-10-5%S	49.2	Water	1
	16.4	Muriate of Potash	2
	31.4	N-pHURIC 15/49	3
	3.0	10-34-0	4
5-5-5-5.4%S	52.0	Water	1
	23.5	N-pHURIC 15/49	2
	9.7	Sulfate of Potash	3
	14.8	10-34-0	4
5-5-5-5.3%S	48.7	Water	1
	9.7	Phosphoric Acid 52%	2
	33.4	N-pHURIC 15/49	3
	8.2	Muriate of Potash	4

continued

Table 15. Continued.

N-P-K-pHURIC®s	Percent by Weight	Ingredients	Sequence of Addition
5-10-5-5.3%S	39.1	Water	1
	19.3	Phosphoric Acid 52%	2
	33.4	N-pHURIC 15/49	3
	8.2	Muriate of Potash	4
8-4-4-8.5%S	35.0	Water	1
	45.5	N-pHURIC 15/49	2
	7.7	Sulfate of Potash	3
	11.8	10-34-0	4
10-2-10-3%S	34.7	Water	1
	16.4	Muriate of Potash	2
	33.7	N-pHURIC 28/27	3
	5.9	10-34-0	4
10-2-10-3.3%S	43.5	Water	1
	16.4	Muriate of Potash	2
	13.8	Urea 46%	3
	5.9	10-34-0	4
	20.4	N-pHURIC 15/49	5
12-8-4-3.1%S	35.3	Water	1
	23.6	10-34-0	2
	34.5	N-pHURIC 28/27	3
	6.6	Muriate of Potash	4
12-8-4-3.8%S	35.1	Water	1
	15.4	Phosphoric Acid 52%	2
	42.9	N-pHURIC 28/27	3
	6.6	Muriate of Potash	4
7.5-26-0-8%S	50.0	N-pHURIC 15/49	No special sequence
	50.0	Phosphoric Acid 52%	
10-16-0-11%S	68.7	N-pHURIC 15/49	No special sequence
	31.3	Phosphoric Acid 52%	
11.6-11.6-0-12.4%S	77.6	N-pHURIC 15/49	No special sequence
	22.4	Phosphoric Acid 52%	
13-6.5-0-14%S	87.4	N-pHURIC 15/49	No special sequence
	12.6	Phosphoric Acid 52%	

Note: Plastic equipment and tanks are recommended when mixing these compounds with muriate of potash. The 10-34-0 is ammonium polyphosphate.

UREA SULFURIC ACID AS A MAINTENANCE STRATEGY

The presence of carbonates, bicarbonates, and silicates in the irrigation water not only reduce fertilizer efficiency by forming insoluble precipitates, the precipitates may also restrict the passage of water out of the system. Moreover, these precipitates may also destroy irrigation equipment.

Urea sulfuric acid decomposes bicarbonates and carbonates in irrigation water in addition to adding urea and sulfate to the soil. In addition, acidifying the irrigation water to pH 6.5 greatly enhances chlorine's ability to kill bacteria (see Chapter 12, "Drip System Maintenance"). The reaction of urea sulfuric acid in irrigation water is as follows:

$$\underset{\text{Urea Sulfuric Acid}}{CO(NH_2)_2{\bullet}H_2SO_4} + \underset{\text{Bicarbonate}}{2HCO_3^-} \longrightarrow \underset{\text{Urea}}{CO(NH_2)_2} + \underset{\text{Sulfate}}{SO_4^{2-}} + \underset{\text{Carbon Dioxide}}{2CO_2} + \underset{\text{Water}}{2H_2O}$$

Urea sulfuric acid contains varying amounts of acid depending on the specific formulation. This form of acid is considered safer than mineral acid forms; nevertheless, care should be taken when these compounds are used with aluminum, cement, or transite irrigation pipelines since adverse reactions may occur when the water pH falls below 4.5.

CHAPTER 11. PLANT AND SOIL TESTING

HIGHLIGHTS

- Conventional, infrequent laboratory plant and soil testing practices were historically designed to confirm visual diagnoses or determine problems of previously low-yielding crops. In other words, they diagnosed problems after the plant had suffered a yield decrease.
- Progressive nutrient management strategies include frequent testing throughout the season to monitor the nutrient status of the crops, thus providing a mechanism to prevent nutrient problems and to enhance crop quality.
- New capabilities of microirrigation systems allow farmers to "spoon feed" nutrients to plants (i.e., providing fertilizers in small doses throughout the season as opposed to all of the fertilizer once or twice during the growing season).
- New technologies allow growers to conduct some soil, plant, and water analyses on-farm.
- Many of the old rules which applied to traditional fertility management no longer apply to fertigation.
- Several different types of soil and plant testing procedures are available, each of which provides different indices for interpretation.

NEW ATTITUDES ABOUT NUTRIENT MANAGEMENT

Standard (not fertigation) fertilization procedures rely on large, occasional applications of fertilizer to a crop. There is little or no ability to "spoon-feed" the crop, so there is little need to make frequent observation of soil and plant fertility. The fertilizer which is applied consists of fairly standard formulations. The formulation and application strategies often depend on estimated amounts of annual fertilizer leaching, residual nitrogen, and the time for conversion from urea to ammonium, or from ammonium to nitrate. The time scale of thinking is seasonal or a few times per season, rather than daily or weekly.

Fertigation opens up a whole new world in terms of *control* of nutrient availability, in particular if it is accompanied by drip/microirrigation. Soil solution nutrient levels can be micro-managed in time and amount. The time scale of management can become weekly or even daily. If an immediate nitrogen response is needed, a special blend of nitrate fertilizer can be applied, and a nutrient response can be seen that same day. In areas of drip/microirrigation with high value crops (vegetables, trees, and vines), the fertilizer companies now sell much more sophisticated and varied assortments of fertilizer blends than previously.

This excellent ability to *control* nutrient levels has necessitated better (more accurate and rapid) *feedback* of information on the nutrient status in the soil, plant, and water. Many farmers are now experiencing an evolution in how they view the tools for nutrient analysis, how they interpret those analysis results, and how they choose the fertilizer mixes they use to respond to their interpretations. Of particular interest are new "Quick Tests" for on-farm testing of soil or plant tissue.

Another reason for the heightened interest in improved diagnostic techniques includes the changing markets for some crops which require new levels of crop quality. For example, the "value added" lettuce which is pre-shredded appears to require better control of some nutrient levels immediately before harvest.

Stages in Improving Nutrient Management

There appear to be several stages in farmer acceptance of new nutrient diagnostic procedures.

Stage 1 occurs when growers initially adopt drip/microirrigation systems. Fertigation in any form may be a new experience. Initial farmer efforts focus on just getting the irrigation system and fertigation hardware to function properly. Fertilizer practices are usually based on traditional experiences, with infrequent soil/plant testing and visual diagnosis of nutrient deficiencies. Commercial laboratories are the major, if not sole, source of information and testing.

Stage 2 occurs as the growers become more proficient in managing irrigation and cultivation practices under drip irrigation. Attention sometimes turns to new fertigation and nutrient management strategies, including more frequent plant testing and fertilizer applications. Soil testing will be done annually. Commercial laboratories remain the major, if not sole, source of testing. Many growers will purchase simple quick test equipment for pH and EC during this stage.

Stage 3 involves a more sophisticated examination of total fertility (soil, plant, and water), along with detailed record keeping. The fertility analysis will be conducted at least weekly on many fields, and correlations will be developed between stages of growth, nutrient levels, nutrient ratios, and yield. Growers continue to use commercial laboratories, but may also make much more extensive use of in-field testing equipment. The need is for more immediate access to the nutrient information to aid in decision-making.

At Stage 3, growers realize that it is difficult to correlate results from various test procedures. It is also unclear exactly what levels are adequate at various stages of growth. The rapid and vigorous growth associated with drip/microirrigation seems to bring different nutrient "adequacy" levels than does conventional farming. New guides for nutrient interpretations are being developed to assist growers with these decisions.

CATEGORIES OF TESTS

There are five typical categories of nutrient tests. These include the following:

1. Soil sample
2. Soil solution
3. Plant tissue
4. Plant sap
5. Irrigation water

Each of these tests may be conducted with different equipment and procedures, which may provide different results. Furthermore, the results of each test will provide a different type of insight to crop nutrition needs. Table 16 provides a comparison of each test.

Table 16. Differences between nutrient test categories.

	Category of Nutrient Test				
	Soil	Soil Solution	Plant Tissue	Plant Sap	Irrigation Water
Typical timing	Preplant, and if deficiency symptoms arise	Weekly	Several times during growing season	Several times during growing season	Once or twice during growing season
What is determined	Total fertilizer need for the season	Availability of nutrients at that moment	Are levels of specific nutrients sufficient for that growth stage?	Are levels of specific nutrients sufficient for that growth stage?	Nutrient contribution by water; potential toxic elements such as boron, chloride
Special observations	Potential for problems in fertility or infiltration. Examine nutrient ratios in the soil	Release rates of various fertilizers	Nutrient ratios in the plant itself (DRIS)	Nutrient ratios in the plant itself (DRIS)	Permeability hazards
Typical procedure	Laboratory	Field	Laboratory	Field	Laboratory or field

These categories will be described in more detail later in this chapter.

DESCRIBING AMOUNTS OF NUTRIENTS

GENERAL GUIDELINES

The amount of a substance, such as nitrogen, can be expressed in several ways. One should pay close attention to the units of measurement used to describe the quantity of nitrogen as (1) a test result, and (2) a recommendation for fertilizer applications. Major descriptions for nitrogen are as follows:

1. Total N quantifies all of the nitrogen which is in the NH_4, NO_3, and organic N forms.
2. NH_4 refers to the amount of ammonium only.
3. NH_4-N represents how much nitrogen is in the ammonium form, but measured in terms of an equivalent amount of elemental nitrogen as N.
4. NO_3 refers to the amount of nitrate only.
5. NO_3-N represents how much nitrogen is in the nitrate form, but measured in terms of an equivalent amount of elemental nitrogen as N.

In other words, if a test indicates "ppm NO_3," *this is not equivalent* to "ppm NO_3-N."

Table 17. Conversions of relevant nutrient units of measurement.

Unit		Multiplier		Result
Any compound (ppm)	×	2.0	=	same compound pounds/acre
N (ppm) (i.e., elemental N)	×	4.42	=	NO_3 (ppm) if the nitrogen analyzed is nitrate.
NO_3-N (ppm) (i.e., elemental N)	×	4.42	=	NO_3 (ppm)
N (ppm)	×	1.29	=	NH_4 (ppm) if the nitrogen analyzed is ammonium
NH_4-N (ppm) (i.e., elemental N)	×	1.29	=	NH_4 (ppm)
NH_4 (ppm)	×	.78	=	NH_4-N (ppm)
NO_3 (ppm)	×	.226	=	NO_3-N (ppm)
NO_3 (ppm)	×	0.452	=	NO3-N pounds/acre
NH_4 (ppm)	×	1.556	=	NH_4-N pounds/acre
Organic Matter (%)	×	10.0	=	pounds available N/acre/year [1]
Organic Matter (%)	×	20.0	=	pounds available N/acre/year [2]
P_2O_5 (ppm)	×	.67	=	PO_4 (ppm)
P_2O_5 (ppm)	×	.44	=	PO_4 - P (ppm)
PO_4-P (ppm)	×	3.07	=	PO_4 (ppm)
K_2O (ppm)	×	.83	=	K (ppm)
NO_3-N (%)	×	71.4	=	NO_3-N meq/100 g of soil
K (%)	×	25.57	=	K meq/100 g of soil
Ca (%)	×	49.9	=	Ca meq/100 g of soil
Mg (%)	×	82.24	=	Mg meq/100 g of soil
NH_4-N (%)	×	71.4	=	NH_4-N meq/100 g of soil

[1] Assumes organic matter contains 5% nitrogen and mineralization rate of 1% per year (mineralization rates depend on environmental factors and are not constant).

[2] Assumes organic matter contains 5% nitrogen and mineralization rate of 2% per year.

For example, assume there are 50 pounds NO_3-N /acre. This can be converted as follows:

GIVEN: 50 pounds NO_3-N /acre

FIND: Pounds NO_3/acre

SOLUTION: $$= \frac{50 \text{ pounds } NO_3\text{-N}}{\text{acre}} \times \frac{4.42 \text{ pounds } NO_3}{1 \text{ pounds } NO_3\text{-N}}$$

$$= \frac{221 \text{ pounds } NO_3}{\text{acre}}$$

MILLIEQUIVALENT CONVERSIONS

While nutrient quantities are often discussed in terms of weight (e.g., ppm), nutrients in fertilizers, soil, and water are often quantified in reference to their electrical charges, such as "milliequivalents per liter" (meq/L). The table and example below explains how to convert from *weight* measurements to *charge* measurements.

Table 18. Common constituents of fertilizer, soil, and water with their weights and electrical charges.

Element Name or Ion Name	Symbol	Electrical Charge	Atomic Weight, gm	Milliequivalent Weight, mgm
CATIONS (have positive charge)				
Ammonium	NH_4	+1	17.0	17.0
Calcium	Ca	+2	40.1	20.1
Hydrogen	H	+1	1.0	1.0
Magnesium	Mg	+2	24.3	12.2
Sodium	Na	+1	23.0	23.0
Potassium	K	+1	39.1	39.1
ANIONS (have negative charge)				
Bicarbonate	HCO_3	-1	61.0	61.0
Carbonate	CO_3	-2	60.0	30.0
Nitrate	NO_3	-1	62.0	62.0
Phosphate	PO_4	-3	95.0	31.7
Sulfate	SO_4	-2	96.1	48.1

To convert from ppm of a substance in a water sample to milliequivalent per liter:

$$\text{meq/L} = \frac{\text{ppm}}{\text{meq wt. in mgm}}$$

For example, assume 120 ppm of carbonate in a water sample:

Given: 120 ppm carbonate
Carbonate meq wt. = 30 mgm (from Table 18)

Find: Carbonate (meq/L)

Solution:

$$= \frac{120 \text{ ppm}}{30 \text{ mgm}}$$

$$= 4.0 \text{ meq/L}$$

Soil Sample Testing

Highlights

- Soil sample testing generally measures what is available and/or potentially available for plant uptake throughout the season.
- New technologies are available for quick tests of some soil parameters.
- Variations exist in the ways various laboratories express results; distinctions must be made between units (ppm vs. milliequivalents) and the form of the nutrient tested (total nitrogen vs. nitrate-nitrogen).
- These distinctions must be clear because they will affect the interpretation and recommendation based on the test.

Standard Laboratory Procedures

Soil sample testing is commonly used to determine the nutrient supplying power of the soil before planting (i.e., what is available or *potentially* available for plant uptake). The importance of soil testing is acknowledged by growers utilizing both traditional fertilizer application techniques and by those fertigating through microirrigation systems. Laboratories have traditionally conducted soil tests after the farmer or laboratory personnel collect the sample, but new technologies have recently allowed growers to conduct some rapid soil tests on the farm.

Laboratories routinely test for several factors: pH, electrical conductivity (EC), cation exchange capacity (CEC), macronutrients (i.e., N, P, K), secondary nutrients (S, Ca, Mg), and micronutrients (i.e., Zn, Fe, Cu, B, Mn, Mo, Cl). Costs for laboratory analysis range from $50 to $300, depending on the extent of analysis.

Laboratory testing and sampling provide two big advantages for growers:

- A *good* laboratory will provide results which are more accurate than some "quick tests."
- Growers can contract out all of the work, so it is done on a consistent basis without requiring special management time.

Currently, the standards for soil testing and analysis techniques used by laboratories in California are found in the *California Fertilizer Association Handbook of Soil Analysis Methods.* Not all California laboratories use these recommended methods. While the majority of labs operating in California provide high quality analysis, variations in results and recommendations will exist, depending on the techniques utilized. In 1994, a proficiency testing program was initiated for soil and plant samples for agricultural laboratories in the Western United States. The program involved sending samples to 104 labs for pH, EC, and nutrient analyses. Overall, soil salinity and pH results were highly reproducible among soils and labs. Soil phosphorus and plant nitrate results showed high variability (Miller and Kotuby-Amacher, 1995).

This proficiency testing program shows that while it is not critical to be well versed in all the various analysis techniques employed by these labs, knowing that the variation exists can be useful when evaluating laboratory test data and recommendations.

ON-FARM QUICK TESTS

Various technologies exist which enable growers to monitor the nutrient status in the soil using on-farm procedures. For some simple tests, in-field "Quick Tests" are becoming more popular for use by individual farm employees for the following reasons:

- They provide immediate results.
- It is easy to take a quick look at the field's problem areas.
- Some tests can be run quite inexpensively, even considering the employee time and purchase price of the test equipment.

"Quick Tests" allow farmers to measure such factors as pH, EC, NO_3, P, and K. More elaborate tests allow farmers to analyze SO_4-S, organic matter, Zn, Cu, Mn, and Fe using on-farm facilities. Some "Quick Tests" do not offer quantitative results; rather, the results are relative (i.e., potassium is high, medium, or low). The more elaborate tests may be too involved for what the farmer wants. Quick Test equipment is described in Appendix C.

Techniques range in complexity, accuracy, and costs. However, they all require that a liquid sample be extracted or derived from the solid soil sample. Once the liquid sample is obtained, the actual nutrient concentration may be measured using a variety of techniques. For NO_3, these techniques may include:

1. A nitrate-selective electrode meter.
2. A paper "test strip" which is color-sensitive.
3. A colorimetric or spectophotometric method to compare sample colors with a color dial.

The real questions for quick tests are these:

1. How does one obtain the liquid sample? For example, what liquid must be mixed with the soil sample, in what ratio, and how is the liquid separated from the sample?
2. How does one translate the concentrations given by the quick test into equivalent concentrations obtained from conventional soil analysis procedures?

A study by Hartz et al. (1994) showed excellent correlations (r^2 = .87) between NO_3-N concentrations as measured by a specific quick test procedure and standard laboratory analysis, *without* the need to oven dry soil samples for the quick test. These quick test procedures are described in Appendix C.

SOIL SAMPLING

Regardless of whether a sample is tested in a lab or on-farm, the results and interpretation can be no better than the quality of the sampling. Obviously, fertility levels will vary throughout a field depending on irrigation distribution uniformity, existence of diseases, and soil types.

A real question with sampling involves the desirability of collecting a "representative" sample, which is generally recommended in books and laboratories. The argument for a "representative" sample is that the field will be fertigated uniformly, and therefore many samples from a field should be taken and mixed together. From the mixture of sub-samples, a portion of the mixture (composite sample) is analyzed.

It is probable that *fertigation* sampling will eventually evolve to have more similarity with *irrigation* system evaluations. In irrigation system evaluations, a primary consideration is how water distribution varies across a field, and why it varies. In irrigation, it has been found that a "representative" sample of emitter discharge, for example, hides other problems which are associated with non-uniformity.

The following list outlines general rules for soil sampling (Miller and Donahue, 1990).

1. The sample should represent one uniform area or soil condition within a field.
2. Avoid mixing sub samples from unusual areas. Sample and treat unusual areas separately, if they are large enough.
3. Sample row crop fields in the middle away from rows to avoid sampling old fertilizer bands.
4. Collect samples in the late summer or fall following summer crops, but not immediately at harvest time. For multiple cropping and year-around crops in warm climates, try to sample as near to pre-seeding time as is practical. Unless the soil dries during a long period, continual plant growth (even weeds) constantly changes the nutrient status of such soils. The best sample is one taken as soon before planting as will permit time for the soil analysis and your preparation for fertilizer purchase and addition to the field.
5. Subdivide fields into sampling units based on their differences in recent cropping or fertilization histories.
6. Subdivide large uniform fields into smaller units, probably not to exceed about 15 acres, and collect a composite sample from each unit.
7. Adjust sampling depth to fit (i) laboratory suggestions, (ii) the crops to be grown, and iii) the cultivation practices used.
8. Sample annually until the field needs and crop responses are understood. Then sample every two or three years.

Some consultants recommend analyzing several soil samples at various depths to give an indication as to the *trends* in nutrient movement. This is especially important for growers fine-tuning their nutrient management strategies while fertigating with microirrigation systems.

Soil Sample Nitrogen

Laboratories generally test for parts per million (ppm) nitrogen in the nitrate form (NO_3-N). Laboratories may also test for ppm of total inorganic nitrogen ($NO_3 + NH_4$). Sometimes labs express nutrients in terms of milliequivalents per 100 grams of soil (meq/100 g). Labs which provide output results in (meq/100 g) generally provide the conversion to ppm, percent, or pounds/acre. These differences must be noted when interpreting test results and making a recommendation. Also, some labs may test for organic matter which can be converted (approximately) to an amount of available nitrogen (Table 17). For simplicity it is best to convert all forms of nitrogen to the equivalent elemental N which equals the form of nitrogen reported by the grade of a fertilizer.

Nitrogen may be held in crop residues (commonly referred to as residual N) which would not be shown in a soil analysis. Crop residues from a heavily fertilized broccoli crop in which two-thirds of the residue remains in the field may conservatively add 100 pounds of nitrogen to the soil once the residue decomposes (Hartz et al., 1994). Residue from a crop such as lettuce, in which a lesser portion remains in the field, would obviously add less nitrogen to the soil. Also, nitrogen can be added to a field through the irrigation water if there is NO_3 in the water. All nitrogen sources must be considered when planning a fertigation strategy.

Soil Sample Phosphorus

Ideally, soil sample testing methods should replicate what the plant roots experience (i.e., an extractant should not remove highly insoluble forms of P, because a plant root can not use this form of P). Soil testing methods generally depend on a soil's pH and attempt to remove only "plant-available phosphorus." In the Western United States, the Olsen or Mehlich methods are most commonly used and these measure the phosphate ion held by the soil in a readily available form.

Laboratories report phosphorus results in terms of phosphate-phosphorus (PO_4-P) in units of ppm or meq/100 g. Most soil phosphorus tests are run in commercial laboratories, although some procedures allow farmers to test P on-farm. It is best to convert all phosphorus measurements to the same form as elemental P. Note that fertilizer grades report the % P_2O_5 which can be multiplied by 0.44 (see Table 17) to obtain the equivalent elemental P concentration of the fertilizer.

Soil Sample Potassium

Traditional soil sample potassium testing methods generally measure exchangeable potassium which has been extracted from the soil through an exchanging process involving another cation (i.e., NH_4^+). These results would indicate what is *potentially* available for plant uptake; such availability hinges on cation exchange (K^+ being released from the CEC into the solution). Chapter 8, "Other Nutrient Processes" describes other details of potassium equilibrium.

Total *potential* available soil K^+ does not accurately depict what is *currently* available to plants. Rather, measuring the "Rate of Potassium Release" better indicates what the farmer should apply to make up the difference of what the plant needs and what the soil supplies.

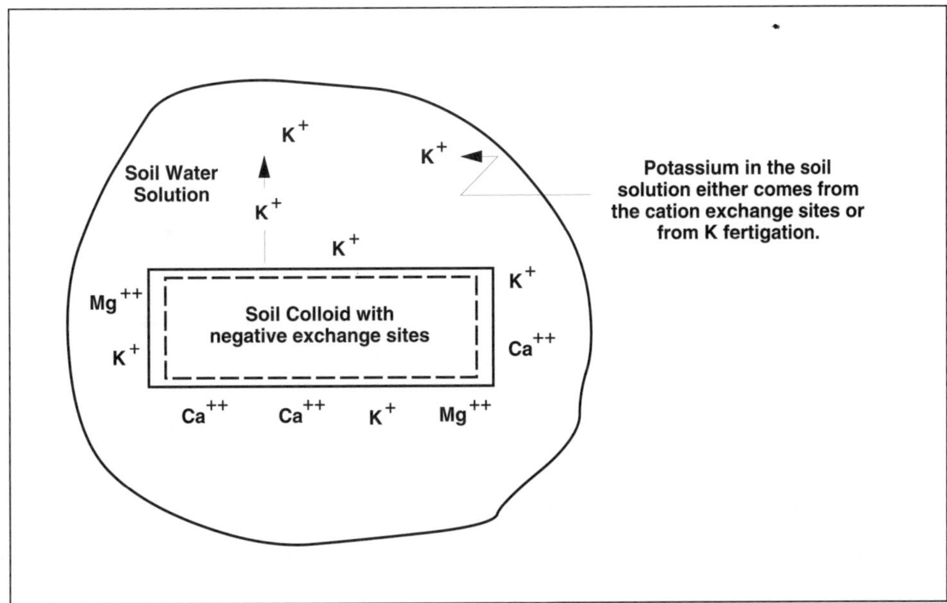

Figure 47. Fertigation must often supplement the slow natural release of potassium from the cation exchange sites.

The Union Oil Company (Unocal) has developed a special soil testing technique which measures the ability of the soil to supply potassium ions to the soil solution. It involves evaluating the soil solution several times over a period of several days to determine the K+ release rate from the exchange sites to the soil solution. Although this is not the most common method for measuring K^+, it is gaining in popularity and several commercial laboratories are able to perform the Unocal potassium test.

By determining the rate of K^+ release by a soil on a per day basis, incremental amounts of K^+ can be applied to make up the difference between what the soil is releasing and what the plant uses.

It is best to use the elemental form of K for all work. Note that fertilizer grades report the % K_2O which can be multiplied by 0.83 (see Table 17) to give the equivalent elemental K concentration in the fertilizer.

SOIL SECONDARY AND MICRONUTRIENTS

The metal micronutrients (copper, iron, manganese and zinc) are most commonly evaluated by laboratories using the DTPA chelate; results are given in ppm. The DTPA test is widely used and specifically considers the complications associated with high lime content soils. In California, deficiencies of these metal micronutrients are most common on high lime soils.

INTERPRETATIONS FROM SOIL SAMPLE TESTS

Soil sampling is used primarily to estimate annual fertilizer needs and to diagnose special problems such as toxicities and soil reclamation requirements.

Plants vary in their nutrient requirements, so it is impossible to publish absolute, quantitative standards to apply for every crop. In addition, factors other than nutrient *quantity*, such as pH, soil aeration, soil type, microbial activity, and temperature affect nutrient *availability* for plant uptake.

Commercial soil testing laboratories generally offer recommendations in conjunction with soil analyses. Such recommendations are often based on data which have been derived for various crops in specific regions. These recommendations are used as guidelines and express a likelihood of a positive yield response to a given amount of fertilizer. Commercial laboratories are generally reluctant to disclose the foundations for recommendations, so farmers are limited in their ability to develop similar recommendations outside of their own experience.

Even though it is difficult to locate published recommended soil nutrient levels which are sufficient for specific crops, certain rules regarding nutrient *balance* can be taken into consideration. Many authors agree that nutrient balance within a soil must be considered in addition to total nutrient quantity in a soil. Along these lines, *general* guidelines are given regarding nutrient balance (Table 19).

Table 19. General rules for interpreting soil test results. (Tisdale et al., 1985)

Soil Nutrient	Rule
NO_3-N	Deficient if NO_3-N < 10 ppm Generally sufficient if NO_3-N > 20 ppm
Ca	Ca (meq) should occupy 65 – 75% of CEC If Ca/Mg < 2/1, then Ca deficiency may occur
Mg	Mg (meq) should occupy 10 – 15% of CEC If Ca/Mg > 20/1, then Mg deficiency may occur
K	K (meq) should occupy 2.5 – 7.0% of CEC

Soil sample Ca : Mg ratios are commonly calculated. If the Ca : Mg ratio in the soil falls below 2 : 1, plant growth is often adversely affected. This is particularly the case for tomato growers, where low calcium levels in the plant will lead to blossom end rot. Generally the *desired* soil Ca : Mg ratio is at least 5 : 1.

P : Zn ratios have recently been examined by several growers. In cases where P : Zn ratios differ from "normal" ranges, growers might conclude that zinc levels are deficient. After further examination of the soil test values, many growers are concluding that the soil has a very high level of phosphorus. The general experience in these situations is that applying zinc by itself has had mixed results. Growers have found that they can cut back on further phosphate fertilizer for that cropping cycle, when combined with the addition of zinc supplements, this seems to improve the P : Zn *balance* and improve crop growth.

SOIL SOLUTION TESTING

HIGHLIGHTS

- The water held in the soil is referred to as the soil solution.
- Nutrients in solution are considered immediately available for plant uptake.
- Any procedure which analyzes water can also analyze the soil solution.

GENERAL

Soil solution testing is done weekly or more frequently to monitor the available nutrients during the growing season. Therefore, it is used quite differently from standard "soil sample testing," and guidelines for what constitutes a "sufficient" nutrient level in the soil solution is very different from a "sufficiency" level for annual nutrient planning.

Plants use nutrients from the water which is held in the soil; this water is referred to as the "soil solution." Testing the soil solution generally indicates what is immediately available to the plants via mass flow. This type of analysis can be helpful on drip-irrigated systems because fertigating with these systems allows the farmer to make small applications of nutrients for immediate crop demands. Also, this type of analysis avoids complications associated with soil testing (i.e., drying the soil, extracting the nutrients, etc.) because the procedures require only testing the water (the water generally doesn't require any preparation before testing). Generally, soil solution testing is used in conjunction with plant analysis.

Analyzing nutrients in the soil solution consists of simple water analysis and can be done either in a lab or on-farm. Laboratory test costs range from $30 to 100. Laboratories have the ability to analyze all nutrients, pH, and EC in the soil solution. On-farm tests of soil solution can measure NO_3, P, K, Ca, Mg, Cl, SO_4, pH and EC depending on the available equipment. Most on-farm solution testing has been limited to Quick Tests for pH, EC, and NO_3^-.

SOIL SOLUTION – TESTING PROCEDURES

Soil Solution Access Tubes (SSAT) extract the soil solution and are placed in the field with the tip in the active root zone. They consist of a porous ceramic cup (similar to a tensiometer) which is connected to a hollow tube; the tube is capped off with a cork. A hand vacuum pump is then connected to a hose in the cork, and a vacuum of 60 centibars is created in the tube. This causes the soil solution to flow into the tube through the ceramic tip. The tube remains under vacuum for a period of time; the length of time depends on soil moisture and texture, and ranges from two hours to two days. At this time, the cork is removed, and a vacuum pump with a long tube (e.g., a syringe) is used to remove the collected soil water from the bottom of the tube. Manufacturer's information for these tubes is included in Table 20.

Table 20. Manufacturer's information regarding soil solution access tubes.

Manufacturer	Part	Price/Notes
Irrometer P.O. Box 2424 Riverside, CA 92516 909-689-1701	Access tube	$7.60 – $ 9.60 depending on length; greater than 72" in length can be ordered for slightly more money
	Suction line	$6.60, connects tube with syringe to collect the sample
	Syringe	$2.80, draws water from the tube
	Vacuum pump (hand)	$51.25, applies suction to the tube
Soilmoisture Equipment Corp. P.O. Box 30025 Santa Barbara, CA 93105 805-964-3525	Soil water sampler	$39 – 58, depending on length
	Hand pump with gauge	$100, required to put tube under suction
	Extraction kit	$42, required to extract solution from tube
Soil Test Company P.O. Box 8004 Lake Bluff, IL 60044 708-295-9400	Lysimeters	$44 – 60 for access tubes, depending on length
	Service kit	$112 for hand pump and dial gauge; required equipment in addition to the access tubes

This method is best suited for drip irrigation, in which the soil moisture content is maintained fairly high. Because the moisture content is fairly high, the access tubes can have a relatively small diameter (1" or so) and still have enough ceramic contact area to collect sufficient soil solution. A minimum of 4 or 5 SSAT's should be installed per management unit for accuracy, especially if a grower is just beginning to use them. The SSAT's will only collect the soil solution from their immediate surroundings, so different placements (by depth or offset from the plant or water source) will provide different results. Nitrate in the soil solution tends to concentrate towards the edges of the wetted zone of an emitter (Hartz et al., 1994).

The soil water can be analyzed on-farm or by a lab for a variety of factors, depending on the needs and equipment. Nitrate in the soil solution is most commonly tested by nitrate selective meters. Appendix B gives manufacturer information on quick test products.

Soil Solution Testing – Interpretations

This method is appropriate for detecting *sufficiency* levels of soil nitrogen. Soil solution nitrate concentrations exceeding 50 – 75 ppm NO_3-N indicate sufficient levels of N during the early half of the season for most crops (when crop uptake of N is less and residual N in the soil is high) (Hartz et al., 1994). However, because this is a relatively new technique, considerably more work will be necessary before excellent guidelines can be published. It is hypothesized by the authors of this book that the soil solution sufficiency levels under drip irrigation fertigation will be several times greater than sufficiency levels for traditional fertilizer and irrigation regimes, because (i) "traditional" practices often have a greater wetted root zone, and (ii) "traditional" practices often have banded zones of some fertilizers, such as phosphorus, around which the availability is much higher than for the root zone as a whole.

The method of examining soil solutions can be useful in minimizing nitrogen fertigation and maximizing the use of residual nitrogen during this period, because the soil water solution concentrations tell exactly what has been converted and what remains in the soil solution. Hartz et al. (1994) recommend that solution testing be conducted in addition to tissue testing later in the season when crops approach periods of high nutrient uptake. Table 21 gives *general* information regarding solution sufficiency ranges. These values should be used as guidelines only. Specific sufficiency ranges for the soil solution will vary depending on crop, soil type, and percentage of the soil wetted by the irrigation method.

Table 21. General interpretations of nutrients in the soil solution.
(Hartz et al., 1994; Tisdale et al., 1985)

Nutrient	Sufficiency Level	Notes
NO_3-N	> 50 – 75 ppm	Generally considered sufficient during early half of the season
K	20 – 60 ppm is generally adequate	Suggested solution balance: K (ppm) = 0.10 × Ca (ppm)
Ca	unclear	Suggested solution balance: Ca (ppm) = 10 × K (ppm)
Mg	24 ppm	

If the nitrogen content of the soil water is 25 ppm and the plant needs 50 ppm in the soil water (see Table 21), then 25 ppm nitrogen should be added in the irrigation water. The solution can be tested to determine if the plant is utilizing the nutrient. Such a strategy can be used to determine nutrient uptake efficiency. The fertilizer injection rate can be adjusted to insure that nutrient supply is meeting the crop's nutrient demand. Note that a decrease in soil solution NO_3 does not necessarily indicate plant uptake. Rather, such results might indicate that leaching has occurred from over-irrigation or from a poor irrigation distribution uniformity (DU).

The problem with soil solution testing is that the results will be highly variable depending on the tubes' depth and location relative to the emitters and edge of the wetted pattern. This technique may be more suited for monitoring nutritional *trends*, rather than determining absolute sufficiency levels.

PLANT TISSUE ANALYSIS

HIGHLIGHTS

- Good fertilizer strategies include *on-going nutrient monitoring* of crops.
- This strategy allows farmers to detect nutrition problems before they decrease yields.
- Fresh plant tissue (sap) testing can be done by labs or farmers and allows for frequent, inexpensive testing.
- Time of sampling and plant part sampled are *critical* for meaningful tissue analyses, because nutrient concentrations change throughout the plant and throughout the season.
- Proper interpretation depends on the tissue sampled being the same as the tissue used as the standard for comparison in making the nutrient recommendations.

GENERAL

Jones et al. (1991) makes a distinction between plant testing and tissue testing. Plant (leaf) testing consists of an analysis of the total elemental content of the leaf. "Tissue testing" extracts only a portion of the nutrients from a specific plant part (i.e., petiole). Plant testing has traditionally occurred in the laboratory and been done at the end of the cropping season to help diagnose nutritional problems in low yielding plants. Laboratory costs for leaf and petiole tests range from $20 to 100; generally these tests are run to determine the primary nutrients along with whatever micronutrients are of interest to the grower.

Tissue testing is the technique which will be discussed in this book. Tissue testing is similar to soil water solution testing, in that it is intended for very frequent monitoring of nutrient conditions. Whereas soil solution testing provides information on what is available in the soil, tissue testing shows what is actually used.

As mentioned earlier, those using fertigation in microirrigation systems have sought improved methods for nutrient monitoring due to their new capability to "spoon feed" nutrients to plants. Relatively simple tissue tests provide growers with additional ways to receive nutrition information faster and at a lower cost than traditional leaf testing. This has allowed them to conduct more tests throughout a growing season—thus allowing them to *monitor* the nutrient status in the field. Tissue testing, unlike soil solution monitoring, is already very common in commercial irrigated agriculture. However, it is typically only used a few times per season on most crops. On high value crops there is a definite trend towards more frequent testing.

Past cropping history is the best guide to anticipate what problems the crop may encounter during the current production cycle. Previous low nutrient conditions or other limiting factors may persist from season to season. This is especially true of nutrient problems such as low levels of soil phosphorus or zinc, or high levels of boron or salt. Most growers have found that keeping careful records of soil test nutrient levels, irrigation water nutrient concentrations (particularly nitrates), plant leaf tissue levels, petiole concentrations, and yields gives them an excellent picture of what to expect next season with the same crop. Computer spreadsheets or databases allow them to manage this information quickly and conveniently.

Frequent tissue sampling allows the grower to make mid-season fertigation adjustments to resupply the nutrients for the currently growing crop. Such adjustments can potentially help the plant before a yield reduction occurs. This strategy requires rapid turnaround time: on-farm tests and certain laboratories facilitate the rapid turnaround time. Some laboratories now provide one or two day turn around on tissue testing. Many growers find that it is most convenient to contract with a laboratory to take the samples on a regular basis. This service is generally convenient for the grower because the grower does not have to remember the testing and sampling schedule.

Tissue analysis allows the grower to obtain a picture of the plant's performance which may not be apparent from visual observations. Once a deficiency shows up visually, yield has already been reduced appreciably. Petiole testing is done by growers every week or two to assess the nutrient movement into the crop. The nutrients which are most commonly monitored are nitrogen and phosphorus and less often potassium. Monitoring of petiole nutrients has been compared to counting the number of people presently using an escalator to the next floor of a shopping mall. The number only gives you the current capacity of the escalator or of the petiole conductive system. It does not tell you how much of the nutrient has or has not accumulated in other parts of the plant. Nevertheless, when the escalator is nearly full, the system is working at near the optimum.

Plant Tissue Sampling Procedures

The nutrient concentrations change with each different leaf over time, and the nutrient concentrations will differ from lower to upper leaves. For nutrients which are immobile within the plant (i.e., Ca, S, Fe, Zn, Mn, B), nutrient concentrations will decrease in new leaves if the availability of a nutrient in the soil is marginal to low. For nutrients which are mobile within a plant (i.e., N, Mg, Cu, K, P), older leaves will usually have a lower concentration of the nutrient than the newer leaves because the plant translocates a mobile nutrient to new tissue if there is a deficiency. Older leaves produce enzymes which break down the old leaf tissue and export the usable nutrients to the newer leaves to sustain plant growth.

Therefore, it is critical that the stage of growth and plant part be considered when collecting tissue samples. Specifically, these factors must be matched with those which were used to derive the interpretative standards. For example, if a farmer wants to use U.C. sufficiency guidelines for tissue analysis, then the grower must use the same plant part and stage of growth which U.C. used. Whether whole leaves or only petioles are sampled, the most commonly sampled tissue is from the most recently developed, fully mature leaf.

The frequency of plant nutrient testing is highly variable depending on the grower's situation. Some growers only test once during the season. Others will test three to five times per season, with an interval between samples ranging from 7 – 14 days.

Tisdale et al. (1985) list several general factors regarding plant testing:

1. It is ideal to follow the uptake of nutrients through the season by testing five or six times. Nutrient levels should be higher in the early season when the plant is not in stress. [Note, however, that the five to six times rule was written in a non-fertigation situation; fertigation allows better control and may require more frequent feedback to optimize that control.]
2. The plants' greatest need for nutrients generally comes at the time when they are preparing to make seed (i.e., flowering stage). If the field is to be checked only once a season to determine the adequacy of the fertilization program, this will be the time (although the information may be too late for use that season).

3. Comparison of plants in a field is helpful. Plant samples from deficient areas should be compared with plants from normal areas.
4. Plants vary. Sub-samples should be taken from several plants, or growers can test 10 to 15 plants and average the results.

TISSUE TEST INTERPRETATION – CRITICAL LEVEL APPROACH

Interpretation of tissue test results is perhaps the most difficult step in plant testing. Several methods for interpretation exist and all have distinct advantages and disadvantages.

A critical level is the concentration in which an element is deemed either insufficient (below) or sufficient (above). These levels are often defined as the critical nutrient concentration corresponding to 90% yield (Sumner and Boswell, 1981). Critical level values are often established from the outcomes of single-factor experiments (Sumner, 1990). In these experiments, researchers keep all factors at an "optimum" level and vary a single factor (such as N application rate). Therefore, critical values may not be transferable to crops grown under different conditions. In addition, critical levels are dependent on the time of year and plant part tested, as nutrient concentrations are known to change throughout the growing season. This eliminates the flexibility in sampling time that is needed for microirrigation fertigation. The trend is moving away from critical levels to sufficiency *ranges* for tissue analysis interpretation.

TISSUE TEST INTERPRETATION – SUFFICIENCY RANGE APPROACH

Sufficiency ranges are perhaps the most common plant tissue interpretation method used in California. Sufficiency ranges establish a specified range below which, nutrients are considered deficient, and above which, they are considered in excess. These ranges are developed from either single-factor experiments or from analysis of large population data through field surveys. As with the critical value approach, this method requires that the plant species, part of plant sampled, and timing of sample, are done as specified by the procedure developed. Most often, this is defined by a certain growth stage and requires not only correct diagnosis of crop age and plant part in the field, but also sufficiency level standards for many growing stages and many crops. Table 22 contains widely used U.C. sufficiency ranges to use for interpretation of tissue results. This table is commonly used when interpreting tissue test results. Growers should be sure that the nutrients tested (i.e., NO_3-N, PO_4-P, K) are consistent with those used in this table. In addition, growers must be sure that the *units of measurement* are consistent with those used in this table.

Table 22. Plant analysis guide for sampling time, plant part, and nutrient levels of vegetable crops based on samples from California field experiments. (Lorenz and Tyler, 1983)

Crop	Time of sampling	Plant part	Nutrient[1]	Nutrient Level — Deficient	Nutrient Level — Sufficient
Asparagus	Midgrowth of fern	4" tip section of new fern branch	NO_3-N, ppm	100	500
			PO_4-P, ppm	800	1,600
			K, %	1	3
Bean (bush snap)	Midgrowth	Petiole of 4th leaf from tip	NO_3-N, ppm	2,000	3,000
			PO_4-P, ppm	1,000	2,000
			K, %	3	5
	Early bloom	Petiole of 4th leaf from tip	NO_3-N, ppm	1,000	1,500
			PO_4-P, ppm	800	1,500
			K, %	2	4
Broccoli	Midgrowth	Midrib of young, mature leaf	NO_3-N, ppm	7,000	9,000
			PO_4-P, ppm	2,500	4,000
			K, %	3	5
	First buds	Midrib of young, mature leaf	NO_3-N, ppm	5,000	7,000
			PO_4-P, ppm	2,500	4,000
			K, %	2	4
Brussel Sprouts	Midgrowth	Midrib of young, mature leaf	NO_3-N, ppm	5,000	7,000
			PO_4-P, ppm	2,000	3,500
			K, %	3	5
	Late growth	Midrib of young, mature leaf	NO_3-N, ppm	2,000	3,000
			PO_4-P, ppm	1,000	3,000
			K, %	2	4
Cabbage	At heading	Mid rib of wrapper leaf	NO_3-N, ppm	5,000	7,000
			PO_4-P, ppm	2,500	3,500
			K, %	2	4
(Chinese)	At heading	Mid rib of wrapper leaf	NO_3-N, ppm	8,000	10,000
			PO_4-P, ppm	2,000	3,000
			K, %	4	7
Cantaloupe	Early growth (short runners)	Petiole of 6th leaf from growing tip	NO_3-N, ppm	8,000	12,000
			PO_4-P, ppm	2,000	3,000
			K, %	4	6
	Early fruit	Petiole of 6th leaf from growing tip	NO_3-N, ppm	5,000	8,000
			PO_4-P, ppm	1,500	2,500
			K, %	3	5
	First mature fruit	Petiole of 6th leaf from growing tip	NO_3-N, ppm	2,000	3,000
			PO_4-P, ppm	1,000	2,000
			K, %	3	4

continued

Table 22. Continued.

Crop	Time of sampling	Plant part	Nutrient[1]	Nutrient Level — Deficient	Nutrient Level — Sufficient
(muskmelon)	Early growth	Blade of 6th leaf from growing tip	NO_3-N, ppm	2,000	3,000
			PO_4-P, ppm	1,500	2,300
			K, %	1	3
	Early fruit	Blade of 6th leaf from growing tip	NO_3-N, ppm	1,000	1,500
			PO_4-P, ppm	1,300	1,700
			K, %	1	2
	First mature fruit	Blade of 6th leaf from growing tip	NO_3-N, ppm	500	800
			PO_4-P, ppm	1,000	1,500
			K, %	1	2
Carrot	Midgrowth	Petiole of young, mature leaf	NO_3-N, ppm	5,000	7,500
			PO_4-P, ppm	2,000	3,000
			K, %	4	6
Cauliflower	Buttoning	Midrib of young, mature leaf	NO_3-N, ppm	5,000	7,000
			PO_4-P, ppm	2,500	3,500
			K, %	2	4
Celery	Midgrowth	Petiole of newest fully elongated leaf	NO_3-N, ppm	5,000	7,000
			PO_4-P, ppm	2,500	3,000
			K, %	4	7
	Near maturity	Petiole of newest fully elongated leaf	NO_3-N, ppm	4,000	6,000
			PO_4-P, ppm	2,000	3,000
			K, %	3	5
Cucumber (pickling)	Early fruit set	Petiole of 6th leaf from tip	NO_3-N, ppm	5,000	7,500
			PO_4-P, ppm	1,500	2,500
			K, %	3	5
(slicing)	Early harvest period	Petiole of 6th leaf from tip	NO_3-N, ppm	5,000	7,500
			PO_4-P, ppm	1,500	2,500
			K, %	4	7
Eggplant	At first harvest	Petiole of young, mature leaf	NO_3-N, ppm	5,000	7,500
			PO_4-P, ppm	2,000	3,000
			K, %	4	7
Lettuce	At heading	Midrib of wrapper leaf	NO_3-N, ppm	4,000	6,000
			PO_4-P, ppm	2,000	3,000
			K, %	2	4
	At harvest	Midrib of wrapper leaf	NO_3-N, ppm	3,000	5,000
			PO_4-P, ppm	1,500	2,500
			K, %	2	3

continued

Table 22. Continued.

Crop	Time of sampling	Plant part	Nutrient[1]	Nutrient Level: Deficient	Nutrient Level: Sufficient
Pepper (chili)	Early growth 1st bloom	Petiole of young, mature leaf	NO_3-N, ppm	5,000	7,000
			PO_4-P, ppm	2,000	2,500
			K, %	3	5
	Early fruit set	Petiole of young, mature leaf	NO_3-N, ppm	1,000	1,500
			PO_4-P, ppm	1,500	2,000
			K, %	2	4
	Fruits, full size	Petiole of young, mature leaf	NO_3-N, ppm	750	1,000
			PO_4-P, ppm	1,500	2,000
			K, %	2	3
	Early growth 1st bloom	Blade of young, mature leaf	NO_3-N, ppm	1,500	2,000
			PO_4-P, ppm	1,500	2,000
			K, %	3	5
	Early fruit set	Blade of young, mature leaf	NO_3-N, ppm	500	800
			PO_4-P, ppm	1,500	2,000
			K, %	2	4
Pepper (sweet)	Early growth 1st flower	Petiole of young, mature leaf	NO_3-N, ppm	8,000	10,000
			PO_4-P, ppm	2,000	3,000
			K, %	4	6
	Early fruit set 1" diam.	Petiole of young, mature leaf	NO_3-N, ppm	3,000	5,000
			PO_4-P, ppm	1,500	2,500
			K, %	3	5
	Fruit 3/4 size	Petiole of young, mature leaf	NO_3-N, ppm	2,000	3,000
			PO_4-P, ppm	1,200	2,000
			K, %	2	4
	Early growth 1st flowers	Blade of young, mature leaf	NO_3-N, ppm	2,000	3,000
			PO_4-P, ppm	1,800	2,500
			K, %	3	5
	Early fruit set, 1" diam.	Blade of young, mature leaf	NO_3-N, ppm	1,500	2,000
			PO_4-P, ppm	1,500	2,000
			K, %	2	4
Potatoes	Early season	Petiole of 4th leaf from growing tip	NO_3-N, ppm	8,000	12,000
			PO_4-P, ppm	1,200	2,000
			K, %	9	11
	Midseason	Petiole of 4th leaf from growing tip	NO_3-N, ppm	6,000	9,000
			PO_4-P, ppm	800	1,600
			K, %	7	9

continued

Table 22. Continued.

Crop	Time of sampling	Plant part	Nutrient[1]	Nutrient Level: Deficient	Nutrient Level: Sufficient
	Late season	Petiole of 4th leaf from growing tip	NO_3-N, ppm	3,000	5,000
			PO_4-P, ppm	500	1,000
			K, %	4	6
Spinach	Midgrowth	Petiole of young, mature leaf	NO_3-N, ppm	4,000	6,000
			PO_4-P, ppm	2,000	3,000
			K, %	2	4
Summer squash (zucchini)	Early bloom	Petiole of young, mature leaf	NO_3-N, ppm	12,000	15,000
			PO_4-P, ppm	4,000	6,000
			K, %	6	10
Sweet corn	Tasseling	Midrib of 1st leaf above primary ear	NO_3-N, ppm	500	1,000
			PO_4-P, ppm	500	1,000
			K, %	2	4
Sweet potato	Midgrowth	Petiole of 6th leaf from the growing tip	NO_3-N, ppm	1,500	2,500
			PO_4-P, ppm	1,000	2,000
			K, %	3	5
Tomato (cherry)	Early fruit set	Petiole of 4th leaf from growing tip	NO_3-N, ppm	8,000	10,000
			PO_4-P, ppm	2,000	3,000
			K, %	4	7
	Fruit 1/2" diam.	Petiole of 4th leaf from growing tip	NO_3-N, ppm	5,000	7,000
			PO_4-P, ppm	2,000	3,000
			K, %	3	5
	At first harvest	Petiole of 4th leaf from growing tip	NO_3-N, ppm	1,000	2,000
			PO_4-P, ppm	2,000	3,000
			K, %	2	4
Tomato (processing and determinant, fresh market)	Early bloom	Petiole of 4th leaf from growing tip	NO_3-N, ppm	8,000	12,000
			PO_4-P, ppm	2,000	3,000
			K, %	3	6
	Fruit 1" diameter	Petiole of 4th leaf from growing tip	NO_3-N, ppm	4,000	6,000
			PO_4-P, ppm	1,500	2,500
			K, %	2	4
	First color	Petiole of 4th leaf from growing tip	NO_3-N, ppm	2,000	3,000
			PO_4-P, ppm	1,000	2,000
			K, %	1	3
Tomato (fresh market non-determinant, type)	Early bloom	Petiole of 4th leaf from growing tip	NO_3-N, ppm	10,000	14,000
			PO_4-P, ppm	2,500	3,000
			K, %	400	7

continued

Table 22. Continued.

Crop	Time of sampling	Plant part	Nutrient[1]	Nutrient Level: Deficient	Nutrient Level: Sufficient
	Fruit 1" diameter	Petiole of 4th leaf	NO_3-N, ppm	8,000	12,000
		from growing tip	PO_4-P, ppm	2,500	3,000
			K, %	3	5
	Full ripe fruit	Petiole of 4th leaf	NO_3-N, ppm	4,000	6,000
		from growing tip	PO_4-P, ppm	2,000	2,500
			K, %	2	4
Watermelon	Early fruit	Petiole of 6th leaf	NO_3-N, ppm	5,000	7,500
		from growing tip	PO_4-P, ppm	1,500	2,500
			K, %	3	5

1 2% Acetic acid soluble NO_3-N and total K (dry-weight basis).

Table 23 differs from the previous table because the following table uses TOTAL NUTRIENTS as the interpretive parameter. For example, for nitrogen, total nitrogen includes organic (proteins and amino acids) and inorganic forms (ammonium and nitrate). Organic nitrogen held in the plant tissue represents, by far, the greatest portion of total nitrogen, and organic nitrogen does not vary significantly in a day. Consequently, measurements of total nitrogen are not as variable over the course of a day as those for inorganic nitrogen.

Table 23. Total nutrient analyses for diagnosis of the total nutrient level of vegetable crops. (Lorenz and Tyler, 1983)

Crop	Time of sampling	Plant part	Nutrient	Nutrient Level: Deficient (% dry weight)	Nutrient Level: Sufficient (% dry weight)
Asparagus	Early fern growth	4" tip section of	N	4.00	5.00
	(May – June)	new fern branch	P	0.20	0.40
			K	2.00	4.00
	Mature fern	4" tip section of	N	3.00	4.00
	(July – Sept.)	new fern branch	P	0.20	0.40
			K	1.00	3.00
Asparagus	Early fern growth	4" tip section of	N	4.00	5.00
Bean (bush snap)	Full bloom	Petiole: recent	N	1.50	2.25
		fully exposed	P	0.15	0.30
		trifoliate leaf	K	1.00	2.50
	Full bloom	Blade: recent	N	1.25	2.25
		fully exposed	P	0.25	0.40
		trifoliate leaf	K	0.75	1.50
		leaf	P	0.20	0.30
			K	1.50	2.25

continued

Table 23. Continued.

Crop	Time of sampling	Plant part	Nutrient	Nutrient Level — Deficient (% dry weight)	Nutrient Level — Sufficient (% dry weight)
Bean (lima)	Full bloom	Oldest trifoliate	N	2.50	3.50
Cantaloupe (muskmelon)	Early growth (short runners)	Petiole of 6th leaf from growing tip	N	2.50	3.50
			P	0.30	0.60
			K	4.00	6.00
	Early fruit	Petiole of 6th leaf from growing tip	N	2.00	3.00
			P	0.20	0.35
			K	3.00	5.00
	1st mature fruit	Petiole of 6th leaf from growing tip	N	1.50	2.00
			P	0.15	0.30
			K	2.00	4.00
Celery	Midgrowth	Petiole	N	1.00	1.50
			P	0.25	0.55
			K	4.00	5.00
Garlic	Early season (pre-bulbing)	Newest fully elongated leaf	N	4.00	5.00
			P	0.20	0.30
			K	3.00	4.00
	Midseason (bulbing)	Newest fully elongated leaf	N	3.00	4.00
			P	0.20	0.30
			K	2.00	3.00
	Late season (post bulbing)	Newest fully elongated leaf	N	2.00	3.00
			P	0.20	0.30
			K	1.00	2.00
Lettuce	At heading	Leaves	N	1.50	3.00
			P	0.20	0.35
			K	2.50	5.00
	Nearly mature	Leaves	N	1.25	2.50
			P	0.15	0.30
			K	2.50	5.00
Onion	Midgrowth	Tallest leaf	N	2.00	2.50
			P	0.10	0.20
			K	2.00	2.50
Pepper (sweet)	Full bloom	Blade and petiole	N	2.00	3.50
			P	0.15	0.25
			K	1.50	2.50
	Full bloom fruit 3/4 size	Blade and petiole	N	1.50	2.50
			P	0.12	0.20
			K	1.00	2.00

continued

Table 23. Continued.

Crop	Time of sampling	Plant part	Nutrient	Nutrient Level Deficient	Nutrient Level Sufficient
				% dry weight	
Potato	Early plants 12" tall	Petiole of 4th leaf from tip	N	2.50	3.50
			P	0.20	0.30
			K	9.00	11.00
	Midseason	Petiole of 4th leaf from tip	N	2.25	2.75
			P	0.10	0.20
			K	7.00	9.00
	Late nearly mature	Petiole of 4th leaf from tip	N	1.50	2.25
			P	0.08	0.15
			K	4.00	6.00
	Early plants 12" tall	Blade of 4th leaf from tip	N	4.00	6.00
			P	0.30	0.60
			K	3.50	5.00
	Midseason	Blade of 4th leaf from tip	N	3.00	5.00
			P	0.20	0.40
			K	2.50	3.50
	Late nearly mature	Blade of 4th leaf from tip	N	2.00	4.00
			P	0.10	0.20
			K	1.50	2.50
Southern pea (cowpea)	Full bloom	Blade and petiole	N	2.00	3.50
			P	0.20	0.30
			K	1.00	2.00
Spinach	Midgrowth	Mature leaf blades and petiole	N	2.00	4.00
			P	0.20	0.40
			K	3.00	6.00
	At harvest	Mature leaf blades and petiole	N	1.50	3.00
			P	0.20	0.35
			K	2.00	5.00
Sweet corn	Tasseling	6th leaf from base of plant	N	2.75	3.50
			P	0.18	0.28
			K	1.75	2.25
	Silking	Leaf opposite 1st ear	N	1.50	2.00
			P	0.20	0.30
			K	1.00	2.00
Tomato	Flowering	Leaf blade and petiole	N	2.00	3.00
			P	0.20	0.35
			K	2.50	4.00
	First ripe fruit	Leaf blade and petiole	N	1.50	2.50
			P	0.15	0.25
			K	1.50	3.00

A common concern of California growers is that many currently published recommendations of critical levels and sufficiency ranges for California crops may not be appropriate for the increased yields produced under microirrigation systems. Many growers begin by using published guidelines, but they eventually adjust the standards upward as crop yields increase. Other growers feel that the sufficiency ranges are acceptable, but the pounds of nutrient uptake per acre increases with larger yields. Several testing labs have begun to calibrate new levels for crops for specific areas by using their customer database. This has proven to be beneficial for many crops, but much work is still needed.

Table 24 illustrates a sample laboratory petiole analysis report, along with the sufficiency levels recommended by the laboratory.

Table 24. Sample laboratory petiole analysis for citrus.

	Nutrient Concentrations [1]				
	N	P	K	Zn	Mn
Test ID	(%)	(%)	(%)	(ppm)	(ppm)
1	2.2	0.13	1.36	23	49
2	2.2	0.12	0.92	31	30
3	2.1	0.13	1.18	28	30
4	2.4	0.14	1.01	22	32
5	2.4	0.13	1.07	21	27

1 Sample data taken from actual laboratory analysis on citrus.

Laboratory Recommendations:

N The optimum nitrogen range for Navels and Valencias is 2.4 – 2.6% and for Satsuma is 2.2 – 2.4%. Satsuma produce best at a nitrogen level of 2.4 – 2.5%. Adjust your nitrogen program according to N levels, tree vigor and crop size.

P The optimum phosphorus range is 0.12 – 0.16%.

K The optimum potassium range is 0.7 – 1.09%.

Zn Zinc levels should be maintained above 25 ppm.

Mn Manganese levels are normal between 25 and 200 ppm.

TISSUE TESTING INTERPRETATION – NUTRIENT RATIO/PRODUCT APPROACH

Nutrient ratios have been touted as a method to overcome the problems of tissue age and nutrient interactions. The idea is that even though nutrient concentrations change with time, ratios and products of nutrients remain fairly constant throughout the growing season. Walworth and Sumner (1987) concluded that the percentage of N, P, and K varies greatly over different ages of the plant. On the other hand, the nutrient ratios (e.g., N : P, N : K and K : P) showed much less variation for different growth stages.

Nutrient concentrations in plants, (expressed as a percent of dry matter) decline over time for N, P, K, S, Cu, Zn, and B, while Ca and Mn increase, and Mg increases or stays the same (Tisdale et al., 1985). Again, these relationships are a function of whether the nutrients are mobile or immobile within the plant. To compare two nutrients which both increase over time, one should use ratios (i.e., N : P, N : K, N : S). To compare one nutrient which increases with another which decreases over time, one should use products (i.e., (N × Ca) or (K × Ca)).

Table 25. Criteria for selecting a nutrient ratio or product.

Change in concentrations of two nutrients over time	Nutrient combination to examine
Both increase	A : B
Both decrease	A : B
One increases, one decreases	A × B

Interpreting results in terms of nutrient ratios or products reduces the influence of plant age on the results. Even so, the changing *rates* of different nutrient accumulation in plant tissues causes variations in the ratios. For example, if nutrient A increases faster than nutrient B, the ratio will change also. The issue of unequal rate changes provides a disadvantage in using the nutrient ratio approach. Nevertheless, these variations in ratios with plant age or products are significantly less than those of absolute concentrations.

Several key ratios have already been identified by growers as potentially indicative of nutritional status. Jones et al. (1991) specifies optimal ratio ranges for many different crops. The details of their results are too large to incorporate in this text.

N : P ratios in plant tissue, in particular, are becoming increasingly important in managing these two nutrients. The ratio of N : P is approximately 10 for most healthy plants. The ratio between N : K is approximately 1.0 for most healthy plants (Tisdale et al., 1985).

Of course, one must look beyond the simple ratios. If the optimum ratio of N : P for tomatoes is 10 and the sample indicates a level of 8, it is impossible to tell if N is too high, P is too low, or if N is too high and P is too low. When other ratios of the same nutrients are included, they can help to determine which nutrient is either low or high. When coupled with sufficiency range data, the determination may be more easily made.

Growers are beginning to look at high yielding crops when establishing nutrient ratios. This gives a greater indication of *optimum* nutrient ratios. Only a few growers and laboratories currently use nutrient ratios as part of their diagnostic program.

Table 26 computes the ratios of some key nutrients which were analyzed in Table 24.

Table 26. Nutrient ratios derived from actual laboratory petiole analyses results.

Test ID	N : P	N : K	P : Zn	Zn : Mn
1	16.9	1.6	56.5	0.5
2	18.3	2.4	38.7	1.0
3	16.2	1.8	46.4	0.9
4	17.1	2.4	63.6	0.7
5	18.5	2.2	61.9	0.8

It can be seen from the above table that there are substantial differences in ratio values between different tests. Although specific information for citrus regarding optimum ratio balances is difficult to find, these ratios or products, when combined with yield information from the different tests, could provide valuable insight into nutrient management. Presumably, the optimum ratio balances will be more uniform as the crop yield or quality approaches near maximum economic production levels.

Tissue Testing Interpretation – Diagnosis and Recommendation Integrated System (DRIS) Approach

The DRIS method was developed by Beaufils in the early 1970's, in South Africa. DRIS goes one step further than nutrient ratios or products, alone. DRIS *ranks* the deficient nutrients, therefore including a recommendation as to which nutrient should be added first.

The DRIS method can be applied to both soil and plant testing and is an extension of the nutrient ratios concept. It purportedly overcomes the problems associated with plant age and nutrient interactions, while being able to *rank* the nutrients in the order of deficiency.

Three steps are needed to utilize the DRIS method: determining the standard nutrient ratios for peak yields (or norms), making a diagnosis, and calibrating the norms and diagnoses.

Through analyzing the nutrient status of high and low yielding crops, researchers use statistics to determine the "norms" (the *ideal* nutrient ratios) for the particular crop (Walworth and Sumner, 1987). These procedures are normally done by researchers and laboratories. In addition, researchers make recommendations according to a crop's *deviation* from the norm. Finally, they determine whether the fertilizer recommendation promoted a yield increase.

Several researchers have compared traditional plant analysis interpretations to the DRIS method of making correct diagnoses (Table 27). Kelling and Schulte (1988) claim that sufficiency range and DRIS indicated similar diagnoses in a Wisconsin plant analysis program. Although in some cases, "DRIS has pointed to problems that may have been overlooked previously. Farmer and consultant feedback has been extremely positive." (Kelling and Schulte, 1988)

Table 27. Comparison of successes and failures of fertilizer recommendations depending on the interpretation method (Critical Value vs. DRIS) for corn. (Kelling and Schulte, 1988)

		Successful Diagnoses		Unsuccessful Diagnoses	
Diagnostic Method	Nutrient Applied	Number	Yield responses (bu/acre)	Number	Yield responses (bu/acre)
DRIS					
	N	47	+ 980	8	- 5
	P	40	+ 1500	6	- 40
	K	79	+ 2300	10	- 90
Critical Value					
	N	31	+ 510	9	- 70
	P	35	+ 1100	8	- 30
	K	67	+ 2170	17	- 170

Table 27 indicates that DRIS influenced more successful and less unsuccessful diagnoses than did the critical value method. Similarly, the positive yield response with a correct diagnosis was greater with DRIS than the critical value method. Except for P, the negative response with an incorrect diagnosis was less with the DRIS method.

In organic and mineral soils in Florida, DRIS correctly diagnosed a probable K response in lettuce when the sufficiency ranges did not make such a diagnosis (Sanchez et al., 1991). Similarly, DRIS norms improved the nutritional diagnosis of greenhouse tomato plants over critical level methods (Caron and Parent, 1989).

One reported disadvantage of the DRIS method is that it may indicate a nutrient limitation for a nutrient that is low relative to others even though the nutrient may not limit yield (Beverly et al., 1984 and Sanchez et al., 1991). It appears that the DRIS method will often determine that a nutrient is not in balance with the others, yet an addition of this nutrient may not increase yields. Several techniques may overcome this problem. Kelling and Schulte (1988) discuss a method used in Wisconsin to overcome interpretation problems with DRIS results by using *ranges* in which nutrients are considered in balance. If nutrients lie within this range, no application is made.

Individual growers can approximate the DRIS process by analyzing leaves for total nutrient content of the various essential plant nutrients. They can use as their norms, the best producing plants in the field. The nutrient concentrations can be entered into any spreadsheet on a computer. The computer can be used to calculate the nutrient ratios or products. It will not make any difference whether the grower enters percentages or ppm concentrations, as long as the data for any one nutrient is always entered in the same units of measure.

The authors believe that some form of DRIS has potential for use in determining the nutritional status of California crops grown using microirrigation systems. The ability of microirrigation systems to deliver fertilizers in a timely fashion, and the increased yields associated with better crop management allowed by microirrigation, fit well with the DRIS method's reported ability to determine the nutrient most limiting yield, ranking of subsequent nutrients in order of deficiency, and greatly reducing the effects of growing stage on nutritional status. The fact that DRIS norms are also calculated using data from high yielding crop populations also favors the adaptation of this method to microirrigation systems.

Table 28. Citations for published DRIS norms. (Adapted from Walworth and Sumner, 1987)

Crop	Author	Source
Alfalfa	Kelling et al. (1983)	Proc. Wisc. Fert. Aglime Pest Management Conference 22:239 – 243.
Alfalfa	Walworth et al. (1986)	Agron. J. 78:1046 – 1052.
Corn	Sumner (1981)	Soil Sci. Soc. Am. J. 45:87 – 90.
Corn	Escano et al. (1981)	Soil Sci. Soc. Am. J. 45:1140 – 1144.
Corn	Elwali et al. (1985)	Agron. J. 77:506 – 508.
Citrus	Beverly et al. (1984)	J. Am. Soc. Hort. Sci. 109:649 – 654.
Citrus	Sumner (1985)	Food and Fertil. Tech. Center, Ext. Bull. 231. Taipai, Taiwan.
Lettuce	Sanchez et al. (1991)	Hor. Sci. 26(3):274 – 276.
Oats	Chojnacki (1984)	Proc. VI Int. Colloq. Opt. Plant Nutr. 1:139 – 148.
Peaches	Sumner (1985)	Food and Fertil. Tech. Center, Ext. Bull. 231. Taipai, Taiwan.
Pineapples	Langenegger and Smith (1978)	Proc. 8th In. Colloq. Plant Anal. Fertil. Prob. DSIR Information Series No. 134, Wellington, N.Z.
Poplars	Leech and Kim (1981)	For. Chron. 17 – 21.
Potatoes	Meldal-Johnsen (1980)	J. Plant Nutrit. 2:569 – 576.
Rubber	Beaufils (1957)	Fertilite 3:27 – 38.
Soybeans	Sumner (1977e)	Agron. J. 69:226 – 230.
Soybeans	Hallmark (1984 – 85)	J. Fertil. Iss. 1:104 – 109;2:66 – 73.
Sugarcane	Beaufils (1976)	Proc. S. Afr. Sugar Tech. Assoc. 50:118 – 124.
Sugarcane	Elwali (1983, 1984)	Agron. J. 75:79 – 83;76:466 – 470.
Sunflower	Grove and Sumner (1982)	Fertil. Res. 3:367 – 378.
Tea	Lee (1980)	Crop Prod. 9:207 – 209.
Tomatoes	Caron and Parent (1989)	Can. J. Plant Sci. 69:1027 – 1035
Wheat	Sumner (1981)	Soil Sci. Soc. Am. J. 45:87 – 90.

Several authors have published norms for various crops (Table 28). These values are not necessarily transferable from region to region, because there are other environmental factors which limit yields. Nevertheless, they can be used as guidelines. To determine the extent of transferability of DRIS norms, one has to consider the sample size and geographic diversity (the "widely" accepted corn norms had 8,000 samples from all over the world). Also, the samples should be diverse in their plant stage of growth. This would ensure that the norms are not dependent on crop stage. In cases where norms vary by region, one could average the values for practical application of DRIS.

PLANT SAP TESTING

Analyzing fresh plant sap is a relatively new sampling/diagnostic technique, but it is being recommended by researchers for use on some vegetable crops in California. The samples are often tested in an office or farm building, as the process is temperature sensitive. The petioles are rinsed with distilled water and then crushed in a garlic press. The sap is placed on the end of a meter electrode and the reading is displayed digitally (Hartz et al., 1994). Appendix C explains the specific process for sap analysis using the Cardy Meter®, the most popular field meter used at the time of this publication. The Cardy Meter® electrode lasts for 250 – 400 tests, with replacement electrodes available from the manufacturer. The primary advantage is the simplicity and quickness of this test, as compared to traditional tissue testing.

It is important to note that nitrate concentrations in the plant sap are not the only indication of crop health. For example, sap nitrate levels may be low but the plants may be doing fine. This can be attributed to the fact that plants take up ammonium also, which would not be indicated in a sap nitrate test. If the sap tests are high and soil solution tests are high, this may indicate that there is over fertilization of nitrogen. If the sap tests are low and the solution tests are high, this might indicate that another ion (i.e., Cl^-) might be interfering with the nitrate uptake or some nutrient (such as molybdenum deficiency) may be limiting the plant's ability to use the nitrate. Several consultants agree that a weakness of sap testing is the tendency to *ignore* other nutritional problems.

Sap and tissue can also be tested using test strips and/or other extraction procedures. Obviously manufacturer's directions should be followed depending on the device used.

The concepts of critical levels, sufficiency ranges, nutrient ratios, and DRIS can potentially be applied to plant sap measurements as well as to tissue analysis. Table 29 provides information on sufficiency ranges for both tissue and plant sap testing.

Figure 48. Nitrate nitrogen readings from the Cardy Meter®.

Table 29. Tissue and sap petiole NO_3-N sufficiency value ranges for drip-irrigated vegetables. (Hartz et al., 1994)

Crop	Growth stage	——Petiole NO_3-N concentration (ppm)—— Dry tissue	Fresh sap
Broccoli	Midgrowth	10,000 – 20,000	1,000 – 1,600
	Button formation	8,000 – 15,000	800 – 1,200
	Pre-harvest	5,000 – 8,000	600 – 1,000
Cantaloupe	Early flower	12,000 – 15,000	1,000 – 1,200
	Fruit bulking	8,000 – 10,000	800 – 1,000
	First Harvest	4,000 – 6,000	700 – 800
Celery	Midgrowth	7,000 – 10,000	300 – 500
	Pre-harvest	6,000 – 8,000	300 – 400
Lettuce	Early head formation	7,000 – 12,000	400 – 600
	Pre-harvest	6,000 – 8,000	300 – 500
Pepper	Vegetative growth	7,000 – 10,000	900 – 1,200
	Early flower/fruit	5,000 – 8,000	700 – 1,000
	Fruit bulking	5,000 – 8,000	700 – 1,000
	Pre-harvest	5,000 – 7,000	700 – 900
Tomato	Vegetative growth	10,000 – 14,000	700 – 900
	Early flower/fruit	9,000 – 12,000	600 – 800
	Fruit bulking	6,000 – 8,000	500 – 700
	Pre-harvest	4,000 – 6,000	400 – 600
Watermelon	Early flowering	12,000 – 15,000	900 – 1,100
	Fruit bulking	8,000 – 15,000	700 – 900
	Fruit harvest	5,000 – 8,000	500 – 700

IRRIGATION WATER TESTING

In terms of plant nutrition, irrigation water testing is important as it pertains to nutrients added to the field via the irrigation water (primarily NO_3). Such information is important when determining the amount of fertilizer to add to a field. Table 30 summarizes relevant water quality conversions.

Table 30. Conversion of water analysis of ions to pounds material per acre-foot of water. (California Fertilizer Association, 1980)

Ions	Mg/L (ppm)[1](A)	Pounds/AF of water	Conversion Factor[2](B)	Equivalent to Material(C)
Ca	20.0	54.4	6.8	136 pounds CaCO3
Mg	12.2	33.2	9.4	115 pounds MgCO3
Mg	12.2	33.2	13.5	165 pounds MgSO4
Na	23.0	62.6	6.9	159 pounds NaCl
K	39.1	106.4	3.3	128 pounds K_2O
K	39.1	106.4	5.2	203 pounds KCl
CO_3	30.0	81.6	—	—
HCO_3	61.0	165.9	—	—
Cl	35.5	96.6	—	—
SO_4	48.0	130.6	0.9	43 pounds S
SO_4	48.0	130.6	4.9	234 pounds gypsum

1 To convert milligrams per liter (mg/L) to pounds/acre-foot, multiply by 2.72.
2 $A \times B = C$

HOW MUCH NUTRIENT TO APPLY

The previous sections described the different types of testing which are currently available to improve nutrient management. The scope of these tests has dramatically increased in the last several years. Nevertheless, none of these testing capabilities matter if a grower cannot translate test results into a fertilizer recommendation.

Depending on the method used to establish a fertilizer recommendation by a laboratory, nitrogen recommendations can easily vary threefold. Discrepancies among various laboratories' recommendations occur for various reasons—some factors affect the quantity of nutrient available for plant uptake, and other factors affect the quantity that plants require. For example, the percentage of leaching and the nutrient conversion rates (i.e., mineralization) affect the quantity of nutrient *available* for the plant. The environment, timing of fertilizer applications, and irrigation timing affect crop growth, and consequently affect the quantity of nutrient *required.* In addition, the concepts of net and gross are intermingled in terms of nutrient recommendations, and traditional recommendations assumed poor fertilizer efficiencies. It was common to apply twice the fertilizer which was actually needed by the crop. In addition, yield goals vary between researchers; these variations will effect the fertilizer recommendations. The authors believe that the data regarding nutrient management will become more transferable as the concepts change with the increase in "spoon feeding" fertigation.

To determine a fertigation schedule, growers should consider the following:

- How much nutrient is supplied by the soil and water?
- How much nutrient does the plant require?
- When does the plant require the nutrient?
- What are the plant's *present* and *historical* tissue/sap nutrient levels?

The grower's goal would be to supply the difference of what the plant needs and what is already available for the plant through the irrigation water and the soil.

One strategy may be to consider the nutrients which are removed from the harvested portion of the crop (Table 31). Growers may then get a general idea of the quantity of nutrient "mined" from the soil and removed from the field during a growing season. Growers should be aware that these are average values, and they will vary depending on region, cultural practices, and crop cultivars. The downfall of this strategy is that growth stage information is not included, therefore the timing of fertilizer application must be approximated. In addition, Table 31 only refers to the nutrients contained in the harvested portion. Additional nutrients are required for the leaves and roots. If the soil and water do not have enough nutrients to satisfy the requirements of the non-harvested portion, extra fertilizer will be needed in excess of what this table suggests. Adjustments must also be made to account for losses (volatilization or leaching) and immobilization.

Table 31. Quantity of nutrients contained in the harvested portion of various crops. (California Fertilizer Association, 1980)

Crop	Yield	Nutrients Removed in Harvested Portion of Crop (pounds/acre)		
		N	P_2O_5	K_2O
VEGETABLE CROPS				
Asparagus	3,000 pounds	95	50	120
Beans (snap)	10,000 pounds	175	40	200
Broccoli	18,000 pounds	80	30	75
Cabbage	35 t	270	65	250
Celery	75 t	280	165	750
Lettuce	20 t	95	30	200
Potatoes (Irish)	500 cwt	270	100	550
Squash	10 t	85	20	120
Sweet Potatoes	15 t	155	70	315
Tomatoes	30 t	180	50	340
FRUIT AND NUT CROPS				
Almonds (in shell)	3,000 pounds	200	75	250
Apples	15 t	120	55	215
Cantaloupes	30 t	220	70	400
Grapes	15 t	125	45	195
Oranges	30 t	265	55	330
Peaches	15 t	95	40	120
Pears	15 t	85	25	95
Prunes	15 t	90	30	130

Another strategy may be to consider the net nutrient uptake by plants (Table 32). This information details nitrogen uptake by growth stage. Knowing how uptake patterns vary by stage of growth is extremely helpful in establishing a fertigation schedule. Unfortunately, some of this available data is limited in its application to fertigation. Most of the data pertains only to nitrogen, much of the available data does not apply to crops grown in California under microirrigation, and the absolute values will vary depending on factors listed earlier. In addition, this data reflects what the plant needs (net), not what should be applied (gross). Growers will need to factor in the nutrients supplied from the soil and water, along with irrigation system efficiencies, before they can determine how much nutrient to apply. Nevertheless, the data may be helpful when determining a fertilizer schedule. Table 32 provides net nitrogen uptake for several crops. Again, growers should recognize that published values will vary.

Table 32. Nitrogen uptake (net) by growth stage for various crops. (Adapted from Doerge et al., 1991)

Crop	Days after planting	Growth Stage	Daily N Uptake (pounds N/ac/day)	Cumulative N Uptake (pounds N/ac)
Cabbage	0	Planting	0.00	0
	20		0.00	0
	30		0.00	0
	40	4 – 6 Leaves	0.24	1
	46		0.76	4
	48		1.05	6
	62	10 – 12 Leaves	4.17	41
	77	Folding	4.89	120
	91	Head Formation	2.73	169
	108		2.58	213
Cantaloupe	0	Planting	0.00	0
	10		0.00	0
	24	3 – 4 Leaves	0.00	0
	32		0.22	1
	37		0.76	3
	40	2 in. melons	1.00	6
	48		0.85	12
	56		1.23	22
	72		5.36	64
	78		0.00	100
	82		0.00	110
	88	5 in. melons	0.87	117
Cauliflower	0	Planting	n/a	0
	10		n/a	0
	20		n/a	0
	30		n/a	0
	36		n/a	0
	39		n/a	2
	42	4 – 6 Leaves	n/a	5
	55		n/a	28
	70	10 – 12 Leaves	n/a	66
	85	Buttoning	n/a	165
	97	Curd Formation	n/a	215
	112		n/a	340

continued

Table 32. Continued.

Crop	Days after planting	Growth Stage	Daily N Uptake (pounds N/ac/day)	Cumulative N Uptake (pounds N/ac)
Corn (sweet)	0	Planting	0.00	0
	10		0.00	0
	20		0.00	0
	28		0.00	0
	34		0.11	17
	38		0.41	1
	40	3 – 4 Leaves	0.71	3
	46		1.92	12
	52	8 Leaves	2.52	25
	60		4.09	51
	72	Silking	3.12	104
	80		1.92	3
	94	Harvest	4.38	170
Cotton (Upland and Pima)	0	Planting	n/a	0
	10		n/a	0
	20		n/a	1
	30		n/a	4
	40	1st Square	n/a	8
	45		n/a	12
	51		n/a	20
	65	1st Boll	n/a	55
	80		n/a	99
	93	Peak Bloom	n/a	156
	108	1st Open Boll	n/a	193
	160	Pre Harvest	n/a	248
Lettuce (head)	0	Planting	0.00	0
	10		0.00	0
	20		0.00	0
	30		0.00	0
	40	4 – 6 Leaves	0.00	0
	50		0.03	0
	62	10 Leaves	0.24	2
	66		0.35	3
	81	Folding	0.65	11
	98	Heading	2.04	31
	112		3.00	70
	127	Harvest	2.67	111
Potatoes	0	Planting	0.00	0
	10		0.00	0
	20		0.16	1
	40	6 – 8 Leaves	0.75	9
	50		1.50	20
	60		3.00	43
	74	Tuber Development	2.88	91
	87		4.70	129
	101		4.83	217
	122		-1.77	218
	137	Harvest	-1.19	198
Watermelon	0	Planting	0.00	0
	30	3 – 4 Leaves	1.64	16
	44	3 in. Melons	6.38	67
	58	10 in. Melons	7.18	181
	72	Final Harvest	3.00	241

Finally, growers may want to ignore the "pounds per acre" strategy. Instead, they may want to approach fertilizer scheduling from a ppm basis, incorporating tissue, sap and soil solution test results throughout the season. In fact, many growers approach fertigation scheduling exactly in this way (Chapter 16, "Grower Drip Fertigation Experiences"). Growers attempt to maintain tissue concentrations within sufficiency ranges, and adjust the concentrations of nutrients applied until the desired levels are achieved. Growers may use this same method with fresh sap nutrient concentrations or with soil solution concentrations. In either case, growers are able to compare tissue ppm with chemical injection rate ppm and are able to correlate what is being applied to what is being used by the plant (or what is remaining in the soil solution). Therein lies one advantage of on-farm quick tests—a rapid turn around allows growers to turn the test results into a fertilizer recommendation immediately. In addition, by making only small adjustments in the injection concentrations, there is little chance for the equilibrium to be significantly affected. Growers must be aware that the concentration in ppm in the soil solution is different from the ppm in the dry soil, because the first is based on the weight of water and the second is based on the weight of soil.

Because the fine tuning of fertigation through microirrigation systems is relatively new, agronomists are still trying to overcome the previously mentioned problems with making fertilizer recommendations *specific* to fertigation. Previous rules, which applied to irrigated agriculture for many years, simply may not be appropriate with progressive nutrient management strategies. Nevertheless, some researchers have established *general* guidelines for fertigation schedules for several crops (Table 33). This information should be helpful to growers in establishing a base fertigation schedule. Unfortunately, this information is limited to nitrogen—the authors believe that this type of information will be expanded to include other nutrients in the future. In addition, growers may want to deviate from these guidelines depending on irrigation efficiencies, yields, and cultural practices. Again, the Chapter 16, "Grower Drip Fertigation Experiences" provides information regarding actual fertigation practices and offers limited information on fertigation schedules for a few crops.

Table 33. Weekly nitrogen fertigation estimate of (gross) requirements of vegetable crops under California conditions. (Hartz et al., 1994)

Crop	Growth Stage	Approximate nitrogen fertilizer requirement (pounds N/ac/day)	Approximate nitrogen fertilizer requirement (pounds N/ac/wk)
Broccoli	Early growth	0.71 – 2.14	5 – 15
	Midseason	1.42 – 2.86	10 – 20
	Button formation	2.14 – 4.29	15 – 30
	Head development	1.42 – 2.86	10 – 20
Cucumber	Vegetative growth	0.71 – 1.42	5 – 10
	Early flowering/fruit set	1.42 – 2.86	10 – 20
	Fruit bulking	1.42 – 2.14	10 – 15
	First harvest	0.71 – 1.42	5 – 10
Lettuce	Early growth	0.71 – 1.42	5 – 10
	Cupping	1.42 – 2.86	10 – 20
	Head filling	2.14 – 4.29	15 – 30
Melon	Vegetative growth	0.71 – 1.42	5 – 10
	Early flowering/fruit set	1.42 – 2.86	10 – 20
	Fruit bulking	1.42 – 2.14	10 – 15
	First harvest	0.71 – 1.42	5 – 10
Pepper	Vegetative growth	0.71 – 1.42	5 – 10
	Early flowering/fruit set	2.14 – 4.29	15 – 30
	Fruit bulking	2.14 – 2.86	15 – 20
	First harvest	0.71 – 1.42	5 – 10
Squash	Vegetative growth	0.71 – 1.42	5 – 10
	Early flowering	1.42 – 2.86	10 – 20
	First harvest	0.71 – 1.42	5 – 10
Tomato	Vegetative growth	0.71 – 1.42	5 – 10
	Early flowering/fruit set	2.14 – 2.86	15 – 20
	Fruit bulking	1.42 – 2.14	10 – 15
	First harvest	0.71 – 1.42	5 – 10

Note: Upper threshold values represent fertigation needs in low residual N soils and/or under high temperature (rapid growth) conditions. Higher values are for soils with low residual nitrogen or conditions of rapid growth under favorable environmental conditions.

Conclusions

New irrigation systems favor improved nutrient monitoring to decrease the likelihood of nutritionally related yield declines. The ability of these systems to spoon feed nutrients allows farmers to apply nutrients as they are required by the crops. This strategy requires that growers monitor nutrient status throughout the growing season in order to maximize the potential benefits of this strategy.

It is becoming clear that relative amounts of nutrients in plants, soil, and solution are important, and that progressive nutrient strategies require looking at nutritional balances in addition to absolute amounts. Although quantitative data is hard to find regarding specific balances, there are guidelines. Each grower has the ability to compare balance differences between high and low yielding plants to establish their own guidelines for nutritional levels. In addition, nutrient interactions should be considered by those monitoring fertility regimes.

It is also becoming clear that "old rules" of fertilizing may not apply to fertigating through microirrigation systems. The new conditions promote higher yields through more precise applications. Many farmers believe that previously accepted sufficiency range guidelines may not be adequate for crops intensively grown under fertigation and microirrigation.
Good record keeping seems to be a key factor in fine-tuning new fertigation strategies. Since these technologies are new, and many "old rules" may not apply to modern fertilizing techniques, growers are forced to examine their own nutrient trends of high yielding plants in order to determine the new rules and guidelines for fertigating with microirrigation systems.

The Soil Science Department at California Polytechnic State University is beginning to develop a data base of plant nutrient balance ratios for near optimum economic yield levels for many California crops. Growers interested in participating in this service may contact the Soil Science Department at California Polytechnic State University, San Luis Obispo, CA 93407, tel. (805) 756-2261 or FAX (805) 756-5412.

CHAPTER 12.
DRIP SYSTEM MAINTENANCE

INTRODUCTION

Chemical injection is essential for sustaining drip irrigation in arid environments long-term. Treatment for avoiding plugging and permeability problems is usually necessary unless municipal water is used. The specific treatment must be customized for the water quality of each system or region.

Table 34 provides some guidelines to potential plugging problems when water of various qualities is used with microirrigation. These guidelines were developed for emission devices having flow rates in the 0.5 – 2.0 GPH range.

Table 34. Guidelines for potential of water plugging.
(Adapted from Bucks and Nakayama, 1980)

	Hazard Level		
Type of Problem	Low	Moderate	Severe
PHYSICAL			
Suspended solids	<50 ppm	50 – 100 ppm	>100 ppm
CHEMICAL			
pH	<7.0	7.0 - 8.0	>8.0
Salt	<500 ppm	500 - 2000 ppm	> 2000 ppm
Bicarbonate		<100 ppm	
Manganese[1]	<0.1 ppm	0.1 – 1.5 ppm	> 1.5 ppm
Total iron[1]	<0.2 ppm	0.2 – 1.5 ppm	>1.5 ppm
Hydrogen sulfide	<0.2 ppm	0.2 – 2.0 ppm	> 2.0 ppm
BIOLOGICAL			
Bacterial population	<2,642/gal	2,642 – 13,210/gal	13,210/gal
	<10,000/L	10,000 – 50,000/L	>50,000/L

1 When testing for iron and manganese, acidify the water sample to a pH of 3.5 immediately after taking it.

There are many causes of plugging in drip systems which require chemical injection into the water. These include:

- Bacterial growth
- Algae which originated in surface water supplies. The algae has little subsequent growth in the hoses because there is no sunlight available for its growth. However, in some cases the algal growth is strong enough that it grows at the outlet of the emitters, and can plug them.
- Iron and manganese oxides and sulfides
- Calcium and magnesium precipitation
- Root intrusion into buried emitters

The solutions for these plugging problems are found later in this chapter, after chemical additives have been discussed.

Chemicals Used for Plugging Prevention

Chlorine

Either liquid or gas may be used. The most common form of chlorine used depends on the region. In some areas, chlorine gas injection is illegal. Sodium hypochlorite (NaOCl) solution (household bleach) is readily available, and is generally legal to inject. Calcium hypochlorite ($Ca(OCl)_2$) is a powder or pellets which must be dissolved in water to form a stock solution. Calcium hypochlorite may give problems with calcium precipitation in the irrigation water.

Equipment is available commercially to make potassium hypochlorite in the field by electrolyzing a KCl solution to produce a 0.8% hypochlorite solution from 3% KCl. This form of hypochlorite is new on the market, and is not available in solid form. Manufacturers of equipment to produce compounds such as KOCl may be reluctant to advertise the compound as an antibacterial agent, as that would initiate EPA registration. By selling such new compounds as fertilizer, EPA involvement is minimized.

Chlorine has different chemical properties, depending on its concentration. At low concentrations (1 – 5 ppm), chlorine acts as a bactericide and an oxidizer of iron, and at higher concentrations (100 – 1000 ppm) it acts as an oxidizing agent for organic materials, and can be used to "burn up" or disintegrate organic materials (Abbott, 1985).

When chlorine is injected into water, the *free chlorine* is composed of two compounds: hypochlorous acid (HOCl) and hypochlorite (OCl^-). The reaction is shown below for chlorine gas.

Cl_2	+	H_20	——>	HOCl	+	H^+	+	Cl^-
Chlorine		Water		Hypochlorous Acid		Acid		Chloride

The HOCl then exists in a pH-dependent equilibrium as follows:

HOCl	<——>	H^+	+	OCl^-
Hypochlorous Acid		Acid		Hypochlorite

The relative percentage of HOCl vs. OCl^- varies with the water pH. A low pH (i.e,. high H^+ concentration) pushes the reaction from the OCl^- side to the HOCL direction. Hypochlorous acid (HOCl) is 40 to 80 times *more powerful* as a biocide than hypochlorite (Boswell, 1990). For effective chlorine treatment, alkaline water should be acidified to a pH of 6.5 so that hypochlorous acid (HOCl) predominates. This chlorine reacts with anything organic.

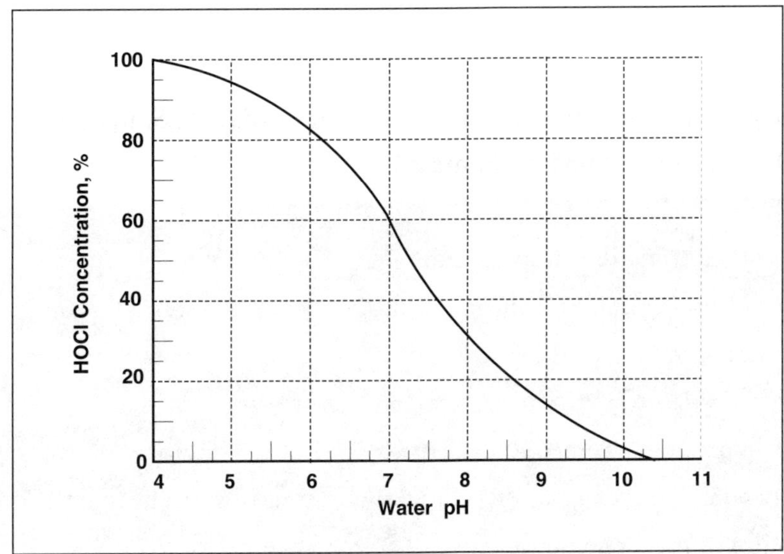

Figure 49. Hypochlorous acid concentration as a function of water pH. (Boswell, 1990)

"Free" (residual) chlorine is easy to measure with swimming pool test kits. "Total" chlorine tests analyze for free chlorine plus other chlorine compounds. Total chlorine test results are not as meaningful for microirrigation as those measuring free chlorine.

When using liquid sodium hypochlorite for injection, use the following formula:

$$\text{Chem. soln. Injection rate} = \frac{K \times (\text{desired ppm of free chlorine}) \times (\text{System flow})}{\%\ OCl}$$

Where:

- Solution injection rate is in LPH (metric) or GPH (U.S. meas.)
- K = 0.36 (metric) or .006 (U.S. meas.)
- Desired ppm of free chlorine is the desired concentration of hypochlorite (OCl) and hypochlorous acid (HOCl) in the irrigation water, expressed as parts per million (ppm)
- System Flow is in LPS (metric) or GPM (U.S. meas.)

When using chlorine gas, use the following formula:

Chlorine gas inj. Rate = KC x (desired ppm of free chlorine) x (System Flow)

Where:

- Chlorine gas injection rate is kg/day (metric) or lbs./day (U.S. meas.)
- KC is .06 (metric) or .0085 (U.S. meas.)

Notes of caution for chlorine injection include:

- The water acidification and chlorine injection should be done at two different injection ports; mixing acid and chlorine liquid in the same tank will produce highly toxic chlorine gas. *Never store acids and chlorine together.*
- Chlorine injection combined with herbicides or pesticides may reduce the effectiveness of herbicides or pesticides because the chlorine attacks the organic composition of these chemicals.
- *Always add chlorine supplies (liquid or dry) to water*, not vice versa.
- Chlorine gas is illegal to inject in irrigation systems in many areas.

CAUTION *Never mix or store acid and chlorine together.*

Sodium hypochlorite solutions (5 – 15%) tend to deteriorate with time. During six months storage, a 10% solution loses 20 – 50% of its useful chlorine (Driscoll, 1986). Tanks of the solution should be shielded from the sun to reduce such decomposition.

ACIDS

As noted above, acid treatments enhance the effectiveness of chlorine. In many cases, acids are sufficient by themselves to eliminate small slimy bacteria. Typical acids are:

- Sulfuric acid
- Phosphoric acid
- Urea-sulfuric (e.g., N-pHURIC®) nitrogen fertilizers that provide both acid and fertilizer
- Citric acid (which is handled separately in this section)

Long-term or exclusive acidic nitrogen fertilizer use for bacterial control must be applied with caution because (i) nitrogen should not be applied to some crops near harvest, and (ii) the pH of the soil may be adversely affected by additional acid. Besides affecting the availability of various micronutrients (and increasing the problem of aluminum toxicity), a low soil pH under emitters will inhibit nitrification.

Extreme care should be taken when injecting phosphoric acid, because of the possibilities of calcium phosphate precipitation. Calcium phosphate *cannot* be dissolved with subsequent acid injections, whereas calcium carbonate can be dissolved with acid. In general, if there is more than 50 ppm calcium in the water, phosphoric acid should not be injected. Phosphoric acid injections are usually designed to drop the irrigation water pH to 3.0 for 30 – 60 minutes. This very low pH helps to minimize calcium phosphate precipitation.

Low pH solutions can damage irrigation system hardware. In general, corrosion accelerates rapidly as the pH drops below 5.5. There can also be localized corrosion around the injection point in a pipeline system if the acid is not located properly. Injection ports should be designed to protrude into the center of the pipeline to ensure adequate mixing of the acids with the water.

The amount of acid required to drop the pH can only be determined through tests with samples of the water. A titration curve (relationship between water pH and the amount of acid added) is unique for each water source and type of acid. Since water quality often changes throughout the year, titrations should be run occasionally. Figure 50 illustrates titration curves for two hypothetical water samples. The samples started with different pH's, and it took different amounts of acid to change the pH to 5.0 for each.

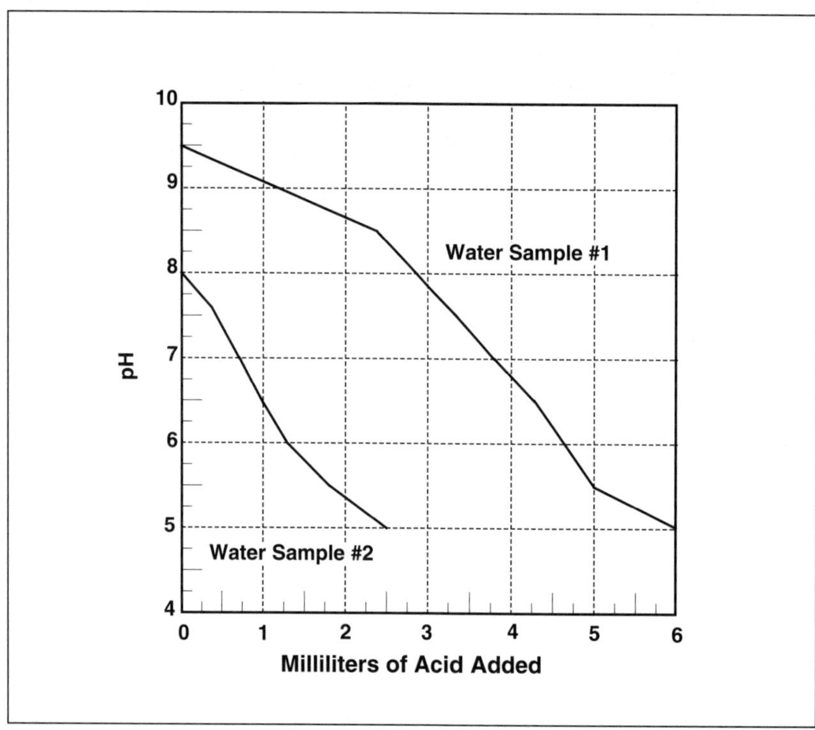

Figure 50. Acid titration curves for two different water samples.

A grower can develop a titration curve in the field. At least 10 gallons of irrigation water should be used, along with a good graduated cylinder. The irrigation water should be stirred to ensure complete mixing of the acid before the pH is read.

Table 35. Quantities of N-pHURIC® required to neutralize 90% of the bicarbonate in an acre-foot of irrigation water. (Unocal, 1993)

Bicarbonates in Water (mg/L)	N-pHURIC 28/27 (gallons)	N-pHURIC 15/49 (gallons)	N-pHURIC 10-55 (gallons)
50	31	16	14
100	61	32	28
200	122	63	56
400	244	126	112

Acids require special materials in pre-filters and injection pump gaskets. Check with the manufacturers of the equipment to ensure compatibility.

CAUTION *Always add acid to water. Do not add water to acid.* All acids can be dangerous to handle, and appropriate eye protection and clothing should be worn.

OVERVIEW OF SYNTHETIC COMPOUNDS

Compounds such as polyphosphates, phosphonates, polymaleic acids (PMA), and polyacrylic acid (referred to as PAA or PAM) have been researched and used for many years in municipal and cooling tower applications (Weijnen and van Rosmalen, 1985; Robinson et al., 1990). Some of the compounds are noted for their scale removing properties, others for the prevention of scale formation, and others for sequestration of iron and/or manganese. There appears to be no question about the fact that all of these compounds have been successful in various applications; there have also been conflicting and inconsistent results with these compounds.

Some of these compounds are sold for prevention of scale and sequestration of iron and manganese in drip irrigation systems. Sequestration is the reaction of a given metal ion to keep it in solution without removing that metal ion from the system. This is most commonly accomplished by neutralizing the metal ion's electrical charge.

These compounds can offer some substantial benefits due to their ease of handling (as opposed to acid, for example) and the lack of registration problems (they are often registered for drinking water applications). However, the water chemistry in drip systems may differ significantly from that of a cooling tower or some municipal systems. Previous research which concentrated on corrosion inhibition characteristics of a polymer may not be helpful in understanding how these polymers will work in some drip systems. Furthermore, some of the polymers being sold for drip irrigation have different molecular weights and other functional groups attached to polymer backbones than those which have been referenced in published research.

There is still uncertainty regarding exactly which compounds should be used, for what conditions, and at what concentrations. There has been almost no in-depth independent research or confirmation of manufacturer/vendor claims. The sections below indicate one of the difficulties of conducting such research—a wide range of compounds is available. Therefore, the sections below have been written more in the format of a "literature search" rather than as providing concrete guidelines. It appears from the sections below that the best drip irrigation cleaning/plugging formulations might well be a mixture of various polymers and compounds, each with a different function. Such a broad-spectrum formulation might be satisfactory for the three major clogging problems:

- Small slimy bacterial growth
- Iron and manganese bacteria
- Calcium and magnesium precipitations

Some such commercial broad-spectrum formulations have already been developed.

POLYPHOSPHATES

Polyphosphates have been used in municipal water treatment for over 60 years. However, it is very important to recognize that there are major differences between various compounds which are commonly called "polyphosphates." Holm and Schock (1991) report the following:

> Polyphosphates usually contain a range of linear chain polyphosphates having different numbers of orthophosphate (PO_4^{3-}) groups that are held together by P-O-P linkages. The average chain length, as measured by the number of phosphorus atoms per chain, is reported to range from approximately 4.5 upwards to 18 in commercial glassy phosphates.

For various reasons relating to the structural chemistry of the different polyphosphate ions in solution, polyphosphates vary in the strength of the complexes they form with different metals. The order of preference for complexing metals can also vary among polyphosphate compounds. In other words, one compound might bind metals in the order $Ca^{2+} > Mg^{2+} > Fe^{2+}$ (strongest first), whereas another compound might bind in the order $Fe^{2+} > Ca^{2+} > Mg^{2+}$. The concentration of a complex formed in solution is a function of the concentration of both the metal ion and the polyphosphate, as well as the stability constant for the complex, so comparisons of one chemical with another are somewhat complicated and depend on specific water chemistry conditions.

Boffardi (1992) notes the following:

> Both the crystalline and glassy condensed phosphates have a linear chain structure with a -P-O-P- backbone. When the polymeric n group equals 0, the orthophosphate species exists (Na_3PO_4); with n equal to 1, the pyrophosphate composition exists ($Na_4P_2O_7$); with n equal to 2, tripolyphosphate species exist ($Na_5P_3O_{10}$). As the number of polymeric groups increases and the $Na_2O : P_2O_5$ ratio approaches 1 : 1, a family of glassy condensed phosphates (polyphosphates) are formed. Commercial glassy polyphosphates used in potable water systems have an n value of about 4 to 20, and more typically 12 to 14 with a 1.1 : 1 $Na_2O : P_2O_5$ ratio. Therefore, the term "sodium hexametaphosphate" is a misnomer. The term implies six repeating units, where in actuality it contains 12 to 14 repeating units due to the $Na_2O : P_2O_5$ ratio of 1.1 : 1.

One of the earliest polyphosphates to be developed was Calgon® ("Calcium-be-gone"), which is sodium hexametaphosphate, a short chain polyphosphate. Its most common usage in fertigation is associated with anhydrous ammonia injection in hard water. (Note that anhydrous and aqua ammonia should never be injected into drip or sprinkler systems.) Recommended rates of Calgon® injection to prevent calcium precipitation during injection of anhydrous ammonia are included in

Chapter 9, "Solubility and Compatibility." Table 36 provides an approximate rate of injection of rosestone (Calgon®) to counteract the "liming effect" of injecting ammonia into irrigation water. Water hardness can be expressed as the sum of the calcium and magnesium in parts per million.

A sample water hardness problem (expressed as ppm $CaCO_3$) can be approximately computed as follows:

GIVEN: Water quality Ca = 40 ppm, Mg = 35 ppm

FIND: Water hardness ($CaCO_3$, ppm)

FORMULA: Hardness $= (2.5 \times Ca) + (4.1 \times Mg)$

SOLUTION: Hardness $= 2.5\,(40) + 4.1(35)$

$= 244$ ppm $CaCO_3$

Table 36. Equivalent rosestone injection to counter the "liming effect" of ammonia fertigation. (Mullinier, 1974)

Irrigation Water Hardness (ppm)	Maximum Ammonia Application per 1000 GPM of Irrigation Water (pounds/hour)	Rosestone required per 1000 GPM of Irrigation Water (pounds/hour)
35 – 120	37	.3
120 – 170	37	.4
170 – 425	37	.6
425 – 850	37	.8

Arceneaux (1974) reported that glassy polyphosphates were used extensively even 20 years ago in the chemical treatment of wells effectively to disperse iron hydroxide, iron oxide, manganese hydroxide, silts, and clays. He stated that an important advantage in using the polyphosphates is that the chemicals are safe to handle. Dollar (1992) noted very successful cleanup of rusty water problems in Sutherlin, Oregon city water lines. Dart (1983) noted that typically polyphosphates should be applied 3 to 5 minutes before chlorine or air contact, and cannot remove more than 1 to 2 mg/L of iron content. The dosage should be two to five times the iron content.

Robinson et al. (1990) completed a fairly broad study of some polyphosphates on behalf of the American Water Works Association. They concentrated on iron and manganese sequestration in their research. That study was funded because they noted inconsistencies in previous research reports and various vendor claims regarding polyphosphates for treatment of municipal water systems. Their conclusions pertaining to treatment of drinking water and rehabilitation of clogged city water pipes, were as follows.

1. Polyphosphates prevent the oxidation of manganese by chlorine. No significant differences were seen between several polyphosphates tested, in their performance for sequestering manganese.
2. Calcium interfered with treatment by polyphosphates, because of calcium phosphate precipitation.
3. Polyphosphates sequester iron. Chlorine should not be added ahead of polyphosphates, because the iron would oxidize and precipitate. Significant differences were found between the sequestering ability of polyphosphates for iron.
4. Chlorine must be added at least 30 seconds *after* the polyphosphates.
5. Some polyphosphates lose their potency with long storage times.
6. Different water suppliers report quite different successes of polyphosphate use.
7. Iron is oxidized from ferrous to ferric during sequestration, but manganese is not oxidized by the polyphosphates.

For irrigators, the mention of calcium phosphate precipitation by Robinson et al. (1990) seems inconsistent with years of successful use of polyphosphates for prevention of calcium precipitation problems. The previous quotes from Boffardi (1992) and from Holm and Schock (1991) indicate that there are major differences between properties of various polyphosphates. It should be noted that Robinson et al. (1990) used polyphosphates with 14 or less repeating units.

One of the more popular polyphosphate mixes sold for drip irrigation systems (Hornack, 1994a) uses a special polymerized linear long-chain polyphosphate. Such long-chain polyphosphates have around 25 repeating units. The formula is:

$$H_2K_6Na_{53}P_{51}O_{157}$$

This longer-chain polymer is claimed to use only 1% of the material, as compared to orthophosphate, to remove calcium or magnesium. It is recommended at rates of 2 ppm to control 400 ppm of hardness.

The very long-chain linear polyphosphates also claimed to hold 17 times more iron or manganese in solution than does hexametaphosphate (Hornack, 1994b). Such very long-chain polyphosphates are created by blending soda ash and phosphate at much higher temperatures (> 1200 degrees F) than is used for shorter-chain polyphosphates. According to Hornack (1994b), very long chain polyphosphates are needed to handle water hardness (calcium and magnesium), whereas shorter chain polyphosphates are more suited to sequestering iron and manganese. Long chain polyphosphates are also supposed to be more durable than shorter chain polyphosphates, in that they withstand shear in pumps and valves.

PHOSPHONATES

Phosphonates have been used in the municipal water industries as a scale inhibitor. They are often combined with phosphate and with zinc or molybdate ions for properties as corrosion inhibitors (Weijnen and van Rosmalen, 1985). Structurally, phosphonates differ from polyphosphates by a direct bond formation between the phosphorus and carbon atoms in contrast to the oxygen-phosphorus linkage. The bonds in the phosphonate molecules create greater hydrolytic stability, reducing the problem of reversion that exists with polyphosphates. These are also used to control corrosion and scale deposits (Boffardi, 1984).

Such scale inhibitors are adsorbed at the crystal surface. For the adsorption of phosphonate inhibitors at low inhibitor concentration levels on gypsum and barite it has been reported that only a small percentage of the crystal surface needs to be covered with inhibitor ions to achieve a drastic growth retardation. Also, phosphonate inhibitor effectiveness drastically increases with rising pH (Weijnen and van Rosmalen, 1985).

A specific brand of phosphonate, ESI-50, was reported by Schwankl and Prichard (1990) to reduce calcium precipitation problems in drip systems if applied continuously or during the last 2 hours of irrigation, when applied at 4 ppm.

POLYELECTROLYTES

Cooper et al. (1979) reported that during the mid-sixties various polymers including polyacrylates and polymethacrylates, used successfully in cooling water systems, were plant tested under a variety of conditions but, in general, results were disappointing. To match the performance of polyphosphates considerably greater dosage levels had to be used.

Weijnen and van Rosmalen (1985) studied polyelectrolytes which were polycarboxylates with a molecular structure based on either polymaleic acid (PMA) or polyacrylic acid (PAA) in relation to boiler water and municipal water treatments. In addition to the carboxylic acid groups originating from the monomers, other functional groups were attached to the polymer backbone to study their influence on the inhibitor performance. The PMA compounds had average molecular weights ranging between 1000 and 1500. Polyacrylates were found to be less effective for gypsum growth inhibition than polymaleates in the same concentration, at pH values greater than 7.

In drip irrigation, a polymaleic anhydride anionic terpolymer is currently sold for the purpose of sequestering iron and manganese. Iron and manganese ions attach to these polymers in the water, and pass through the irrigation system rather than oxidizing and precipitating out in the lines and emitters. In addition, these compounds are claimed to provide de-scaling of certain precipitates, including calcium phosphates and calcium carbonates. The scale can be in either the hoses, emitters, or in the soil. Some of these polymers are registered for irrigation injection as a soil amendment. The authors have no independent, replicated test results to verify or deny these claims. The information is presented to the readers because the authors anticipate that the use of such polymers will increase in the future as more is known about their applications in irrigation.

Bromide Materials

Various bromide materials (e.g., potassium bromide) have been used in post-harvest vegetable washing. At the time of this publication, it is unclear if bromides will be used in the future in drip and microirrigation systems.

Copper Sulfate

During the early days of drip irrigation, copper sulfate was occasionally used to prevent bacterial growth in hoses. Chlorine was found to be more effective, less expensive, and less problematic in regards to plant toxicity and aluminum corrosion. However, it is now included in some commercial drip line cleaning mixtures together with citric acid to reduce small slimy bacterial growth.

For the treatment of algae in reservoirs, copper sulfate is still used. However, even at high concentrations (30 ppm), copper sulfate will not be completely effective because the algal spores can exclude the copper sulfate. An effective option to reduce copper sulfate material requirements may be a mixture of citric acid (60%) with copper sulfate (40%) (Hornack, 1994b).

Specific Plugging Problems and Their Solutions

Small Slimy Bacteria

Some slimy bacteria grow on the interior walls of the hose and emitters. Clay particles in the water, *which are too small to remove with filtration*, stick to the slimy bacteria and exacerbate the problem by providing nutrients for additional bacterial growth.

Chlorine may be applied continuously at low dosages for effective prevention of bacterial growth. The free chlorine should be measured at the ends of hoses; for continuous application, a concentration of about 0.5 – 1.0 ppm free available chlorine should persist at the hose ends. Some growers apply chlorine weekly (instead of continuously) at concentrations of 5 – 10 ppm (at the hose ends), toward the end of an irrigation set.

Recommendations for exact chlorine dosages and injection frequencies are quite variable. This is probably due to differences in the organic matter and nutrient contents of the irrigation water, plus differences in water pH. The best rule of thumb for sufficiency is to reduce the chemical applications to the smallest amount which will maintain the cleanliness of the insides of the hose ends.

Superchlorination (free chlorine concentrations of 200 to 500 ppm at hose ends) is sometimes used if a system is already badly plugged.

CAUTION Superchlorination may damage or kill a healthy crop.

If this practice is used, the following special procedures must be followed.

1. First, thoroughly flush the hoses.
2. With the system shut-off, inject the hoses with chlorine and allow to sit overnight.
3. The next day, thoroughly flush the system.

It may be necessary for this process to be repeated several times.

Many superchlorination recommendations assume that it is possible to allow the chemical to sit in the hose for a period of time to gain contact time. This is only possible though, to some degree, with very flat irrigation systems. However, for systems with sloping hoses, this is impossible because the chemical will drain out of the hoses after the system is shut off.

A mixture of copper sulfate and citric acid may also be used to kill bacteria. Some growers achieve very successful control with acidic fertilizers, such as urea-sulfuric.

Iron and Manganese Bacteria

Very small iron and manganese concentrations (less than 0.2 ppm) are sufficient to provide healthy bacterial growth. The iron bacterial growth generally looks reddish, whereas the manganese bacterial growth is more black in color. These bacteria oxidize iron or manganese from the irrigation water as an energy source. There are various types of bacterial strains which combine to form long, slimy, sheath-like strands which can completely plug a drip system in a matter of a couple of weeks. The reddish color from iron bacteria should not be confused with other bacteria which have a by-product that looks like grains of rust.

In the western U.S., these bacteria are almost always associated with well water; neighboring wells can have completely different characteristics. These bacteria are notoriously difficult to kill, partly because they may reside in the well itself. A periodic acid/chlorine treatment of the well is sometimes sufficient to minimize this type of bacterial problem.

One solution is first to pump well water into a reservoir and aerate (i.e., oxidize) it. The iron and manganese then precipitate out in the reservoir before being picked up by a booster pump for the irrigation system.

Long chain linear polyphosphates and polymaleic acid can effectively sequester iron and manganese so that they stay suspended in the water and move through the irrigation system.

Chlorine gas injection can be used prior to a fine media or disc filter to precipitate the iron from the water and then trap it in the filter but not manganese; its reaction is much slower. It is essential that the mixing be thorough. Chlorine is injected at a rate of 1.4 parts chlorine for each 1.0 part iron in the water. The following formula (Rain Bird International, 1990) can be used:

$$\text{Chlorine gas needed (pounds/acre-foot)} = 3.8 \times \text{Fe concentration (ppm)}$$

The chlorine will also kill the bacteria which oxidize the iron and manganese, thus eliminating the sheath-like growths. However, as mentioned above, the iron will precipitate out of the water with the chlorine injection, so it must be removed with a filter.

Iron and manganese oxides can form without bacterial action, but generally the bacterial formation is dominant.

Iron and Manganese Sulfides

Dissolved iron and manganese in the presence of sulfides can form a black, insoluble precipitant. A combination of greater than 0.6 ppm iron and greater than 2.0 ppm total sulfides present in the water will generally create an iron sulfide sludge (Huntmacher et al., 1994). Iron is the major problem for plugging because the manganese is toxic to many crops in very low concentrations and may damage the crops by toxicity before the plugging becomes a problem. The problem with sulfides is almost exclusively associated with well water, and the well casing is generally fouled up rapidly. Once the water has reached the ground surface, a combination of aeration, acidification, and chlorination can precipitate out any remaining sulfide.

Calcium and Magnesium Carbonate Precipitation

Carbonate precipitation occurs in the emitters themselves, and can plug the emitters. This precipitation is easily identified by placing a drop of muriatic (hydrochloric) acid on a plugged emitter. If the material fizzes, it is a carbonate.

The most popular solution has historically been acid injection to remove the carbonates from the water in the form of CO_2 gas before the precipitation occurs. Many types of acid are used, including sulfuric and phosphoric acids (white phosphoric is preferred over green phosphoric because it has fewer impurities). Additionally, there are several types of gas injectors, including various SO_2 injectors. Dropping the pH to 6.5 will prevent carbonate precipitation.

Recently, there has been interest in active magnetic water treatment (which causes the carbonates to form aragonite that passes through the emitters as microscopic aragonite particles).

Very long chain linear polyphosphates, injected at rates of 1 to 2 ppm, have been effective in the elimination of calcium and magnesium precipitation problems. They also have the advantage of being safer to inject than most acids.

Injection of the acid or polyphosphate during the last one third of the irrigation set time is often sufficient to prevent precipitation problems.

Root Intrusion

As more permanent subsurface drip (PSSD) systems are installed, attention is given to preventing roots from growing into the buried emitters. For annual crops such as lettuce, root intrusion is commonly minimized by avoiding water stress during the growing season, and then killing the roots (and plants) by injecting acid at the end of the season.

For some trees and vines, an active program of chemigation with Trifluralin 5® is the only well understood prevention. This herbicide kills root tips without killing the plants, themselves. Some emitters have the chemical Treflan® impregnated into them. These emitters provide more protection, when new, than do standard emitters. Eventually, the Treflan® discharge from the emitters will decrease in concentration so that other means must be used to inhibit root intrusion.

Schwankl et al. (1994) report that periodic injections of chlorine or acid products can destroy intruding root hairs. Mention is made of a potentially successful trial with weekly chlorine injections resulting in 10 ppm free chlorine concentrations in a subsurface drip system on walnuts. They note the problem of controlling root intrusion is due to a flush of root growth on permanent crops that may occur in the late fall or early spring, even though the plant canopy may be dormant and the irrigation system shut down. University of California researchers are examining the possibility of using copper sulfate at toxic levels selectively to kill tree roots around the emitters (the copper may "fix" onto adjacent soil particles), but the results are not yet known.

AN IMPORTANT NOTE:

Any chemicals mentioned in this publication are listed in the context of observed practices by farmers, rather than as recommendations by the authors or their organizations. Local health/ safety regulations related to chemical injection must always be followed. In the absence of local or state regulations, adhere to USEPA standards.

CHAPTER 13. INFILTRATION PROBLEMS

Water infiltration rates in California commonly decrease during the growing season of row crops as the soil surface becomes more compacted. Certain types of salt in the irrigation water can also cause decreases in water permeability of soils.

Four *water quality* conditions contribute to water infiltration problems:

1. High Adjusted Sodium Adsorption Ratio (adj. R_{Na}) values
2. Low EC of the irrigation water, EC_w
3. High ratios of (Magnesium/Calcium) or of ((Magnesium + Sodium)/Calcium)
4. Fertigation with monovalent cations such as ammonium (NH_4^+) in conditions of low EC_w

The initial water quality can often be satisfactorily amended through chemigation.

Sulfur containing fertilizers may provide special benefit for improved furrow fertigation. Calcium polysulfide, lime sulfur, ammonium polysulfide, and ammonium thiosulfate acidify the surface of high lime soils when they are water run. This helps to keep the soil surface open and allows for good water infiltration. Otherwise, high lime soils (especially when irrigated with water high in lime) tend to become encrusted with lime and clog the soil pores, thus slowing water infiltration.

This chapter will provide a description of the problems and solutions.

HIGH ADJUSTED SODIUM ADSORPTION RATIO

This is the most commonly recognized water quality problem. It contributes to a poor calcium vs. sodium ratio on the soil exchange sites. High percentages of sodium on the exchange sites are bad for water penetration; high percentages of calcium are good. A high soil ESP (Exchangeable Sodium Percentage of greater than 15%) is often the result of irrigating with poor quality water.

Most water quality analysis from agricultural laboratories include a calculated value of "Adjusted Sodium Adsorption Ratio (adj. R_{Na})." At the moment, there are several methods used to calculate and interpret this ratio. Modifications of this and other permeability indexes continue to be made. The specific formulas used in the calculations should always be checked before interpretation. One formula used for the adjusted R_{Na} is:

$$\text{adj. } R_{Na} = \frac{Na}{\sqrt{\dfrac{Ca_x + Mg}{2}}}$$

where Na and Mg are in milliequivalents/liter (Appendix A) taken from the irrigation water quality analysis, and Ca_x (also in meq/L) is obtained from Table 37. Interpretations of adjusted R_{Na} are made using Table 38.

The "Ca_x" value is used instead of "Ca" because the "Ca_x" accounts for precipitation of some of the calcium with bicarbonates and carbonates in the water. If the calcium precipitates as $CaCO_3$, it essentially ends up as a rock in the soil, and the calcium is not available for replacement of sodium on the soil cation exchange sites.

Table 37. Ca_x values in meq/L for use in calculating R_{Na} of irrigation water. Calcium, carbonate and bicarbonate measured in meq/L. (Westcot and Ayers, 1984)

	Ca_x for salinity of applied water ratio (column headings are EC_w - dS/m)						
$\frac{HCO_3 + CO_3}{Ca}$	0.1	0.3	0.5	1.0	1.5	3.0	6.0
.05	13.2	13.92	14.40	15.26	15.91	17.28	19.06
.15	6.34	6.69	6.92	7.34	7.65	8.31	9.17
.25	4.51	4.76	4.92	5.22	5.44	5.91	6.52
.50	2.84	3.00	3.10	3.29	3.43	3.72	4.11
1.00	1.79	1.89	1.96	2.09	2.16	2.35	2.59
2.00	1.13	1.19	1.23	1.31	1.36	1.48	1.63
3.50	0.78	0.82	0.85	0.90	0.94	1.02	1.12
7.00	0.49	0.52	0.53	0.57	0.59	0.64	0.71
20.00	0.24	0.26	0.26	0.28	0.29	0.32	0.35

Table 38. Permeability hazard of irrigation water. (Westcot and Ayers, 1984)

	Degree of restriction on use depending on EC_w -(dS/m)		
adj. R_{Na}	None	Slight – Moderate	Severe
0 – 3	≥ 0.7	0.7 – 0.2	<0.2
3 – 6	≥ 1.2	1.2 – 0.3	<0.3
6 – 12	≥ 1.9	1.9 – 0.5	<0.5
12 – 20	≥ 2.9	2.9 – 1.3	<1.3
20 – 40	≥ 5.0	5.0 – 2.9	< 2.9

An example calculation of the Adjusted Sodium Adsorption Ratio follows:

GIVEN: The sample of water quality below

Measurement	Concentration (meq/liter)
Sodium (Na)	2.5
Calcium (Ca)	1.0
Magnesium (Mg)	0.5
Bicarbonate (HCO_3)	2.0
Carbonate (CO_3)	0.0
Electrical Conductivity (EC), dS/m	1.0

FIND: Adjusted Sodium Adsorption Ratio (adj. R_{Na})

FORMULA:

$$Ca_x : \frac{HCO_3}{Ca}$$

$$\text{adj. } R_{Na} = \frac{Na}{\sqrt{\dfrac{Ca_x + Mg}{2}}}$$

SOLUTION:

$$\frac{HCO_3}{Ca} = \frac{2.0 + 0.0}{1.0} = 2.0$$

$$\text{Since EC} = 1.0 \text{ and } \frac{HCO_3}{Ca} = 2.0$$

Ca_x = 1.31 (from Table 37)

$$\text{adj. } R_{Na} = \frac{2.5}{\sqrt{\dfrac{1.31 + 0.5}{2}}} = \frac{2.5}{.95}$$

$$\text{adj. } R_{Na} = 2.6$$

From Table 38 (adj. R_{Na} = 2.6 and EC_w = 1.0), the degree of restriction on use is classified as "None."

There are two variations of this water quality problem:

1. The water has a high percentage of carbonates and bicarbonates (CO_3 and HCO_3). A relatively simple solution is to remove the bicarbonates and carbonates from the water by

acidifying the water. This is most commonly done with SO_2 generators (which create sulfurous acid, H_2SO_3) or direct injection of sulfuric acid into the water. Sometimes this water treatment can be sufficiently effective to eliminate the need for annual soil amendments, once the soil has been initially reclaimed.

2. The water has a low percentage of CO_3 and HCO_3, and the problem is due to high sodium only. In this case, water treatment alone is sometimes successful, but typically annual soil amendment applications of gypsum are needed to supply sufficient calcium.

Water treatment would involve the injection of gypsum into the irrigation water (see the next section), which is economical only if the water is relatively pure. If the water is relatively pure, it does not take large amounts of gypsum to correct the imbalance.

Pure Irrigation Water

Table 39 shows that water permeability problems in soil can be caused by an interrelationship between EC_w and the relative amounts of various ions. For example, a very high adj. R_{Na} of 30 presents no problem for infiltration if the irrigation water is very salty (greater than 5.0 dS/m). On the other hand, if the irrigation water is very pure (EC_w < 0.2 dS/m), there are severe water permeability problems even if there is a high proportion of calcium in the irrigation water. However, some farmers use a rule-of-thumb that if there is more than 1 meq/liter of calcium in the irrigation water, there will be no problem due to "low EC_w."

There is still considerably more to learn about the effects of low EC_w on permeability (Tanji, 1990). In many California soils, the irrigation water from the Sierra Nevada Mountain range and some wells on the east side of the San Joaquin Valley is very pure (EC_w <0.25 dS/m). If this irrigation water is applied to soils of low organic matter, with no ground cover, and no mechanical disturbance of the soil, the water intake rates can be extremely low. These conditions are widespread with furrows on citrus, where clean-cultivation is practiced with herbicides. With microirrigation (drip and microspray) it is particularly difficult because the ground which receives water is rarely disturbed and is often kept clean with herbicides. On the other hand, irrigation with very pure water in other areas of the U.S. has not presented such problems. The differences may be caused by variations in soil parent materials, organic matter percentages, tillage practices, or the existence of frost in the winter. For example, the high levels of mica in some sandy loam soils may make them especially sensitive to water permeability problems.

GYPSUM INJECTION

A common solution to the pure irrigation water problem is to increase the calcium concentration of the water with injection of pure gypsum. Commercial equipment has been readily available for this purpose since the late 1980's.

High quality bulk and sacked gypsum is available from a number of companies, and the quality and form of the gypsum will vary depending upon the natural deposits which are mined by those companies. Most companies will only guarantee 92% gypsum purity, leaving a potential 8% of other compounds such as clay or $CaCO_3$.

There are two forms of gypsum: (i) natural dihydrate gypsum ($CaSO_4 \cdot 2H_2O$, molecular weight 172.2) and the natural anhydrite ($CaSO_4$, molecular weight of 136.1, also called anhydrous calcium sulfate). The dihydrate dissolves more rapidly because of the attached water molecules.

The gypsum machines generally inject a slurry of suspended (not dissolved) particles into the irrigation water, and the objective is to have everything dissolved before it reaches the emitters or microsprayers. With a good grind, manufacturers expect about 90% of the slurry to be dissolved within a minute, with 100% dissolution within five minutes. If hard particles remain in suspension (i.e., are not dissolved) by the time they reach the first emitters or microsprayers, there is a potential for abrasive damage to the equipment. Such damage has been observed with microsprayers, with the greatest damage closest to the pump (where suspended particles did not have sufficient contact time to dissolve) and less damage with an increased distance from the pump (Rivers, 1995).

Since the reports of abrasion are relatively recent, and because gypsum injection is a relatively new practice, the authors recognize that recommendations to avoid abrasion problems will improve as more experience is gained. However, the following tentative guidelines are provided:

- The anhydrite form of gypsum is a very hard crystal. It dissolves, but not as quickly as the dihydrate form. There may be some advantage to purchasing gypsum with a higher percentage of dihydrate.
- The gypsum should have a very fine grind. Some specialists recommend that 98% of the material should pass through a 200 mesh screen, and 100% through a 100 mesh screen. Gypsum is available with a 99.4% passage through a 200 mesh screen. A good grind should minimize problems with dissolving differences between anhydrite and dihydrate forms.

- Always locate the injector prior to the filtration equipment.
- Obtain a laboratory analysis of the material to determine which impurities exist. For example, calcium carbonate impurities are hard and fairly insoluble, and may cause abrasion problems even if the gypsum dissolves. Silicate impurities may also cause abrasion. The presence of some trace elements such as boron, even in low percentages, can cause plant toxicity problems.
- Purchase a good injection machine which provides a thorough mixing.
- Do not inject gypsum at excessively high rates which exceed the solubility limit of gypsum in water. In theory, one could dissolve up to about 2400 ppm of 100% pure dihydrate gypsum (1900 ppm of pure anhydrite). In the field, 1000 ppm of dissolved gypsum is a very high application.

1000 ppm anhydrite
= 1265 ppm dihydrate
= 14.7 meq/L of calcium
= 2720 pounds of pure anhydrite per acre-foot
= 3440 pounds of dihydrate per acre-foot

One can use Table 38 to estimate the benefits of gypsum injection. When gypsum is injected into irrigation water two factors change: (i) the adjusted R_{Na} decreases and (ii) the EC_w increases.

There is a fairly unanimous opinion in the field that gypsum application must be continuous in order to be effective. Occasional injections may show little if any benefits. This is because the gypsum is a water treatment, rather than a soil treatment. Once the treatment is stopped, pure water can quickly reseal a soil surface. In practice, a continuous maintenance injection rate of 2 – 2.5 meq/L is generally sufficient to prevent future pure water infiltration problems (Rivers, 1995).

There is some uncertainty over the dosage of gypsum which is needed for the initial treatments. A field which has been irrigated with pure water for some time will have an established permeability problem. Some field personnel (Hornung, 1994) believe that unless high dosages of gypsum (1200 pounds of dihydrate per acre-foot or higher) are used during the initial treatments, there will be little benefit derived from the gypsum injection. Later, as the infiltration problem decreases, the injection rate can be decreased to more closely match the guidelines of Table 38.

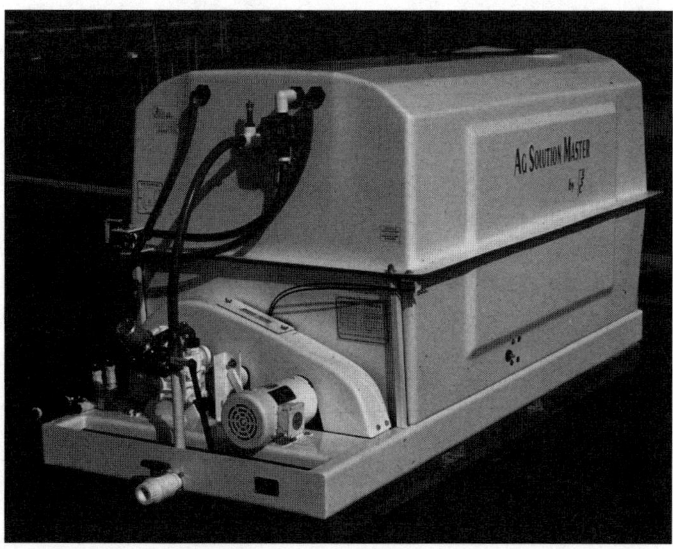

Figure 51. A gypsum injection machine.

HIGH MAGNESIUM/CALCIUM RATIOS

This is another area with relatively little research and reference in textbooks. Ayers and Westcot (1989) stated that field experience shows that if the Mg/Ca ratios of irrigation water (computed as meq/L) exceed 1.0, there is a high probability of water permeability problems. Some field personnel in California believe that to avoid water permeability problems, the water quality should be such that:

$$\frac{Ca}{Na + Mg} \geq \frac{2}{1} \text{ or } \frac{3}{1}$$

This ratio is frequently below this value on the Central Coast of California, where the aquifers are in soils with serpentine parent materials having a high amount of soluble magnesium.

It is interesting to note that in this section, magnesium is considered non-beneficial in terms of water infiltration. That brings up the question of why magnesium is considered a beneficial cation in the calculation of the Adj. R_{Na}.

Obviously, there is still considerable objective and replicated research needed on the subject of water penetration problems and how additions of calcium can be beneficial.

FERTIGATION WITH MONOVALENT CATIONS

Under conditions of low EC_w, any monovalent cation (a positively charged ion with only one charge/ion) resembles sodium in terms of its effects on water permeability. Therefore, in some cases it is preferable to use urea or nitrate fertilizers rather than ammonium-based fertilizers when fertigating. Another option is to increase the EC_w through the injection of gypsum.

POLYMERS

A variety of polymers are sold for the purpose of improving water penetration. Some of these are listed below.

WETTING AGENTS

Wetting agents reduce the surface tension of the water. This helps to prevent the "beading" of water at air-soil-water interfaces. Some wetting agents have very short life spans, and others are ineffective with the water salinities common in agricultural irrigation.

POLYELECTROLYTES

These were described in Chapter 12. They can be effective in dissolving lime and other calcium scales which may be plugging soil pores. A brand new family of polyacrylamides is now available for irrigation injection for purposes of reducing erosion and improving irrigation water infiltration. A special edition of the journal Soil Science (1994) contains ten articles on various aspects of the new polyacrylamides, and is recommended as reference material.

OTHERS

Other compounds are available, which vary from legitimate products to "snake oil." The irrigation industry has had such an abundance of snake oil sales people that many farmers and researchers are very hesitant to look at new polymers, due to unsubstantiated claims.

CHAPTER 14. SOIL PH MODIFICATION

SOIL ACIDIFICATION

In most western U.S. soils, growers have traditionally been more concerned with high pH problems rather than acidity. Nevertheless, there is an increasing number of cases of low soil pH (i.e., acidic) conditions being created, especially under drip irrigation. One of the sources of the hydrogen ions is bacterial oxidation of NH_4^+ from fertilizers (i.e., nitrification):

$$\underset{\text{Ammonium}}{NH_4} + \underset{\text{Oxygen}}{2O_2} \xrightarrow[\text{Nitrifying Bacteria}]{} \underset{\text{Nitrate}}{NO_3} + \underset{\text{Water}}{H_2O} + \underset{\text{Hydrogen (Acid)}}{2H^+}$$

Since the late 1980's acidic fertilizer (e.g., urea sulfuric) injection has become popular with drip systems, because these fertilizers can help prevent plugging in addition to providing nitrogen fertilizer. They have also contributed to the acidification of some soils.

Many western U.S. soils have high levels of free lime ($CaCO_3$), which effectively buffer the pH and keep it well above neutral. Therefore, the problems of soil acidification tend to be regional.

With conventional farming and irrigation, low pH problems have been remedied through liming. Traditional liming involves incorporating a finely ground limestone or other such material into the soil with a disc. This is not possible under drip irrigation, because the hoses would be damaged. Furthermore, under drip irrigation the low pH problem is often localized directly under the emitter and several feet to each side, rather than being spread evenly across the field. Traditional liming techniques have obvious limitations with permanent buried drip systems.

The following sections explain traditional concepts of soil liming for raising the soil pH, and then summarize techniques available for drip irrigation.

BENEFITS OF LIMING

There are several benefits of liming acidic soils:

1. Decrease the potential for certain nutrient toxicities at lower pH.
2. If calcium or magnesium carbonate are added as liming agents, Ca and Mg will be added for crop nutrition as a secondary benefit.
3. Lime improves phosphorus availability in acidic soils by reducing the solubility of Fe and Al, thereby prohibiting their complexing with phosphorus.
4. Lime makes plant uptake of potassium more efficient. Since Ca will be prevalent in limed soils, plants will use more calcium rather than using potassium in excess of their needs.
5. Lime improves the microbial environment in acid soils. Liming acidic soils increases Mo availability.
6. Finally, liming an acidic soil decreases the plant uptake of potentially toxic heavy metals (i.e., Cd, Cu, Pb, Ni, and Zn).

NEUTRALIZING MATERIALS

If a compound is able to neutralize soil acidity (i.e., neutralize the H^+ ions and raise the pH), then it is considered to be a liming agent. The most common "liming" reaction is as follows:

$$\underset{\text{Acidic Soil}}{2H^+ + \text{Soil}} + \underset{\text{Lime}}{CaCO_3} \longrightarrow \underset{\text{Neutral Soil}}{Ca^{2+} + \text{Soil}} + \underset{\text{Water}}{H_2O} + \underset{\text{Carbon Dioxide}}{CO_2}$$

The presence of calcium in a compound does not guarantee that it can change the pH. For example, gypsum ($CaSO_4$) has no effect on the soil pH.

Calcium carbonate ($CaCO_3$) is considered the standard for liming compounds, although the only true "Lime" is calcium oxide (CaO). The "calcium carbonate equivalent" refers to the total neutralizing power of a compound *relative to* that of calcium carbonate. Chemically pure limestone is considered to have a calcium carbonate equivalent of 100%; all other liming materials are rated relative to pure limestone. Table 39 lists several liming materials and their calcium carbonate equivalents.

Table 39. Traditional liming materials and their calcium carbonate equivalent. (modified from California Fertilizer Association, 1980)

Name[1]	Chemical Formula	Equivalent $CaCO_3$(%)	Source
Shell meal	$CaCO_3$	95	Natural shell deposits
Limestone	$CaCO_3$	100	Pure form, finely ground
Potassium carbonate	K_2CO_3	72	
Hydrated lime (slacked lime)	$Ca(OH)_2$	120 – 135	Steam after being burned
Lime (burnt lime, or unslacked lime)	CaO	150 – 175	Kiln burned
Dolomite	$CaCO_3 \bullet MgCO_3$	110	Natural deposit
Sugar beet lime[2]	$CaCO_3$	80 – 90	Sugar beet by-product lime
Calcium silicate	$CaSiO_3$	60 – 80	Slag

1 Ground to same degree of fineness and at similar moisture content.
2 From 4% to 10% organic matter.

In addition to the chemical composition of the material, the physical qualities must be considered. The finer the material and the better mixing there is of the material into the soil the quicker the neutralizing effect in the soil.

It should be noted that most of the materials in Table 39 above are solid calcium materials with very low solubilities.

Determining the Liming Schedule

The amount of pH change, the organic matter content, and the soil texture (specifically the clay type which influences the cation exchange capacity) will determine the amount of limestone needed to neutralize soil acidity. These concepts apply whether the grower is applying limestone to raise the pH (Figure 52) or whether small portions are applied to *maintain* the soil pH.

Figure 52 gives information regarding the tons of limestone needed to raise the pH of a given soil texture with a given cation exchange capacity.

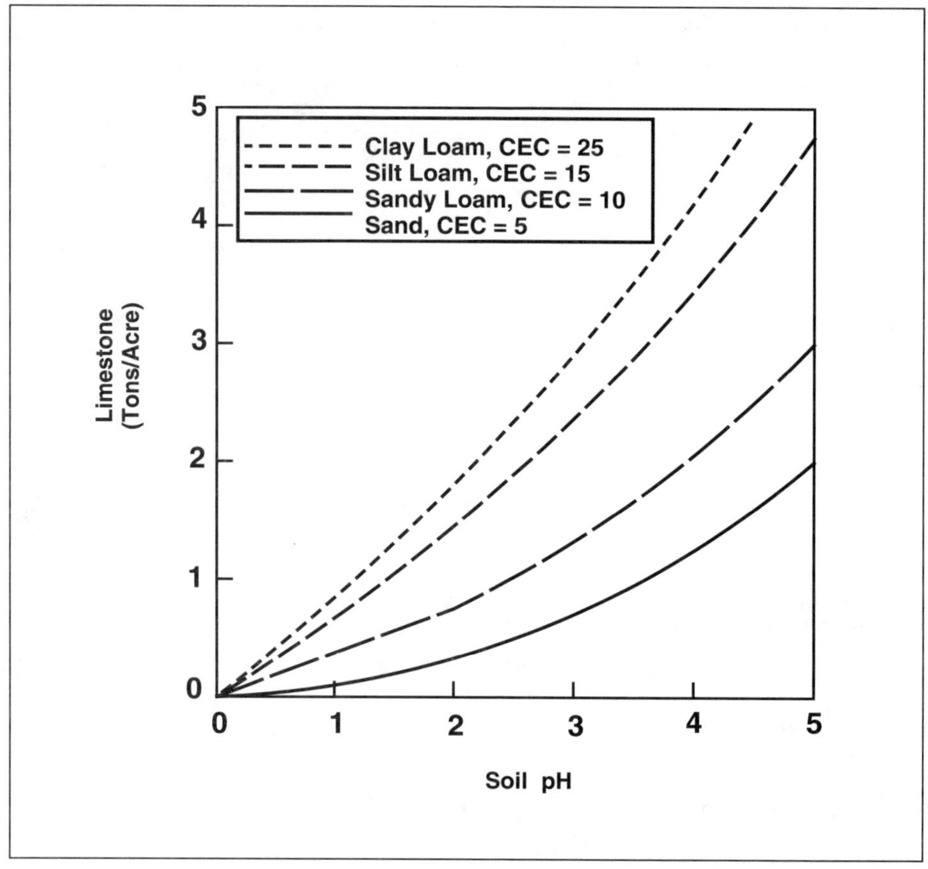

Figure 52. Approximate tons of limestone required to raise the soil pH of a 7-inch layer of soil of four textural classes with typical cation exchange capacities. (Miller and Donahue, 1990)

Various clay types influence a soil's cation exchange and buffering capacity; consequently, the soil texture affects the lime requirement. Laboratories can perform a lime requirement analysis and offer recommendations regarding a lime maintenance schedule.

Each pound of nitrogen from ammonium fertilizers potentially generates soil acidity equal to 7.14 pounds of pure calcium carbonate. The anions released from anhydrous ammonia, aqua ammonia, and urea fertilizers neutralize one half of this acidity and consequently require only 3.57 pounds of calcium carbonate. If the plant takes up the ammonium directly, then this ammonium does not produce acidity in soils. Thus, the fertilizer industry uses an average of one pound of ammonium-nitrogen, producing acidity equal to 5 pounds of pure calcium carbonate. To maintain soil pH, for each 100 pounds of ammonium-nitrogen fertilizer applied, 500 pounds of calcium carbonate equivalent must be applied or be available naturally in the soil (Miller and Donahue, 1990). These of course are *general* guidelines and will vary between soils. A lime requirement analysis should be conducted on the soil.

Liming Material Injection Techniques

Limestone is very insoluble, and it is impossible to dissolve sufficient quantities in irrigation water to change the soil pH appreciably. However, various machines have been developed to mix and inject gypsum and solid fertilizers. Low concentrations of limestone slurries can also be injected by these same machines through irrigation systems.

When injecting powdered limestone through drip and microirrigation systems, growers must use a very pure form which is ground finely enough to pass a 200 mesh sieve. Specially manufactured limestones and hydrated limes may be used, although limestone is preferred because of ease and safety of handling.

At the time of publication, the authors are unaware of good replicated test results for limestone injection through drip and microirrigation systems. It is probable that undissolved limestone particles will damage microsprayer orifices and spray plates—but at the time of this publication it is not known at what concentrations or with what types of limestone it might occur. Also, because the limestone does not dissolve, the depth of its movement into the soil will be limited and it may not be effective in reclaiming soil some distance away from the emitters. It may, however, be effective in preventing a drop in pH if it is applied prior to the application of acidifying fertilizers. These points are conjecture by the authors rather than proven.

As alternatives to calcium liming materials, growers may investigate other compounds. It is possible to inject potassium hydroxide (KOH), but that compound is very expensive and is hazardous to handle. Potassium carbonate (K_2CO_3) is expensive, but is available in 100-pound sacks, and has evidently been used in some areas for pH modification. Potassium carbonate has a high solubility (1.12 Kg/L of cold water). Growers must be cautious with the use of these potassium compounds for use in liming. They do work for immediate correction of small amounts of acidity. When applied at rates adequate to correct overall acidity in a large soil volume, the grower may be supplying excessive amounts of potassium. This has two possible adverse consequences. First, this can create cation competition within plants and decrease the uptake of calcium and magnesium, and probably ammonium. Second, in very pure irrigation water situations the potassium may contribute to water infiltration problems (see Chapter 13).

Chapter 15. Insecticide, Fungicide, and Herbicide Injection (Chemigation)

General

Treatment of irrigation water for pests, diseases, and weeds is more common in the Midwest than in California. Most of the literature gives specific guidelines with regards to linear moves and center pivots. New et al. (1990) and New (1993) are excellent references regarding pesticides used in chemigation through linear moves and center pivots. The data regarding chemical injection of various pesticides through drip irrigation systems is much more limited.

Chemigation Advantages

Injecting crop protection chemicals through the irrigation water of some irrigation systems has many advantages over traditional aerial and ground spraying of these materials:

- Injecting pesticides uses less energy than traditional application techniques (i.e., less fuel, labor, and equipment wear).
- Sometimes there are not additional equipment requirements—injecting pesticides can generally use the same equipment as injecting fertilizers.
- Injecting chemicals is more flexible than traditional applications. Growers do not have to worry about wet fields and equipment entry constraints.
- Injecting chemicals through microirrigation systems allows for split applications in smaller doses rather than one large dose at the beginning of the season. This allows for more prolonged protection against disease and pests throughout the season.
- Many of the hazards associated with traditional applications (in terms of worker protection) are eliminated because chemical concentrations are *significantly less* when applied in the irrigation water. Aerial applications are 140 – 1000 times more concentrated than treated irrigation water. Ground spray applications are 27 – 200 times more concentrated than treated irrigation water (Young, 1988).
- Injection through drip irrigation systems is better able to get the chemical where it is needed, whether it needs to be in the top 2 or top 10 inches of soil (White, 1986).

- Herbicide injection through systems which do not wet the leaves eliminates the "tie up" with organic residue which would occur with traditional applications (Anonymous, 1982a).
- Injecting herbicides through drip systems delays the chemical degradation because the soil remains moist, allowing for long-term protection (Anonymous, 1990).
- Drip system injection into soil avoids problems of drift and air contact.

Chemigation Disadvantages

- Chemical applications take longer than traditional applications. For sprinkler systems, favorable weather conditions (no wind) over a longer period of time is critical and virtually impossible to guarantee. With sprinkler systems, wind significantly affects the irrigation water uniformity, and consequently the chemical application uniformity. Non-uniformity of chemical applications may cause serious problems in a field.
- Chemigation is generally limited to center pivots, linear moves, or drip/ microirrigation systems because these systems do not require personnel in the field to perform the irrigations.
- Sometimes, additional equipment is necessary. Smaller injectors may be necessary because of the lower injection rates compared to fertilizers; often times special agitators are needed.
- Chemical injection requires greater management input for accurate and safe operations in terms of equipment calibration, chemical handling, and proper functioning of safety equipment.
- Chemical injection can increase environmental hazards in terms of water contamination and worker safety. If irrigations are not managed properly (there is runoff), then surface water supplies can be contaminated. If irrigations are not timed properly (there is excessive deep percolation) or backflow prevention devices fail, then groundwater supplies can be contaminated.
- Injecting a preemergent herbicide may require an unnecessary irrigation.

Herbigation

Several factors should be considered when selecting and chemigating herbicides (Bean, 1990).

1. Solubility: Herbicides with greater solubility move deeper with irrigation water.
2. Adsorption: Herbicides are held onto clays and organic matter (adsorption) to varying degrees. If a material is tightly adsorbed it may not be available to kill weeds with deeper roots. If it does not adsorb well, it may move with the water and damage crop roots.

3. Volatility: This factor represents the likelihood of the chemical to transform into gas. If it volatilizes, it its unavailable to kill plants.

Total herbicide movement throughout the soil profile depends on volatility, solubility, and adsorption.

Preemergence herbicides should be applied with enough water to distribute them in the top 2.5 inches of soil. The amount of water will vary depending on the solubility of the herbicide and soil texture. Generally 0.5 inches of water will be required on sandy soils and 0.75 inches of water on fine textured soils (Bean, 1990). Furrow and border strip irrigation methods are incapable of applying such a small amount of water uniformly.

It is also important that herbicides be applied within about 5 days after the last tillage, otherwise some weed seeds may have started to germinate before the herbicide was applied.

With highly volatile herbicides, herbicides should be applied to dry soil. When applied to wet soil, more of the herbicide will be lost through volatilization. Do not apply the herbicides at the end of the set after the soil has been wetted.

Post-emergence herbicides should generally be applied in a low volume of water (< 0.5 inches). If applied with sprinklers, too much water will cause the herbicide to be lost by dripping from the leaf surface (Bean, 1990).

Fungigation

The accuracy of fungigation is not as critical as herbigation because there is less chance of crop damage from high application rates of fungicides. The most important concept with successful fungigation is adequate coverage which is affected by the fungicide formulation, the amount of water applied, and irrigation uniformity (Kaufman and Lee, 1990).

For irrigation systems which wet the leaves, fungigation can be extremely successful for several reasons (Kaufman and Lee, 1990):

1. The fungicide is applied at the time of maximum leaf wetness when the fungus is most active.
2. Nearly complete coverage is achieved due to redistribution of the fungicide on the leaf with successive droplets.
3. There is great reduction of inoculum in the field by complete coverage of plants and trash on the soil surface.

Insectigation

Insectigation (other than control of soil insects) has been most widely used with center pivots. Its use is relatively minor, as of this date, in drip. Changes in chemical registration in the 1980's forced researchers to find alternative ways to treat mites on corn that was irrigated with center pivots. They discovered that LEPA (Low Energy Precision Application) nozzles could successfully apply miticides to corn. With this application, the miticides could be applied to the underside of lower corn leaves, the area where mites feed, by center pivots modified for chemigation. Special LEPA nozzles specific to chemigation were commercially available in 1987. The following text is specific to insectigation through center pivots with LEPA:

> LEPA nozzles should be positioned 6 to 18 inches above the ground to treat the underside of lower corn leaves for spider mites. The miticides should be mixed with non-emulsified cottonseed oils (1.5 to 2 pints/acre) before injection for spider mite control. In many cases, such mixtures with oil improves efficacy and residual control. Oil carriers should contain no emulsifiers (Knutson and Patrick, 1990).

New (1993) recommends that all new center pivots should be purchased with 60- or 80-inch mainline outlets so that irrigation/chemigation applicators can be placed every two crop rows. The crop should be planted in a circular pattern, so that the sprinklers will remain within the same crop row during the pivot rotation.

Insecticides can be injected undiluted or they can be diluted with water or oil. Studies in Texas show that the insecticide formulations best suited for chemigation through center pivots are those insoluble in water and soluble in oil.

Nemagation

Nematicides are used fairly widely in California through drip/microirrigation systems. Historically, the principal use of nematicides was to decrease nematode populations in the soil before planting annual crops. However, with drip systems they are applied through the irrigation system on permanent and row crops after planting and establishment. (Johnson et al., 1987).

Before a systemic nematicide is applied to an orchard or vineyard, the weeds should be killed because the weeds can take up the nematicide before the desired crop does, or before it is moved further down into the root zone. Multiple small applications are typically required for effective control.

Specific Chemicals

The following text offers examples of some specific chemicals which are currently being applied successfully through microirrigation systems. This is not an endorsement of these products, but rather it constitutes observations on their use. The few chemicals discussed below are by no means an exhaustive list, and their availability and use are subject to state and local regulations.

Vapam®

Vapam® (Metam-sodium) can be used as a soil fumigant. This product can also be used for weed and nematode control, and it has been recently used instead of methyl bromide in California in some trials. Some past injection experiments through drip systems probably failed because the emitters and hoses were spaced too far apart (Bryant, 1993). However, that problem is minimized with row crop drip, in which the emitters form a continuous wetted strip down the row.

Some strawberry growers have found that injecting Vapam is less expensive than traditional methyl bromide applications (and probably by the time the reader sees this, illegal) in terms of chemical costs, application, tractor fuel, and labor. Advantages of these applications are also that the plastic does not have to be removed to apply Vapam in the irrigation water via drip systems. Growers in Orange County found that these applications cut the pesticide costs by 50% and increased yields. Growers in Watsonville found that the weeding labor was decreased from $1,000 per acre to $300 per acre (Anonymous, 1985a).

Rates vary depending on the application; labeling instructions should be followed. As an herbicide, it is injected into the top 1 – 2 inches of soil and has been successful in killing broadleaf weeds in Santa Maria when applied at 15 gallons/acre (Anonymous, 1983a). The product has also been used to fight *Sclerotinia* on lettuce, carrot scab, and diseases on broccoli, garlic, and cauliflower. This product has also been successfully applied to strawberries to fight malva and nematodes.

If the product is to be used as an herbicide, smaller doses should be applied to the top 2 inches of soil with sprinklers. If the product is to be used as an insecticide or nematicide, it should be applied to the top 8 – 10 inches of soil (Anonymous, 1983a).

NEMACUR®

Nemacur® has been successfully used in California to combat root-knot and dagger nematodes in vineyards. It works by both contact action in the soil and systemic action in the roots. Chemigating through microirrigation systems allows growers to apply split applications which increases the length of nematode control over a longer period of time. Nemacur® acts as a nematostat, which does not kill the nematodes, but disorients it so it can not complete its life cycle. This product is also being successfully applied on deciduous fruit trees and nut trees. In California, it is being used on citrus according to special local needs (White, 1986).

VYDATE®

Vydate® was one of the first insecticides in the nation to be registered for application through drip systems One of its first applications was on pole tomatoes for nematode applications, with biweekly applications. Vydate® is absorbed into the vascular system of the plant and allows beneficial organisms to grow, whereas foliar applications will kill the beneficials (Anonymous, 1982a). Vydate® is popular with some farmers using buried drip on tomatoes, because they have found it effective in protecting the tape and the crop from wireworms.

It has been used as a nematicide through drip systems on lemons, oranges, and grapefruits. Citrus quality has been improved on nematode infested citrus orchards: color improved, tree vigor enhanced, and fruit size increased. In addition, growers did not see trees being "set back" for the first year of treatment; rather, the trees showed a very quick positive response (Anonymous, 1983b).

FURADAN 4F®

This product is an insecticide/nematicide and has been injected through microirrigation systems on table, wine, and raisin grapes. University researchers recommend applying 3 – 6 quarts per acre before December 1. The spring treatment should be 2 – 3 quarts per acre before May 1. These dates generally correspond with the root flush of the vines. To control nematodes and suppress *Phylloxera*, applications should be 1 – 2 quarts per acre per application over a 30 day period. Once the problem is under control, re-treatment can occur every second or third year thereafter to maintain control. Periodic agitation is suggested to keep Furadan® in suspension (Anonymous, 1986b). Recently, California restricted the application of Furadan® through above ground drip systems because of potential toxicity problems for birds drinking the water. However, it is allowed in specific areas if applied through subsurface drip systems.

Chapter 16.
Grower Drip Fertigation Experiences

The ITRC interviewed a number of growers and fertility specialists throughout California to assess what growers were actually doing in regards to fertigation with specific nutrients, and why they were taking those actions. These next pages offer specific information on a few growers' fertigation practices. They should be interpreted as observations rather than recommendations, and are provided only to give a sample idea of what some successful growers regard as important.

Bell Peppers

One of the main advantages to growing bell peppers with drip irrigation is the ability to control *Phytophthera.* Land previously unable to support bell pepper production under conventional irrigation methods has been brought back into production with subsurface and surface drip tape. The ability to keep the soil moisture content at a desired level, instead of having soil moisture fluctuate from wet to dry, positively affected pepper production. Growers never have to shut down operations because anaerobic conditions do not occur under microirrigation. In addition, the drip irrigated peppers do not experience the moisture stress that furrow irrigated peppers experience between irrigations. This seems to allow for more uniform growth rates without periods of starting and stopping.

Both calcium and potassium are believed to have a significant effect in controlling blossom end rot. One bell pepper farmer growing peppers reduced the cull rate from 45% to 5% by injecting soluble gypsum on the fields.

The growers agreed that potassium is important to grow bigger and better fruit. As the fruit matures, it becomes a "sink for K." The growers also believed that potassium is important in photosynthesis and in nitrogen uptake and synthesis. Some growers use a 10-0-10, others use

KTS to supply potassium. Other growers applied potassium at preplant. Again, there was a general reluctance to disclose specific fertigation practices, but one grower outlined a basic fertigation schedule for bell peppers (Table 40).

Table 40. One farmer's general fertigation strategy on bell peppers.

Time	Action	Notes
Early season	30 gallons of 6-16-0	Only half of this is applied if preplant P is applied
Fruit set	a "shot" of 6-16-0	Supplies P to seeds, fruit, and roots; seems to insure good fruit set
Post fruit set	3 – 5 gallons of AN-20 or CAN-17/acre every other irrigation; 60 – 70 gallons total	Irrigation sets are generally six hours every other day; this is to "pump" plants with N to build thick walls and increase pepper size

Lettuce

Growers seem to be very pleased with fertigating their lettuce crops. In fertigated drip fields, the lettuce growers were able to make several common comments. Overall, they are able to harvest sooner than in traditional (furrow irrigated) fields. Also, fertigated fields tend to have bigger and heavier heads developed in a shorter amount of time. Nitrogen levels seem to be more uniform in the field throughout the season as compared to traditionally irrigated fields. One grower claimed a 20% yield increase with their new fertigation strategies. Many growers believe that they have maximized yields at this point, and that the only room for improvement is the quality.

Plant and Soil Testing

Some lettuce growers conduct weekly petiole analyses, others conduct tissue tests only if a problem arises, or if they are growing a new variety. All of the growers interviewed conducted soil tests at least once per year. Generally, they try to maintain soil levels so that the soil nutrition is not depleted. One grower conducts tissue and soil tests on the farm. The soil test indicates how much preplant fertilizer to apply, and the tissue tests determine the fertigation injection rates. These growers generally use the U.C. sufficiency range guidelines—one grower uses the University of Arizona guidelines (Doerge et al., 1991). Nevertheless, each grower adjusts the published values according to the regional conditions and the variety grown. Another grower uses a laboratory to determine threshold injection rates for the fertilizers in which precipitation will not occur.

Fertilizers for Lettuce

Most of the growers interviewed used AN-20, UAN-32, or a combination of both on their lettuce crops. Both of these products are compatible with 0-10-10 (used by two growers). One grower preplants with 24 pounds of N/acre, but no other growers discussed preplant nitrogen. Several of the growers use UAN-32 at the beginning of the season (one grower applies 16 pounds of UAN-32/acre in two applications directly after thinning). Another grower injects 14 pounds N and 7 pounds of 0-10-10/acre at thinning. Several growers switch to AN-20 later in the season due to its quicker crop response. For many, this switch occurs at head formation. At head formation time, one grower applies 10 pounds N per week for four weeks. Another grower tries to get all of the nitrogen in the ground during the first six weeks of the crop and manipulates the water later in the season to control head thickness. This grower does not want nutrient considerations to interfere with the irrigation practices later in the season. One grower applies 80 pounds total per acre throughout the growing season.

Only a few lettuce growers interviewed inject phosphorus; one applies 30 gallons of 6-16-0 early in the season. This grower feels that it helps build a big frame. Two growers inject 0-10-10 throughout the growing season at a constant rate. They use this product and strategy because the plants require phosphorus early in the season for root development. They also feel that this strategy stores the potassium for future use in head development. They agree that the constant feed strategy is better for the plants and reduces the likelihood of nutrient leaching from winter rains. Other growers apply phosphorus and potassium at preplant.

One grower also applies 25 pounds of KTS/acre. This grower feels it improves the color and head formation.

Several growers inject CAN-17 for the nitrogen and calcium; some inject it only at the end of the season, others inject it about six weeks into the season. They generally agree that this product improves the lettuce color. One farmer injects CAN-17 if harvesting is delayed. The grower believes that it maintains the quality of the crop if it has to stay in the field. Each of these growers inject calcium during the second half of the injection cycle.

Special Practices on Lettuce

One grower injects N-pHURIC® once per week to prevent bacterial accumulation in the lines. Injection rates range from 2 to 3 gallons N-pHURIC per acre per week. Another grower keeps the beds wet for 4 or 5 days in between crops to inhibit root intrusion.

TOMATOES

Tomato growers interviewed produced both processing tomatoes and fresh market tomatoes. The fresh market growers were located on the west side of the San Joaquin Valley and the south end of the Salinas Valley. Processing tomato growers were located in Hollister, Stockton, and Tracy areas.

PLANT AND SOIL TESTING

Plant and soil testing were common practices for the tomato growers interviewed. A few consultants have compiled large amounts of data for tomatoes in these areas. Doubling the average yield per acre was claimed by various growers using drip irrigation. Traditional petiole standards for tomatoes were too low according to all growers contacted.

Frequency of petiole sampling varied among tomato growers. Some growers completed tests weekly, others conducted petiole analyses five times per season. All growers conducted yearly soil tests. In some cases, the farmer completed the analysis on farm; other growers sent out samples to laboratories. One grower contracted with a service to sample and analyze the petiole samples. Generally, nitrogen and potassium concentrations in the plant are the nutrients monitored most carefully throughout the season.

Overall, growers try to keep the nitrate and potassium levels high to maximize growth until the pre-bloom stage. They gave a general fertigation schedule which they use and adjust weekly according to petiole analysis results. Absolute and relative amounts of calcium in the soil were believed to be important to control blossom-end rot. Each grower stressed the importance of good record keeping, and they generally agreed that after correlating test results with physiological changes in the plant for several years, that the frequency of testing can be reduced.

FERTILIZERS FOR TOMATOES

Fertigation regimes varied significantly among growers, but there seems to be several common practices. All of the growers use U.C. sufficiency guidelines as starting points for determining plant nutrition, but they adjust the numbers upward to account for the increased yields. Several of the growers interviewed inject the fertilizers on a ppm basis according to frequent tissue tests throughout the season. All of these growers have seen significant increases in crop quality and quantity as a result of fine-tuning their nutrient management strategies. One grower is able to supply the quota of processed tomatoes on less acreage, and is able to grow additional crops (peppers) on the remaining land.

Many tomato growers apply UAN-32 at transplant, but if the weather is cold at this time, they apply AN-20 instead. Several growers preferred UAN-32 throughout the season because it is not as "hot" and the plant does not see the sudden starts and stops as with AN-20. There were other growers though who preferred AN-20 because it is less expensive and provides a quicker response. These growers generally inject AN-20 continuously at small concentrations, so they do not see problems otherwise associated with this product. One grower injects N-pHURIC® products to supply nitrogen and maintain a low water pH; the product is injected after the filters to prevent hardware damage. The grower cautions about soil micronutrient toxicities, and therefore tests the soil pH regularly to foresee potential toxicities. All of these growers tend to keep the nitrogen levels high until bloom. At bloom time, nitrogen levels are allowed to fall to promote fruiting. They caution about keeping N levels too high at this stage because fruit quality and quantity are adversely affected.

Growers interviewed used several different products to supply phosphorus. One grower injects 10-34-0 to supply P, but others have seen plugging problems with this product. One grower uses a 4-10-10 blend and another grower uses 10-27-0 (an N-pHURIC based phosphate) to supply P. Another grower applies 30 gallons of 6-16-0 two months into the growing season to get phosphorus to the plants early in the growth cycle. This grower also applies this product later in the season because it is believed that it promotes better blooms. One grower prefers KTS to free up the phosphorus already in the soil. Even though this product is more expensive than many other fertilizer products, the grower believes that this strategy is more economical because phosphorus compounds generally become unavailable after they are applied. All growers believe that phosphorus helps root, stem, and leaf structure.

Several growers are satisfied with KTS to supply potassium, and all agree that potassium is particularly important for their tomato crops. The potassium improves fruit solids and wall thickness, in addition to improving fruit color. One grower feels that the potassium improves the fruit ripening, and also believes that low potassium levels lead to streaked and blotchy fruit. All of these factors contribute to better fruit quality and increase the price received for the tomatoes. All growers keep potassium levels high at the end of the growing season.

Growers who have Ca/Mg imbalances in the soil have injected gypsum and have seen positive crop responses. These growers avoid using CAN-17 as their *only* calcium source because this will add too much nitrogen to the crop, although some growers apply CAN-17 later in the season to improve the fruit quality.

No ideal micronutrient solution has been found for fertigation through the drip system. Boron and manganese are applied at first fruit set and again fourteen days later in foliar applications at night by one grower. Manganese and zinc are applied at full fruit set, and another application is made three weeks from harvest. This grower believes that zinc improves nutrient uptake, prevents a wilt, and provides better color.

Many of the growers were reluctant to give details regarding the fertigation schedule, but one grower gave information on the fertigation schedule for tomatoes. This schedule is adjusted according to petiole analyses.

Table 41. One grower's general fertigation schedule for tomatoes.

Time	Fertilizer Application (per acre)
20 – 30 days after transplant	2.5 pounds N from UAN-32 2.5 pounds P from 4-10-10 and 1 gallon KTS
Full bloom	10 gallons UAN-32 10 gallons 4-10-10 and 0.5 gallon KTS
Fruit set	5 gallons UAN-32 5 gallons 4-10-10 and KTS supplement
3 weeks from harvest	2.5 gallons UAN-32 5 gallons 4-10-10 and 2 gallons KTS
2 weeks from harvest	No UAN-32 2.5 gallons 4-10-10 and 1 gallon KTS

This grower injects these products separately and in a specific order during an irrigation set. The grower injects the least mobile nutrient (P or K) first so that it is pushed downward throughout the root zone during the remainder of the set. A typical injection order for a 12-hour set is as follows:

- 1 hour water
- 4 hours of P (or K)
- 1 hour of water
- 5 hours of N
- 1 hour of water

This grower believes that from first bloom to 35 days afterward is the period when most mistakes are made. The grower feels it is important to watch the fruit load per plant growth relationship. The bloom should be barely sticking out of the canopy and the leaf right above the flower. If there is nothing but a mass of yellow flowers, then the plant is being pushed too hard.

Tomato Summary

The growers interviewed have all stressed the importance of fertigation under microirrigation systems in producing higher yielding and better quality crops. The need for increased soil and plant tissue testing in determining crop nutritional needs at these higher yield levels was emphasized by all growers. There was uniform dissatisfaction with currently recommended sufficiency range data; growers commonly use private labs for calibration of petiole analyses. Sufficiency range data was the predominate interpretation technique utilized. Many growers conduct weekly petiole analyses when beginning their fertigation programs to determine which nutrients affect plant and fruit development at various growth stages. Their testing frequency decreases in subsequent years as their proficiency in diagnoses improves.

Grapes

Fertigating vines with a microirrigation system has the same advantages as many other crops. Many of these vineyards use above ground drip systems. With vines, many other chemicals for the treatment of disease are being applied through chemigation. There is an added advantage in the cost savings of these applications. Many growers claim that younger vines are brought into production sooner; this is a distinct advantage specific to vines and orchard crops. While managing fertigation regimes is still being developed and fine tuned, it has produced excellent results for many growers.

Benefits from fertigation practices include the following: an increase in crop tonnage; better fruit quality in terms of uniform size and color; and improved ability to control more precisely the fruit solids content. Increases in tonnage were noted by almost all growers. One grower estimated that a gain of 10% yield increase through keeping water to the vines. Two ways are claimed to build solids in grapes: dry down the vines, thus increasing the sugar concentration in the grape; or keep the vine going and allow it to produce more sugar on its own. The latter strategy produced the 10% yield increase. Improved color and berry solids translate into better wine producing berries which result in a higher price for the grape.

There are some disadvantages in fertigating vineyard crops. The continued application of fertilizers and irrigation water to a restricted root zone leaches nutrients over time. Careful attention must be paid to the water and fertility management of this zone in order to maintain a fertile growing environment.

Plant and Soil Testing

On perennial crops such as vines, close attention must be given to the chemistry of the water source, soil nutritional levels, and plant nutritional levels. Due to the restricted root zone of vineyards grown using restricted flow irrigation systems, the potential for leaching of the nutrients over time, in this zone, is increased. The soil type is also an important factor because lighter, sandy soils with low cation exchange and water holding capacities have an increased potential for leaching.

All growers interviewed conducted yearly water quality testing of all water sources. There were three fundamental reasons for doing so: to anticipate fertilizer solubility problems, to anticipate water penetration problems, and to select the most cost effective means of fertilizer and biological treatment. Annual soil testing was commonly done for several reasons: to evaluate what portion of the crop's nutritional requirement was contributed by the soil; and to evaluate the effects of soil, water, and fertilizer interactions on the chemical and physical properties of the soil. Frequency of tissue sampling varied depending on the vineyard.

Fertilizers for Grapes

The purpose of the post-harvest fertilization is to revive the plant before winter dormancy and to store nutrients in the wood for use in the spring. One grower injects a complete fertilizer (8-8-8) at this time. For post-harvest nutrition, nitrogen and potassium were the most common nutrients used. Most growers run a fairly heavy post-harvest irrigation (24 – 48 hrs.) followed by the injection of nutrients. UAN-32 was used by all growers as a nitrogen source because it is slower acting and relatively inexpensive. Amounts of nitrogen injected for post harvest fertilization ranged from 10 – 30 pounds of N/acre. The amount of N injected usually depended on the harvest size. Potassium was injected in the form of potassium thiosulfate (KTS), potassium chloride, or potassium sulfate. The form of K used depended on the water chemistry, soil chemistry, and physical condition of the soil.

On calcareous soils, potassium sulfate and thiosulfate are more commonly used. The oxidation of the thiosulfate in the fertilizer helps to break down the limestone in the calcareous soils to form gypsum. The gypsum improves soil structure, increases aeration, frees soil calcium, and frees plant available P in the soils.

On non-calcareous soils where the availability of soluble Ca is not a problem, potassium chloride was used. This form of K nutrition is the least expensive, but irrigation water chloride levels must be sufficiently low to avoid Cl toxicities. Growers recommend testing the soil and water for Cl. Once this is done, growers determine the total Cl portion of the potassium chloride to be injected over the course of the year on a ppm basis. Tissue chloride concentrations can then be examined for toxicity problems.

Petiole sampling on mature vines is done at various intervals depending on the grower and grape variety being grown. One grower takes petiole samples once per year and compares the results to other blocks and historic data. Testing usually occurs when the vines are 50 – 100% in bloom. The comparison method allows for determining which blocks are deficient. This grower has no set fertility program but commonly uses UAN-32 as the N source, CAN-17 in low Ca soils, and KTS for potassium.

Another grower who has Cabernet Sauvignon and Chardonnay grapes follows a different program. No fertilizing is done until fruit set, when the first petiole sample is taken (root flush at this time). Depending on the tissue nutrient levels, CAN-17 and KTS are applied. Another petiole sample is taken in May or June during bloom to determine N and K application rates. The last petiole test occurs at véraison (the time the berries begin to color and soften) on the problem blocks only. These results help confirm the fertility program and are used to make needed adjustments. The grower generally injects more K than N and does not fertilize with P at this growth stage.

This grower uses KTS on calcareous soils. KTS is applied at 30 – 40 gallons per acre/year in approximately 10 gallon applications at fruit set, bloom, and véraison. The benefits of the KTS product are generally seen in the color of the grapes, fruit set, grape solids, and berry wall thickness. This particular grower runs field trials and has seen a 30% increase in fruit set on grapes treated with KTS. Non-treated vines have not made color thus forcing the production of a white cabernet wine. This grower believes that too much KTS applied at véraison may cause the vine to produce water berries; thus, KTS is applied in smaller amounts at this growth stage.

A grower in the Stockton area applies 1 gallon of 5-0-12/acre/day at the beginning of the irrigation season. This is done through véraison and adjustments are made depending on the petiole readings. All injections are done on a ppm basis. Petiole concentrations are examined, and how much (if any) of the nutrient being supplied by the water is determined. Then the amount needed in excess of that is determined on a ppm basis, and calibrated for an injection regime. This particular grower uses higher sufficiency levels than those recommended by the U.C. in order to grow an above-average crop. The grower tries to predict when bloom will occur and takes petiole samples at five times: 3 weeks before bloom, 10 days before bloom, at bloom, 3 weeks after bloom, and at véraison. The pre-bloom samples are important in regards to P, Zn, and K levels; the grower believes that Zn is very important for fruit set. Also, the grower believes that zinc is important in preventing "little leaf," (a smaller than normal leaf size) which reduces the plant's photosynthetic abilities. An inexpensive zinc sulfate is applied prior to bloom as a Zn source. If the zinc must be blended with the NK mix, the chelate form is used. This chelate can be 5 times more expensive than the sulfate form.

Phosphoric acid is also injected to meet the P requirements of the vine. Phosphorus is usually applied at bud break to help give the vine more vigor. In this particular case, the water with high Ca and HCO_3 is not suited for a 10-34-0 solution. The post harvest application at this vineyard consists of 20 gallons of a 5-0-12 solution (approximately 10 pounds N/acre and 24 pounds K/acre).

A grower in Livingston who recently converted to drip and grows Malvasia Bianca, Sauvignon Blanc, Barberra, Chardonnay, and Grenache varieties uses a complete 8-8-8 fertilizer. Doses range from 2 – 5 gallons of 8-8-8 per acre/week depending on the irrigation duration. Under this program, second year vines producing one-third of the expected crop reportedly produced crop yields equal to the third and fourth year vines. Previously, the vines were showing K deficiency symptoms which were believed to be caused from leaching and decreased soil pH from the use of N-pHURIC®. The grower is relatively new at drip irrigation and is thinking of using a 20% magnesium sulfate solution to cure Mg deficiencies which have been observed visually.

Growers with permeability problems have also been using the soluble gypsum machines to inject gypsum. The benefits include better soil structure, increased water penetration, improved lateral movement, increased microbial activity, and overall increased availability of the cations in the soils. The soluble gypsum machines can be labor intensive, if numerous water sources are utilized. Gypsum is also dribbled beneath the drip lines by some farms at rates up to one ton of 90% pure gypsum per acre. Two growers also use the pumice left over from the crush as a water penetration aid.

A few growers are experimenting with increased fertilization of young vines in the hope of bringing the vines into production earlier than normal. In these cases, fertilizer application rates, equal to those given to mature vines, are being applied to first and second year vines with positive results. One grower applies 40 pounds of N and 70 pounds of K/acre/year on young vines. Many growers stressed the importance of applying fertilizers on a concentration (ppm) rather than as pounds/acre basis. By doing this, a better correlation can be achieved with the petiole samples.

Other Chemicals for Vines

Other chemicals which are applied through chemigation are Furadan® and Enzone® for the treatment of *Phyloxxera*. Furadan does have some problems with avian life; one solution was to run spaghetti tubing from the emitter to 3 or 4 inches below the soil surface. Chitinous material (crab shell extract) was being used on several ranches to soften the cuticle of *Phyloxxera,* making them easier to be destroyed.

Nemacur® was commonly used in the treatment of nematodes. Nemacur was applied three times per year: prior to bloom, after bloom, and post harvest. Nemacur rates were approximately 1 gallon/acre/year. Trifluralan®, a herbicide, was also being used on vines. N-pHURIC® was a common compound used to control bacteria in the drip lines or to decrease irrigation water pH. N-pHURIC treatments averaged 2 to 3 times per year.

SUMMARY FOR GRAPES

Fertigating vines with microirrigation systems provides advantages and challenges unique to grapes. The ability to bring young vines into productivity at an earlier age is an advantage that appears to have great potential, although long-term effects on vine vigor are not known. Manipulation of berry color and solids is possible with fertigated grapes. Lastly, injecting chemicals to fight diseases and organisms is being experimented with on several farms. The long-term effectiveness of these treatments is not known at this time.

The lack of a wide range of data regarding petiole sufficiency levels for the various varieties of grapes grown in California has made tissue analysis interpretation difficult. Peter Christiansen, Extension Viticulturist at the U.C., Kearny Agricultural Center, has published a bulletin on vineyard tissue testing which helps growers understand some of the problems associated with vineyard tissue testing (Christiansen, 1989).

Growers predominantly use sufficiency range data with vineyard tissue samples. Due to the restricted wetted zone and the problem of leaching and/or mining of the nutrients in this zone, nutrient balance ratio interpretations (or DRIS) may provide unique advantages for interpreting tissue analyses in vineyards. The ability of the DRIS method to recommend nutrients in order of need, and to evaluate nutrient interactions, should prove valuable in predicting nutrient needs as the root zone changes with time.

Soil testing at increased depths might also prove beneficial to determine the nutrient status of soils supporting mature vines. This practice may be especially useful for determining the long range effects of fertigation and irrigation on soil nutrients (i.e., P and K) and of deep leaching of nitrates which are available to deeper roots of crops.

CHAPTER 17. SAMPLE FERTIGATION CALCULATIONS

COST COMPARISONS OF FERTILIZERS

All fertilizers should be evaluated with regard to the cost of the total nutrients in the fertilizer. This can be accomplished by using the following formula:

$$\frac{\text{Dollars}}{\text{Ton of Fertilizer}} \times \frac{100\%\ \text{Fertilizer}}{\_\ \%\ \text{Nutrient}} \times \frac{1\ \text{ton}}{2000\ \text{pounds}} = \frac{\text{Dollar cost}}{\text{Pound Nutrient}}$$

Of course, price is not the only factor when deciding to use one nutrient over another: the constituents of the fertilizers, solubility, nitrogen source, etc. will also play a part in a farmer's decision to use a particular fertilizer.

INJECTION RATE CALIBRATION

RELEVANT CONVERSIONS

One acre-inch of irrigation water = 27,154 gallons of water

One gallon of water = 8.345 pounds of water

One gallon per minute = 0.00221 acre inches of water per hour

$$\text{Inches of water applied} = \frac{(\text{gallons per minute}) \times (\text{hours})}{(450) \times (\text{acres})}$$

$$\text{Gallons fertilizer per hour} = \frac{(\text{acres}) \times (\text{gallons fertilizer per acre})}{\text{hours of set time for irrigation}}$$

To calibrate a liquid fertilizer injection, the flow rate of the liquid fertilizer can be monitored as the time required to fill a quart "jar" with the fertilizer. For better accuracy, one should calibrate an injector by filling a graduated cylinder (one quart = 946 milliliters). The fertilizer flow rate and the time required to fill the quart jar are given below:

Table 42. Guidelines for calibrating injector flow rates. (Anonymous, 1988)

Gallons Fertilizer Injected per Hour	Minutes and Seconds to fill a Quart Jar
1.0	14 minutes + 56 seconds
5.0	2 minutes + 56 seconds
10.0	1 minute + 28 seconds
20.0	0 minute + 44 seconds

DENSITY OF VARIOUS LIQUID FERTILIZERS

In order to calibrate injection rates, the densities of liquid fertilizers must be known. Table 43 gives density information on some popular liquid fertilizers.

Table 43. Densities and grades of various liquid fertilizers.

Fertilizer	Grade	Density (pounds/gallon)
Ammonium Nitrate	21-0-0	10.73
Ammonium Thiosulfate	12-0-0-26S	11.04
Aqua Ammonia	20.5-0-0	7.52
Urea Ammonium Nitrate	32-0-0	11.05
Urea Sulfuric Acid	10-0-0 (10/55)	12.80
	17-0-0 (17/49)	12.80
	28-0-0 (28/27)	11.90

CALCULATION OF FERTILIZER INJECTION RATE

The 1986 *Chemigation Guide* (Anonymous, 1986b) provides a chemigation worksheet for determining the amount of fertilizer to be injected into irrigation water for a center pivot system.

The following guideline has been adapted from the 1986 *Chemigation Guide.*

STEP 1:

Determine the amount of nutrient (usually N) to apply per acre during this fertigation. This amount is often based on the total nutrient need of the crop divided by the total number of fertigation applications. The amount may be based on an estimation of the maximum rate of nutrient uptake by the crop at that specific stage of growth.

For example, assume that you want to inject 20 pounds of N/Acre.

STEP 2:
Choose the type of fertilizer which will be used to meet this nutrient need. This choice should be made based on the price per pound of nutrient and on other agronomic factors related to your soil, crop, water, and management program.

For example, assume that you want to use urea ammonium nitrate (32-0-0) as a liquid fertilizer or urea 46-0-0 as a solid fertilizer.

STEP 3:
Determine the pounds of the particular fertilizer which should be used. Choose *Method A* for solid fertilizers or *Method B* for liquid fertilizers.

Method A. Solid Fertilizers: Determine the pounds of solid fertilizer which must be dissolved into water for injection and the total number of gallons of resulting fertilizer solution. Note that a solid, mixed nutrient fertilizer may not dissolve uniformly at first. Each component has a different dissolution rate and solubility. Thus, it is very important that all of the solid fertilizer be in solution *before* attempting to inject the fertilizer solution into the irrigation line. Calculate as follows:

$$\frac{\text{Pounds of Nutrient}}{\text{Acre}} \times \frac{100\%\ \text{Fertilizer}}{\_\ \%\ \text{Nutrient}} = \frac{\text{Pounds of Fertilizer}}{\text{Acre}}$$

Using our example with solid urea 46-0-0 to supply 20 pounds of N/Acre gives:

$$\frac{20\ \text{Pounds of N}}{\text{Acre}} \times \frac{100\%\ \text{Urea}}{46\%\ \text{N}} = \frac{43.48\ \text{Pounds of Urea}}{\text{Acre}}$$

When dissolving a solid fertilizer it is important to consult the solubility table to obtain an idea of approximately how many gallons of water should be used to dissolve this amount of fertilizer. Table 9 indicates that about 8.34 pounds of urea will dissolve in each gallon of water. Estimate the amount of water as follows:

$$\frac{\text{Pounds of Fertilizer}}{\text{Acre}} \times \frac{1\ \text{Gallon of Water}}{\_\ \text{pounds dissolved}} = \frac{\text{Gallons of Water}}{\text{Acre}}$$

Using our example for urea with a solubility of 8.34 pounds of urea per gallon we calculate:

$$\frac{43.48 \text{ Pounds of Urea}}{\text{Acre}} \times \frac{1 \text{ Gallon of Water}}{8.34 \text{ Pounds of Urea}} = \frac{5.21 \text{ Gallons of Water}}{\text{Acre}}$$

Our estimate suggests that if we try to dissolve all of the fertilizer in a 5 gallon bucket of water that it probably will not all dissolve. We would be better to use a 10 gallon container. Add 6 gallons of water first and then slowly mix in the solid urea fertilizer with mixing until all of it has dissolved. If the fertilizer has a conditioner, allow this to settle to the bottom first. Then siphon off the clear liquid to use for injection into the irrigation water.

Method B. For Liquid Fertilizers: Use the above calculation followed by the conversion of mass to volume (density) so that you will know the volume of liquid to be injected. This general equation is presented below. Note that the fertilizer density value goes on the bottom of the second term. This allows the result to become gallons.

$$\frac{\text{Pounds of Fertilizer}}{\text{Acre}} \times \frac{1 \text{ Gallon of Fertilizer}}{\_ \text{ Pounds}} = \frac{\text{Gallons of Fertilizer}}{\text{Acre}}$$

For example using Urea Ammonium Nitrate 32-0-0 with a density of 11.05 pounds per gallon and the application of 30 pounds of N/Acre gives:

$$\frac{30 \text{ Pounds of N}}{\text{Acre}} \text{ x } \frac{100\% \text{ UAN} -32}{32\% \text{ N}} = \frac{93.75 \text{ Pounds UAN -32}}{\text{Acre}}$$

$$\frac{93.75 \text{ Pounds of UAN-32}}{\text{Acre}} \times \frac{1 \text{ Gallon of UAN - 32}}{11.05 \text{ Pounds UAN-32}} = \frac{8.48 \text{ Gallons of UAN - 32}}{\text{Acre}}$$

STEP 4:

Calculate the total acres to be fertigated with this fertilizer. This would normally be one revolution of a center pivot, or the acres wetted by the connected drip irrigation system, or the area covered by one irrigation set.

For example, assume that you are fertigating 10 acres.

STEP 5:

Multiply the gallons of fertilizer per acre by the total number of acres to be fertigated.
For the example with solid urea dissolved in 6 gallons of water for 10 acres this would be 60 gallons of solution per irrigation.

For the example with Urea Ammonium Nitrate:

$$\frac{8.48 \text{ Gallons of UAN - 32}}{\text{Acre}} \times \frac{10 \text{ Acres}}{\text{Irrigation Set}} = \frac{84.8 \text{ Gallons of UAN - 32}}{\text{Irrigation Set}}$$

STEP 6:

Determine the total time for fertilizer injection. Often this time will equal the total time for the irrigation. However, you may wish to inject the fertilizer only during a portion of the total irrigation. This choice depends on the type of irrigation you use.

For example assume that we are planning on irrigating for 16 hours and to inject with the Urea Ammonium Nitrate during the entire irrigation time.

STEP 7:

Calculate the rate of fertilizer injection into the irrigation system in gallons per hour. The calculation involves dividing the total gallons of fertilizer by the total hours for fertilizer injection. For our example:

$$\frac{84.8 \text{ Gallons of UAN - 32}}{\text{Set}} \times \frac{\text{Sets}}{16 \text{ Hour}} = \frac{5.3 \text{ Gallons of UAN - 32 Fertilizer}}{\text{Hour}}$$

STEP 8:

Check for system capability. At this point it is important to double check that the rate of fertilizer injection which you have just calculated can be delivered by the type of injection system that you use. In this example the injection system must be able to handle approximately 6 gallons of fertilizer injected per hour.

Appendix A.
Units of Salinity Measurement

There are three common methods of describing the amount of salt in a sample. These are described below.

Electrical Conductivity (EC)

This is the most simple method of measurement, and provides a measure of the total amount of salt in a sample. It is a measure of the number of electrically charged particles. The measuring procedure is very simple. An electrical current is applied between two probes in a water sample, and the higher the conductivity of the water, the higher the salinity. The reading is taken from a dial or digital output on the meter. Because this is a single measurement of total salinity, it does not provide information about the various constituents of salt in the sample.

The units of EC are usually decisiemens per meter (dS/m) or millimhos per centimeter (mmho/cm).

$$1 \text{ dS/m} = 1 \text{ mmho/cm}$$

Instruments are now available to provide measurements of the EC of the soil profile without the need for taking soil samples. These nondestructive techniques use probes on or near the surface of the soil which send an electrical current through the soil and measure the conductivity. These meters, mounted on a small vehicle, are being used on large district-level surveys for rapid determination of soil salinity levels. For rapid mapping, the vehicle is sometimes equipped with GPS (Global Positioning System).

Parts Per Million (ppm)

This is a weight (mass) measurement. The following represents a sample problem converting milligrams (mg) per liter to parts per million (ppm).

Given: Total weight of salt in a water sample = 150 mg
Water sample size = 1.5 liters

Find: ppm of total salt in the sample

Formula:

$$\text{ppm} = \frac{\text{milligrams of salt}}{\text{liters of solution}}$$

Solution:

$$\text{ppm} = \frac{150\ \text{mg}}{1.5\ \text{liters}}$$

$$= 100\ \text{ppm}$$

The concentrations of individual salt constituents in a sample are measured by various laboratory devices or procedures. The determination of each individual salt constituent in ppm is a separate test. The measurement of the total ppm of a sample is the sum of all of the individual salt constituents, and is considerably more difficult than the EC determination.

In the western U.S., salinity measurement can also be approximately converted from dS/m to ppm.

Given: Salinity in a river = 1.2 dS/m

Find: Estimate of total ppm of dissolved salt in the river

Formula: 1 dS/m = 700 ppm

Solution:

$$\text{ppm} = 1.2\ \text{dS/m} \times \frac{700\ \text{ppm}}{1\ \text{dS / m}}$$

$$= 840\ \text{ppm}$$

This relationship of 1/700 depends on the specific water sample, as various types of salt have different weights.

MILLIEQUIVALENTS PER LITER (MEQ/L)

The ppm of each salt constituent is a measurement of weight, but many salinity problems are related more to the number of salt electrical charges than salt weight. Therefore, the ppm values are often converted to meq/L for mathematical computations regarding such things as permeability hazard and the amount of soil or water amendments needed to correct a problem.

The formula for determining the meq/L of a particular ion is:

$$\text{meq/L} = \frac{\text{ppm}}{\text{milliequivalent weight in milligrams}}$$

The formula for determining the total number of milliequivalents of a substance is:

$$\text{meq} = \frac{\text{weight of the substance in milligrams}}{\text{milliequivalent weight in milligrams}}$$

Table 44. Common constituents of irrigation waters with their weights and electrical charges.

Element Name	Symbol	Electrical Charge	Atomic Weight gm	Milliequivalent Weight mg
CATIONS				
Ammonium	NH_4	+1	17.0	17.0
Calcium	Ca	+2	40.1	20.1
Hydrogen	H	+1	1.0	1.0
Magnesium	Mg	+2	24.3	12.2
Sodium	Na	+1	23.0	23.0
Potassium	K	+1	39.1	39.1
ANIONS				
Bicarbonate	HCO_3	-1	61.0	61.0
Carbonate	CO_3	-2	60.0	30.0
Nitrate	NO_3	-1	62.0	62.0
Phosphate	PO_4	-3	79.0	26.3
Sulfate	SO_4	-2	96.1	48.1

For example, using the values in Table 44, the meq/L concentration of a water sample with 120 ppm calcium can be calculated as follows:

GIVEN: Water quality = 120 ppm Ca

FIND: Water quality in meq/L

Formulas:

$$\text{meq} = \frac{\text{weight of the substance (mg)}}{\text{milliequivalent weight (mg)}}$$

$$\text{meq/L} = \frac{\text{ppm}}{\text{milliequivalent weight (mg)}}$$

Solution:

$$\text{meq/L} = \frac{(120 \text{ ppm } Ca^{2+})/\text{ liter}}{20 \text{ mg/ milliequivalent weight}}$$

$$= 6 \text{ meq/L of } Ca^{2+}$$

APPENDIX B. QUICK TEST EQUIPMENT

Table 45. Manufacturer information on various quick test equipment.

Manufacturer	Equipment/Price	Tests and Description
Hach Company P.O. Box 389 Loveland, CO 80539 800-227-4224	AgriTrak Portable Laboratory $995	Soil, water, tissue, sap: NO_3 Soil, water: pH Uses portable pH/mV meter; initial reagents available for 50 tests; for sap testing, a 0.1 to 1.0 ml pipette is also required
	NPK Test Kit $385	Soil: NO_3, P, K, pH Water: NO_3, P, pH Initial reagents available for 100 tests for each parameter
	SIW-1 Soil and Water Test Kit $950	Soil: Base Saturation, pH, CEC, , EC, free lime, NO_3, P, K, Ca, Mg, Na, Acidity Water: Ca, Mg, EC, NO_3, pH, P, SAR, Na This kit can also do some tissue tests with additional reagents and altered procedures; 300 tests for each parameter
	Nitrate Pocket Colorimeter $295	Water: NO_3 100 tests
	DREL 2000 Soil and Irrigation Water Laboratory $3,095	Soil: base saturation, CEC, Na, Ca, Mg, free lime, NO_3, pH, P, K, EC, SO_4, acidity Water: alkalinity, Cl, EC, hardness, Fe, NO_3, pH, P, K, Na, SAR, SO_4 100 tests for each parameter; other reagents are available for soil analyses of Zn, Mn, Fe, Cu, organic matter; this kit can also analyze, NO_3-N, PO_4-P, K, SO_4-S in tissue with additional reagents and altered procedures
	N-Trak Test Kit $90	Soil, water: NO_3 50 tests
	P Test Kit $62	Water: PO_4-P 100 tests

continued

Table 45. Continued.

Manufacturer	Equipment/Price	Tests and Description
Spectrum Technologies 12021 S. Aero Dr. Plainfield, IL 60544 800-248-8873	Cardy Meter® $309	Soil, water, tissue, sap: NO_3 Original sensor lasts for 250-400 samples; $65 for replacement sensor; different standards required depending on what is tested
	Cardy Meter $330	Soil, water, tissue, sap: NH_4 Replacement sensor $85; different standards for tissue vs. sap test
	Cardy Meter $299	Soil, water, tissue, sap: K Replacement sensor $65; different standards
	Water Tester $199	Water: dozens of factors Price does not include kits necessary for analysis; this price merely reflects the meter cost
Em Industries 480 Democrat Rd. Gibbstown, NJ 08027-1297 800-222-0342	Reflectoquant Meter $495	Water, tissue, sap, soil solution: NH_4, Cu, Fe, Mn, Mo, NO_3, pH, PO_4 This equipment also requires the use of test strips specific to the nutrient which is sampled; 50 test strips cost approximately $50; test strips for Zn, Cl, Ca, and K will be available soon
Plant Check 4200 Woodville Pike Urbana, OH 43078 513-652-3142	Test Kit $55.65	Tissue: N, P, K; Soil: pH 100 tests for each parameter; uses test strips and color codes; results are *relative*, not quantitative
Irrometer P.O. Box 2424 Riverside, CA 92516 909-689-1701	Soil Solution Access Tubes $16 – 60	Device for extracting the soil solution. Price depends on length of tube and type of pump; price includes suction line and device to draw water from tube
VWR Scientific P.O. Box 6016 Cerritos, CA 90702 800-932-8000	Various chemicals, supplies	
Soil Test Co. P.O. Box 8004 Lake Bluff, IL 60044 708-295-9400	Soil Solution Access Tubes (Lysimeters) $44 – 60	Extracts the soil solution Price depends on length of tube
	Service Kit $112	Hand pump and gauge

continued

Table 45. Continued.

Manufacturer	Equipment/Price	Tests and Description
Soilmoisture Equipment Corp. P.O. Box 30025 Santa Barbara, CA 93105 805-964-3525	Soil water sampler $39 – 58	Price varies depending on length
	Hand pump with gauge $100	Required to put tube under suction
	Stopper assembly, clamp, tube $20	Required
	1000 ml extraction kit $42	Required to extract solution from tube

Appendix C.
Quick Test Procedures

Soil Nitrate Test Strip Procedure

(Hartz et al., 1994)

Equipment

- Aluminum sulfate ($95/100 g of chemical; $0.21/test)
- 1 L volumetric flask with stopper ($35-75)
- 100 ml graduated cylinder ($10-30)
- Weighing dishes ($35 per 500 dishes)
- Balance ($150 – 200 for a triple beam)
- Nitrate strips (approximately $1 per strip)

Preparation of Reagent

1. Weigh 8.55 g of aluminum sulfate ($Al_2(SO_4)_3$) and place in a 1 L volumetric flask.

2. Dilute to 1 L with deionized water. This solution is 0.025 M $Al_2(SO_4)_3$.

3. Stopper the flask and invert twenty times to mix. One liter of 0.025M $Al_2(SO_4)_3$ solution will complete approximately 40 tests.

Procedure

1. Collect a composite soil sample representative of the main root zone of the crop; blend thoroughly in a container. Don't include the top 2 inches of soil because it may be high in N but too dry for active root growth.

2. Fill a volumetrically marked tube or cylinder to the 30 ml level with 0.025 M aluminum sulfate ($Al_2(SO_4)_3$) solution.

3. Add field moist soil to the tube until the liquid level rises to 40 ml; cap tightly and shake vigorously until soil is thoroughly dispersed. Let sit until soil particles settle.

4. When solution is reasonably clear, dip a Merckquant nitrate test strip into the solution, shake off excess solution, and wait 60 seconds. Compare color with the color chart provided.

To minimize the variability inherent in soil sampling, run duplicate tubes for each field soil evaluated.

Interpretation

The test strips are calibrated in parts per million (ppm) ppm NO_3. The approximate conversion to ppm NO_3-N on a dry soil basis is:

$$\frac{\text{ppm } NO_3}{2} = \text{ppm } NO_3\text{-N in dry soil}$$

Soil less than 10 ppm NO_3-N would be considered quite low; levels above 20 ppm NO_3-N have enough available nitrogen to meet immediate crop needs.

Note: Low soil NO_3-N values late in the cropping season may not indicate insufficient nitrogen; it may just indicate highly efficient crop uptake. Tissue testing would be required to confirm low nitrogen status.

Soil Test Correlations

Researchers found that the above mentioned soil NO_3-N quick test was highly correlated ($r^2 = 0.87$) with standard laboratory analyses for a variety of nitrate concentrations. They also concluded that quick tests using nitrate selective electrodes were well correlated to laboratory results (Hartz et al., 1994).

SOIL NITRATE COLORIMETRIC PROCEDURE

Note: This text merely summarizes one particular procedure using the Hach® Test method. Obviously, if Hach Kits are used, the manufacturer's directions should be followed. Depending on the particular kit, different procedures/analyses are available for soil, water, soil solution, or tissue. See Appendix B for details of various products, contacts, and prices.

EQUIPMENT

- All equipment is provided with the particular test kit

SOIL PREPARATION

1. Collect a "representative" soil sample from several locations. Air dry the soil.
2. Use a mortar and pestle to grind the sample. Screen the soil to verify that the diameters of the particles < 2 mm.

EXTRACTION

1. Add 1 level spoonful of dry, sieved soil to a clean bottle.
2. Fill bottle with deionized water to 25 ml mark.
3. Add one nitrate extraction powder pillow to bottle.
4. Cap and shake for 30 seconds. Allow to settle to form a clear liquid soil extract layer.

ANALYSIS

1. Pipette 1 ml of liquid soil extract and place extract in a clean bottle.
2. Fill bottle to 25 ml mark with deionized water.
3. Add nitraver powder pillow to bottle.
4. Cap and shake for 3 minutes.
5. Pour solution into "testing cell" and add nitrite reagent powder pillow.
6. Cap bottle and shake.
7. Analyze using the colorimeter or spectrophotometer.

Sap Nitrate Test with Cardy Meter®

(Hartz et al., 1994)

Equipment

- Cardy Meter®
- Garlic press
- Deionized water
- Ice chest/cooler
- Plastic bags

The following describes the procedure for measuring the nitrate concentration of fresh petiole sap using the "Cardy" nitrate selective electrode meter.

Sample Collection

1. Number of leaves
   At least 20 petioles from different plants throughout a management unit are required for a representative sample; where there are obvious or suspected differences in fertility within a field, separate samples should be collected and analyzed.

2. Leaf age
   As with standard petiole sampling, recently expanded mature leaves should be used for most crops.

3. Time sampling
   There is some disagreement among printed references regarding the best time of day to collect petiole samples. Most sources would agree that most consistent results will be obtained by collecting samples after 8 A.M. but before 2 P.M.

4. Sample storage
   Immediately on collection, leaf blades should be stripped away and petioles put in plastic bags on ice in a cooler until they are analyzed; water loss can occur very rapidly in hot field conditions, leading to inaccurate readings. Once on ice, samples can be held as long as six to eight hours without appreciable change in sap nitrate readings. Always allow petioles to warm up to the temperature of the meter before analysis; once the field heat is removed, petioles in a plastic bag can be held at room temperature for one to two hours without harm. Once sap is pressed from the petioles, it must be analyzed immediately.

USE OF THE CARDY METER®

Although it can be used in the field, the Cardy Meter® is better suited to use indoors. It is sensitive to temperature changes, so frequent recalibration is necessary through the day if used in the field. Also, the readings tend to drift for the first few minutes after it is turned on. From the standpoint of accuracy and efficiency, it is better to collect a number of petiole samples and bring them to a central location for analysis.

1. Calibration
   The Cardy Meter® is calibrated on a set point with a 'standard' solution and an electronic gain function with a 'slope' solution. It should be calibrated with these standard solutions each time it is turned on, and periodically as samples are analyzed. Generally, the calibration is quite stable, even from one day to the next. If the calibration continually drifts badly, it is a sign that the electrode pad may be wearing out; replacement electrodes can be purchased separately and are easily installed.

2. Sap extraction
   Extracting sap from fresh petioles is not a high-tech procedure; whatever you can rig conveniently to press petioles is fine. To ensure maximum accuracy, one should collect as much sap as is reasonably possible from all petioles in the sample, then mix the collected sap before it is analyzed.

3. When to read
   The Cardy electrode takes at least 20 seconds to stabilize each time a new sample is analyzed. Make it a rule to take readings at a standard time (perhaps 30 seconds) after putting each sample on the electrode.

4. NO_3 vs. NO_3-N
   The calibration solutions that come with the Cardy Meter® are listed as ppm NO_3; most reference material on plant and soil nutrition refer to ppm NO_3-N. The conversion is:

$$\text{ppm } NO_3\text{-N} = \frac{\text{ppm } NO_3}{4.43}$$

It should be emphasized that, despite careful attention to these rules, fresh sap analysis with the Cardy Meter® does not have the accuracy of conventional laboratory analysis of dry tissue. The authors believe the usefulness of fresh sap analysis is strengthened when it is used on a regular basis, with good record keeping; this allows a grower to follow the trend of NO_3-N concentrations over the season. Not only does this provide insight on the N dynamics of the crop, it also allows one to pinpoint a suspect value which needs retesting.

SAP TEST CORRELATIONS

Sap tests for a variety of crops, cultivars, growth stages, and locations were conducted using the Cardy Meter®. Conventional laboratory tests were also conducted on the crops. Researchers concluded: "Nitrate concentration of petiole sap, as measured by selective electrode, was highly correlated with conventional laboratory analysis of dry petiole tissue for all crops tested. The relationship was linear across a wide range of nitrate concentrations. Cultivar and site did not significantly influence the relationship in any crop tested. For all crops tested in which recently matured tissue was always sampled, growth stage was not a significant factor." (Hartz et al., 1994)

Table 46. Relationship of NO_3-N concentration of petiole sap and dry petiole tissue for selected crops. (Hartz et al., 1994)

Crop	Regression Equation	r^2
Broccoli	y = 272 + 0.092x	0.84
Celery	y = 21 + 0.038x	0.88
Corn		
pre-silking	y = 197 + 0.055x	0.65
post-silking	y = 104 + 0.150x	0.72
Lettuce	y = 35 + 0.040x	0.77
Pepper	y = 83 + 0.110x	0.89
Tomato	y = 231 + 0.047x	0.83
Watermelon	y = 228 + 0.055x	0.88

APPENDIX D. EPA LABEL IMPROVEMENT PROGRAM

(retyped by ITRC)

Stamped and Dated: MAR 11, 1987

PR NOTICE 87-1

NOTICE TO MANUFACTURERS, FORMULATORS, PRODUCERS
AND REGISTRANTS OF PESTICIDE PRODUCTS

Attention: Persons Responsible for Registration of Pesticide Products

Subject: LABEL IMPROVEMENT PROGRAM FOR PESTICIDES APPLIED THROUGH IRRIGATION SYSTEMS (CHEMIGATION)

This Notice requires registrants of pesticide products registered under FIFRA and applied through irrigation systems to revise the labeling for such products to include additional use directions and other statements described in this Notice.

No end-use pesticide products labeled for agricultural, nursery, turf farm, golf course, or greenhouse uses may be released for shipment by a registrant or producer of that product after April 30, 1988 unless product bears an amended label which complies with the LIP.

I. THE LABEL IMPROVEMENT PROGRAM

On June 5, 1980, the Agency announced the establishment of a Label Improvement Program under which labels of products were to be upgraded, improved, or revised to meet current labeling standards. Notice of this program was issued in the Federal Register and provided to all registrants as PR Notice 80-1. This Notice is issued under that LIP.

Pesticide labels are required to contain directions for use which are necessary for effecting the purpose for which the product is intended and are adequate to protect health and the environment. The label revisions required by this Notice, if adhered to by the users, will decrease environmental risks of pesticide contamination of ground water and will decrease direct human exposure to pesticide-treated irrigation water by providing appropriate use directions and restrictions or prohibitions. Although the Agency has received indirectly only very limited accounts of water source contamination or personal injury resulting from pesticide application through irrigation systems (chemigation), there is potential for such situations due to the increasing popularity of this application method, lack of public awareness that pesticides may be contained in irrigation water, lack of broad-based or uniform regulation by individual states, and the absence of directions for use on pesticide labels. This last factor is due in part to the recent development of this application technology and equipment, and in part to the FIFRA sec. 2(ee)(3) which allows "any method of application not prohibited by the labeling".

The Agency has received accounts of members of the general public intentionally using irrigation water in a variety of ways which could result in direct human exposure to pesticides if the system was being used for chemigation. The required label revisions, particularly those requiring posting to inform persons that irrigation water may contain pesticides, will decrease the likelihood of direct human exposure.

The required labeling will benefit users by providing them use directions for this relatively new method of application or will indicate that certain pesticides or application equipment should not be used and do not have the support of the registrant. In addition there will be uniform label requirements at the Federal level to address environmental concerns for this method of pesticide application. The Agency will also enhance its ability to enforce the requirements under the misuse provisions of FIFRA sec. 12(a)(2)(G) which will encourage compliance by users.

The Agency believes that some of this safety equipment is already in use on more recently installed irrigation systems. Several states have regulations in place requiring similar equipment. There will be an economic impact on users who may have to install additional safety equipment. However, incremental costs of any additional equipment are small compared with the overall costs of the complete irrigation system itself. Another type of cost may be felt by users in the loss of availability of certain pesticides for chemigation because registrants elect to prohibit use rather than add use directions. Costs to registrants will be the cost of revising labels for those who must add the label prohibition or elect to add use directions for chemigation.

In all cases, the Agency believes that incremental costs will be outweighed by the benefits of having comprehensive and appropriate label use directions, of affording greater protection of the environment including ground water, and of greater enforcement capability to ensure compliance.

II. PESTICIDE PRODUCTS TO WHICH THIS NOTICE APPLIES

A. The requirements of this Notice apply to any pesticide product which:

1. May legally be applied through any type of irrigation system including any sprinkler, flood, furrow, drip or greenhouse system (pesticide products whose labels are silent on chemigation, i.e., neither recommend nor prohibit this application method, do legally allow this use); and

2. Is labeled for agricultural uses, nursery use, turf farm uses, golf course uses or greenhouse uses; and

3. Is subject to FIFRA sec. 3 Registration, sec. 5 Experimental Use Permit, sec 18 Emergency Use, or sec. 24(c) Special Local Need regulation.

B. The requirements of this Notice do not apply to any pesticide product which:

1. Is intended solely for residential use (such as indoor, yard or garden);

2. Is intended solely for direct injection into plants;

3. Is intended solely for post harvest application to produce; or

4. Is intended to be applied only as a gas or only as a solid, such as a pellet, tablet, granule, or dust formulation.

C. This notice is not intended to limit chemigation practices to the types of irrigation systems given as examples above to the detriment of developing new technologies, nor is it intended to limit the Agency's concerns to those types of systems named as examples. If chemigation through other types of systems is intended, registrants must submit a detailed description of the system with proposed labeling for Agency review.

D. It is the Agency's intention to maintain as much flexibility as practicable in administering this LIP relative to FIFRA sec. 24(c) registrations. A label prohibition against chemigation of a product under FIFRA sec. 3 will not prohibit states from issuing a sec. 24(c) registration for chemigation of that same product as long as labeling according to this LIP is incorporated into the sec. 24 (c) registration.

III. IF CHEMIGATION IS NOT INTENDED

If it is your intention that your product not be applied through any type of chemigation system, add to your label the following prohibition statement: "Do not apply this product through any type of irrigation system." Skip to Section XI. of this Notice.

If at any time in the future you wish to replace the prohibition statement with the prescribed use directions and precautions in Sections IV. through IX., you must submit an application for amended registration to the appropriate Product Manager in the Registration Division. Applications for amended registration are not subject to the requirements of FIFRA sec. 3(c)(1)(D) pertaining to data compensation procedures.

IV. GENERIC LABEL STATEMENTS REQUIRED FOR CHEMIGATED PRODUCTS

If you intend that your product be applied by any type of chemigation system, all the following generic statements must be included on your label together with the specific requirements from one or more of Sections V. through IX.

A. "Apply this product only through [choose one or more of the following types of systems: sprinkler including center pivot, lateral move, end tow, side (wheel) roll, traveler, big gun, solid set, or hand move; flood (basin); furrow; border; or drip (trickle)] irrigation system(s). Do not apply this product through any other type of irrigation system."

B. "Crop injury, lack of effectiveness, or illegal pesticide residues in the crop can result from nonuniform distribution of treated water."

C. "If you have questions about calibration, you should contact State Extension Service specialists, equipment manufacturers or other experts."

D. "Do not connect an irrigation system (including greenhouse systems) used for pesticide application to a public water system unless the pesticide label-prescribed safety devices for public water systems are in place."

E. "A person knowledgeable of the chemigation system and responsible for its operation, or under the supervision of the responsible person, shall shut the system down and make necessary adjustments should the need arise."

V. LABEL STATEMENTS FOR CHEMIGATED TOXICITY CATEGORY I PRODUCTS

In addition to generic label statements in Section IV., and specific label statements in one or more of Sections VI., VII., VIII., or IX., the labels of the Toxicity Category I products (those with the label signal word DANGER) which allow chemigation must include the statements:

"Posting of areas to be chemigated is required when 1) any part of a treated area is within 300 feet of sensitive areas such as residential areas, labor camps, businesses, day care centers, hospitals, in-patient clinics, nursing homes or any public areas such as schools, parks, playgrounds, or other public facilities not including public roads, or 2) when the chemigated area is open to the public such as golf courses or retail greenhouses."

"Posting must conform to the following requirements. Treated areas shall be posted with signs at all usual points of entry and along likely routes of approach from the listed sensitive areas. When there are no usual points of entry, signs must be posted in the corners of the treated areas and in any other location affording maximum visibility to sensitive areas. The printed side of the sign should face away from the treated area towards the sensitive area. The signs shall be printed in English. Signs must be posted prior to application and must remain posted until foliage has dried and soil surface water has disappeared. Signs may remain in place indefinitely as long as they are composed of materials to prevent deterioration and maintain legibility for the duration of the posting period."

"All words shall consist of letters at least 2 1/2 inches tall, and all letters and the symbol shall be a color which sharply contrasts with their immediate background. At the top of the sign shall be the words KEEP OUT, followed by an octagonal stop sign symbol at least 8 inches in diameter containing the word STOP. Below the symbol shall be the words PESTICIDES IN IRRIGATION WATER." A small-scale illustration of an acceptable sign is attached at the end of this Notice.

Posting required for chemigation does not replace other posting and reentry interval requirements for farm worker safety.

VI. LABEL STATEMENTS FOR CHEMIGATION SYSTEMS CONNECTED TO PUBLIC WATER SYSTEMS

In addition to generic label statements in Section IV., and specific label statements in one or more of Sections VII., VIII., or IX., the labels of pesticide products which allow chemigation through systems connected to public water systems must include the statements:

A. Specific Required Label Statements

The following statements must be used verbatim:

1. "Public water system means a system for the provision to the public of piped water for human consumption if such system has at least 15 service connections or regularly serves an average of at least 25 individuals daily at least 60 days out of the year."

2. "Chemigation systems connected to public water systems must contain a functional, reduced-pressure zone, backflow preventer (RPZ) or the functional equivalent in the water supply line upstream from the point of pesticide introduction. As an option to the RPZ, the water from the public water system should be discharged into a reservoir tank prior to pesticide introduction. There shall be a complete physical break (air gap) between the outlet end of the fill pipe and the top or overflow rim of the reservoir tank of at least twice the inside diameter of the fill pipe."

3. "The pesticide injection pipeline must contain a functional, automatic, quick-closing check valve to prevent the flow of fluid back toward the injection pump."

4. "The pesticide injection pipeline must contain a functional, normally closed, solenoid-operated valve located on the intake side of the injection pump and connected to the system interlock to prevent fluid from being withdrawn from the supply tank when the irrigation system is either automatically or manually shut down."

5. "The system must contain functional interlocking controls to automatically shut off the pesticide injection pump when the water pump motor stops, or in cases where there is no water pump, when the water pressure decreases to the point where pesticide distribution is adversely affected."

6. "Systems must use a metering pump, such as a positive displacement injection pump (e.g., diaphragm pump) effectively designed and constructed of materials that are compatible with pesticides and capable of being fitted with a system interlock."

7. "Do not apply when wind speed favors drift beyond the area intended for treatment."

B. Nonspecific Required Label Statements

The following subjects must be addressed using the registrant's own wording:

1. Indicate whether agitation is or is not recommended, identify the situations when it is recommended in the pesticide supply tank. If it is not always recommended (e.g., when tank mixing with other pesticides or fluid fertilizers).

2. Indicate if the pesticide is to be applied continuously for the duration of the water application. If not, indicate when during the water application the pesticide is to be applied.

3. Provide mixing instructions for dilution of the pesticide in the supply tank.

C. Optional Label Statements

The following subject is not required but may be addressed at the registrant's choosing:

1. Indicate the quantity of water to be applied per acre to achieve efficacy, but not cause runoff or excessive leaching. A range of quantities may be appropriate.

VII. LABEL STATEMENTS FOR SPRINKLER CHEMIGATION

A. Specific Required Label Statements

The following statements must be used verbatim:

1. "The system must contain a functional check valve, vacuum relief valve, and low pressure drain appropriately located on the irrigation pipeline to prevent water source contamination from backflow."

2. "The pesticide injection pipeline must contain a functional, automatic, quick-closing check valve to prevent the flow of fluid back toward the injection pump."

3. "The pesticide injection pipeline must also contain a functional, normally closed, solenoid-operated valve located on the intake side of the injection pump and connected to the system interlock to prevent fluid from being withdrawn from the supply tank when the irrigation system is either automatically or manually shut down."

4. "The system must contain functional interlocking controls to automatically shut off the pesticide injection pump when the water pump motor stops."

5. "The irrigation line or water pump must include a functional pressure switch which will stop the water pump motor when the water pressure decreases to the point where pesticide distribution is adversely affected."

6. "Systems must use a metering pump, such as a positive displacement injection pump (e.g., diaphragm pump) effectively designed and constructed of materials that are compatible with pesticides and capable of being fitted with a system interlock."

7. "Do not apply when wind speed favors drift beyond the area intended for treatment."

B. Nonspecific Required Label Statements

The following subjects must be addressed using the registrant's own wording:

1. Indicate whether agitation is or is not recommended in the pesticide supply tank. If it is not always recommended, identify the situations when it is recommended (e.g., when tank mixing with other pesticides or fluid fertilizers).

2. Indicate if the pesticide is to be applied continuously for the duration of the water application. If not, indicate when during the water application the pesticide is to be applied.

3. Provide mixing instructions for dilution of the pesticide in the supply tank.

C. Optional Label Statements

The following subject is not required, but may be addressed at the registrant's choosing:

1. Indicate the quantity of water to be applied per acre to achieve efficacy, but not cause runoff or excessive leaching. A range of quantities may be appropriate.

VIII. LABEL STATEMENTS FOR FLOOD (BASIN), FURROW AND BORDER CHEMIGATION

A. Specific Required Label Statements

The following statements must be used verbatim:

1. "Systems using a gravity flow pesticide dispensing system must meter the pesticide into the water at the head of the field and downstream of a hydraulic discontinuity such as a drop structure or weir box to decrease potential for water source contamination from backflow if water flow stops."

2. "Systems utilizing a pressurized water and pesticide injection system must meet the following requirements:"

a. "The system must contain a functional check valve, vacuum relief valve, and low pressure drain appropriately located on the irrigation pipeline to prevent water source contamination from backflow."

b. "The pesticide injection pipeline must contain a functional, automatic, quick-closing check valve to prevent the flow of fluid back toward the injection pump."

c. "The pesticide injection pipeline must also contain a functional, normally closed, solenoid-operated valve located on the intake side of the injection pump and connected to the system interlock to prevent fluid from being withdrawn from the supply tank when the irrigation system is either automatically or manually shut down."

d. "The system must contain functional interlocking controls to automatically shut off the pesticide injection pump when the water pump motor stops."

e. "The irrigation line or water pump must include a functional pressure switch which will stop the water pump motor when the water pressure decreases to the point where pesticide distribution is adversely affected."

f. "Systems must use a metering pump, such as a positive displacement injection pump (e.g., diaphragm pump) effectively designed and constructed of materials that are compatible with pesticides and capable of being fitted with a system interlock."

B. Nonspecific Required Label Statements

The following subject must be addressed using the registrant's own wording:

1. Indicate whether or not a pesticide supply tank is recommended for the application. If it is recommended, provide mixing instructions for dilution of the pesticide in the tank. If agitation is not always recommended, identify the situations when it is recommended (e.g., when tank mixing with other pesticides or fluid fertilizers).

2. Indicate if the pesticide is to be applied continuously for the duration of the water application. If not, indicate when during the water application the pesticide is to be applied.

C. Optional Label Statements

The following subject is not required, but may be addressed at the registrant's choosing:

1. Indicate the quantity of water to be applied per acre to achieve efficacy, but not cause a runoff or excessive leaching. A range of quantities may be appropriate.

IX. LABEL STATEMENTS FOR DRIP (TRICKLE) CHEMIGATION

A. Specific Required Label Statements

The following statements must be used verbatim:

1. "The system must contain a functional check valve, vacuum relief valve and low pressure drain appropriately located on the irrigation pipeline to prevent water source contamination from backflow."

2. "The pesticide injection pipeline must contain a functional, automatic, quick-closing check valve to prevent the flow of fluid back toward the injection pump."

3. "The pesticide injection pipeline must also contain a functional, normally closed, solenoid-operated valve located on the intake side of the injection pump and connected to the system interlock to prevent fluid from being withdrawn from the supply tank when the irrigation system is either automatically or manually shut down."

4. "The system must contain functional interlocking controls to automatically shut off the pesticide injection pump when the water pump motor stops."

5. "The irrigation line or water pump must include a functional pressure switch which will stop the water pump motor when the water pressure decreases to the point where pesticide distribution is adversely affected."

6. "Systems must use a metering pump such as a positive displacement injection pump (e.g., diaphragm pump) effectively designed and constructed of materials that are compatible with pesticides and capable of being fitted with a system interlock."

B. Nonspecific Required Label Statements

The following subjects must be addressed using the registrant's own wording:

1. Indicate whether or not a pesticide supply tank is recommended for the application. If it is recommended, provide mixing instructions for dilution of the pesticide in the tank and indicate whether agitation is or is not always recommended in the tank. If agitation is not always recommended, identify the situations when it is recommended (e.g., when tank mixing with other pesticides or fluid fertilizers).

2. Indicate if the pesticide is to be applied continuously for the duration of the water application. If not, indicate when during the water application the pesticide is to be applied.

C. Optional Label Statements

The following subject is not required, but may be addressed at the registrant's choosing:

1. Indicate the quantity of water to be applied per acre (or other measure of surface, volume etc.) to achieve efficacy, but not cause runoff or excessive leaching. A range of quantities may be appropriate.

X. FORMAT OF LABEL INFORMATION

If all information is included on the label itself, the label must meet format requirements of 40 CFR 162.10. If chemigation label requirements are included in supplemental labeling rather than on the pesticide container label, the container label must contain the referral statement:

"CHEMIGATION:

Refer to supplemental labeling entitled (fill-in title) for use directions for chemigation. Do not apply this product through any irrigation system unless the supplemental labeling on chemigation is followed."

Note that supplemental labeling accompanying the pesticide product or referred to on the container label is considered to be labeling under FIFRA and is subject to review and approval by the Agency.

XI. WHAT YOU MUST DO TO COMPLY

If you are the registrant of a pesticide product that is Federally registered under FIFRA sec. 3:

A. First ascertain by examination of each product label whether or not each product is subject to this Notice according to the use patterns, sites, and formulations outlined in Section II. A. and B. above. For products not subject to this Notice no further action is required.

B. If you believe, for reasons other than those listed in Section II. A. and B. that your product should not be subject to the requirements of this Notice, you must submit a copy of the most current approved label and written justification supporting your position for Agency review by July 31, 1987. Claims for relative nontoxic status of a pesticide are not considered sufficient justification to warrant an exclusion from this LIP. Once this information has been submitted to the Agency through the appropriate Product Manager, you need not take any further action to comply with this Notice until a response is received from the Agency.

C. If your product is subject to this Notice you must take one of the following actions:

   1. Revise your product labels in accordance with this Notice by either adding the prohibition statement in Section III. or the generic and specific label statements in Sections IV. through IX. No application for amended registration is required if the wording in Sections III. through IX. is used as given and information required to be provided by the registrant is included. If you choose this option, submit the following to the Agency:

      a. A letter indicating your intent to adopt the label language (either the prohibition or use directions) as specified in this Notice no later than July 31, 1987. A single letter may be submitted listing all affected products.

      b. A copy of the final printed labeling as revised; and a certification, signed by an authorized representative of your company, that the labeling as revised is in compliance with the requirements of this Notice no later than April 30, 1988.

   2. If you wish to modify the required LIP statements in any substantive manner submit an application for amended registration, together with five copies of proposed labeling no later than July 31, 1987. Applications are not subject to the requirements of FIFRA sec 3(c)(1)(D) pertaining to data compensation procedures. Applications must be submitted to the appropriate Product Manager at EPA for all Federally registered products.

If you are or will be the producer of a pesticide product subject to FIFRA sections 5, 18, or 24 (c):

A. There is no need to amend currently approved labeling of products subject to FIFRA sec. 5 or 18.

B. The Agency will not, however, accept labeling submitted later than April 30, 1988 unless that labeling complies with the requirements of this LIP, i.e., either contains the prohibition or appropriate chemigation use directions.

C. Producers of pesticide products subject to FIFRA sec. 24(c) shall revise existing labels as per this LIP by April 30, 1988. Revised labeling and, if appropriate, a letter of certification shall be submitted to the state agency responsible for issuance or renewal of FIFRA sec. 24(c) registrations.

XII. COMPLIANCE DATES

All subject products released for shipment after April 30, 1988 must bear the revised labeling.

Products not in compliance as of this date will be deemed to be misbranded in violation of FIFRA sec. 12(a)(1)(E). The Agency may take enforcement action, issue Notices of Intent to Cancel the product's registration in accordance with FIFRA sec. 6, or both.

Registrants are reminded that:

A. A copy of the revised final printed label and, if appropriate, a certification statement must be submitted to the Agency prior to product distribution under that label. Revised labeling must be submitted by April 30, 1988.

B. It is the responsibility of the registrant to ensure that his distributors (sub-registrants) comply with these requirements within the time frames given.

XIII. FOR FURTHER INFORMATION

If you have questions regarding this Notice, or do not understand what you must do to comply, contact Dr. Thomas Ellwanger, Registration Division, Fungicide-Herbicide Branch at (703)557-1700.

If you have questions concerning the registration or amendment of a specific product, you must contact the Product Manager for the specific product.

The Agency address for correspondence and submission of applications is as follows:

Registration Division (TS-767C)
Environmental Protection Agency
401 M Street, S.W.
Washington, D.C. 20460

(Signature)
Edwin F. Tinsworth, Director
Registration Division

KEEP OUT

PESTICIDES IN IRRIGATION WATER

UNITED STATES ENVIRONMENTAL PROTECTION AGENCY
WASHINGTON, D.C. 20460

OFFICE OF
PREVENTION, PESTICIDES AND
TOXIC SUBSTANCES

Stamped and Dated : APR 11, 1989, AIR DIVISION U.S. EPA, REGION 9
Stamped and Dated: APR 12, 1989, RECEIVED

MEMORANDUM

SUBJECT: Interim Final FIFRA Compliance Program Policy 12.7

FROM: John J. Neylan III, Director *Signed: John J . Neylan*
Policy and Grants Division
Office of Compliance Monitoring

TO: Addressees

Attached is an Interim Final FIFRA Compliance Program Policy 12.7 (chemigation) for immediate use, as well as for your review and comment. Please forward a copy of this policy to each State with which EPA has a FIFRA cooperative enforcement agreement for State review and comment.

This policy addresses the enforcement of the label provisions which were required in PR Notice 87-1. The policy states that EPA will not take an enforcement action under FIFRA section 12(a)(2)(G) against a person for using chemigation equipment which is not specified on the label if it is specified on a current list of comparable systems issued by EPA's Office of Pesticide Programs (OPP). A Chemigation Committee established by OPP has the responsibility for preparing and updating the list of approved, comparable equipment.

The attached

Attachment

UNITED STATES ENVIRONMENTAL PROTECTION AGENCY
WASHINGTON, D.C. 20460

OFFICE OF
PREVENTION, PESTICIDES AND
TOXIC SUBSTANCES

Stamped and Dated: MAR 22, 1989

MEMORANDUM

SUBJECT: List of Alternative Chemigation Safety Equipment

FROM: Anne E. Lindsay, Director *Signed: Anne E. Lindsay*
Registration Division (H7505C)
Office of Pesticide Programs

TO: Phyllis Flaherty, Acting Director
Policy and Grants Division (EN-342)
Office of Compliance Monitoring

Please find attached OPP's approved list of alternative chemigation safety equipment. These devices are offered as alternatives to certain required components of chemigation systems in PR notice 87-1, the Chemigation Label Improvement Program. I am sure you are aware of the urgency to get this information into the hands of State and Federal regulators as soon as possible for the impending growing season. If we can be of assistance in the timely distribution, please let us know. There are several grower groups, equipment manufacturers, committee members, and other interested parties with whom we have been working, so we would appreciate several copies of the final document. I have been advised that there may be some further expansion of this list of alternatives upon further consideration of the Chemigation Committee.

Please contact Dr. Tom Ellwanger (557-1700) on the status of the mailout and if there are questions on the technical aspects of the project.

Attachment

List of Alternative Chemigation Safety Equipment

PR Notice 87-1, the Label Improvement Program for Chemigation, issued March 11, 1987 requires that the labeling of agricultural pesticides intended for application through irrigation systems must include the use of certain types of safety devices to protect ground water from pesticides contamination. As a result of comments and new information received subsequent to issuance, a list of alternative devices to those included in PR Notice 87-1 has been considered and approved for use. In some cases these alternative devices may be less expensive, more reliable, or more available than some of those devices originally required. Be advised that all of the devices originally included in PR Notice 87-1 are still acceptable and that the PR Notice 87-1 is, in its entirety, still in effect. Devices required in PR Notice 87-1 which have no listed alternatives are still required components of all chemigation systems. The original devices as required in PR Notice 87-1 and their corresponding alternatives are listed below:

Original Device

Functional normally closed, solenoid-operated valve located on the intake side of the injection pump.

Alternative Device 1

Functional spring-loaded check valve with a minimum of 10 psi cracking pressure. The valve must prevent irrigation water under operating pressure from entering the pesticide injection line and must prevent leakage from the pesticide supply-tank on system shutdown. This valve must be constructed of pesticidally resistant materials. [Note: This single device can substitute for both the solenoid-operated valve and the functional, automatic, quick closing check valve in the pesticide injection line.]

Alternative Device 2

Functional normally closed hydraulically operated check valve. The control line must be connected to the main water line such that the valve opens only when the main water line is adequately pressurized. This valve must prevent leakage from the pesticide supply tank on system shutdown. The valve must be constructed of pesticidally resistant materials.

Alternative Device 3

Functional vacuum relief valve located in the pesticide injection line between the positive displacement pesticide injection pump and the check valve. This alternative is appropriate for only those chemigation systems using a positive displacement pesticide injection pump and is not for use with venturi injection systems. This valve must be elevated at least 12 inches above the highest fluid level in the pesticide supply tank and must be the highest point in the injection line. The valve must open at 6 inches water vacuum or less and must be spring loaded or otherwise constructed such that it does not leak on closing. It must prevent leakage from the pesticide supply tank on system shutdown. The valve must be constructed of pesticidally resistant materials.

Original Device

Functional main water line check valve and main water line low pressure drain.

Alternative Device 1

Gooseneck pipe loop located in the main water line immediately downstream of the irrigation water pump. The bottom side of the pipe at the loop apex must be at least 24 inches above the highest sprinkler or other type of water emitting device. The loop must contain either a vacuum relief or combination air and vacuum relief valve a the apex of the pipe loop. The pesticide injection port must be located downstream of the apex of the pipe loop and at least 6 inches below the bottom side of the pipe at the loop apex.

Original Device

Positive displacement pesticide injection pump.

Alternative Device 1

Venturi systems including those inserted directly into the main water line, those installed in a bypass system, and those bypass systems boosted with an auxiliary water pump. Booster or auxiliary water pumps must be connected with the system interlock such that they are automatically shut off when the main line irrigation pump stops, or in cases where there is no main line irrigation pump, when the water pressure decreases to the point where pesticide distribution is adversely affected. Venturies must be constructed of pesticidally resistant materials. The line from the pesticide supply tank to the venturi must contain a functional, automatic, quick closing check valve to prevent the flow of liquid back toward the pesticide supply tank. This valve must be located immediately adjacent to the venturi pesticide inlet. This same supply line must also contain either a functional normally closed solenoid-operated valve connected to the system interlock or a functional normally closed hydraulically operated valve which opens only when the main line is adequately pressurized. In bypass systems as an option to placing both valves in the line from the pesticide supply tank, the check valve may be installed in the bypass immediately upstream of the venturi water inlet and either the normally closed solenoid or hydraulically operated valve may be installed immediately downstream of the venturi water outlet.

Original Device

Vacuum relief valve.

Alternative Device 1

Combination and vacuum relief valve.

Appendix E. California Regulations

The following includes California laws specific to chemigation from Title 3, Food and Agriculture, Division 6. Pesticides and Pest Control Operations.

§ 6610. Backflow Prevention.

Each service rig and piece of application equipment that handles pesticides and draws water from an outside source shall be equipped with an air-gap separation, reduced pressure principle backflow prevention device or double check valve assembly. Backflow protection must be acceptable to both the water purveyor and local health department.

§ 6770. Field Reentry After Pesticide Application.

(g) When a pesticide in toxicity category one, or a minimal exposure pesticide listed in section 6790 is being applied through an irrigation system, before the application starts, signs shall be posted that state in English and Spanish "Danger, Pesticides are applied in water through the irrigation system. Do not drink water from the irrigation system." These signs shall be readable at 25 feet, shall be posted at all the usual points of field entry and shall be placed at intervals not exceeding 600 feet around the field. The use of an additional third language is permissible. Unless treatment occurs weekly or at more frequent intervals, signs shall be removed no later than five days after application. As an alternative there shall be on site a sufficient number of persons to prevent unauthorized employees from entering and any employee from drinking from the irrigation system.

REGULATIONS RELATING TO CROSS-CONNECTION

An Excerpt From the

CALIFORNIA ADMINISTRATIVE CODE

TITLE 17 — PUBLIC HEALTH

Effective Date: June 25, 1987

[Seal]

STATE OF CALIFORNIA

DEPARTMENT OF HEALTH SERVICES

PUBLIC WATER SUPPLY BRANCH

TITLE 17 DRINKING WATER SUPPLIES
(Register 87,
No. 23-6-6-87)

GROUP 4. DRINKING WATER SUPPLIES

DETAILED ANALYSIS

Article 1. General

Section

7583.	Definitions
7584.	Responsibility and Scope of Program
7585	Evaluation of Hazard
7586.	User Supervisor

Article 2. Protection of Water System

Section

7601.	Approval of Backflow Preventers
7602.	Construction of Backflow Preventers
7603.	Location of Backflow Preventers
7604.	Type of Protection Required
7605.	Testing and Maintenance of Backflow Preventers

Article 1. General

7583. Definitions.

In addition to the definitions in Section 4010.1 of the Health and Safety Code, the following terms are defined for the purpose of this Chapter:

(a) "Approved Water Supply" is a water supply whose potability is regulated by a state or local health agency.

(b) "Auxiliary Water Supply" is any water supply other than that received from a public water system.

(c) "Air-gap Separation (AG)" is a physical break between the supply line and a receiving vessel.

(d) "AWWA Standard is an official standard developed and approved by the American Water Works Association (AWWA).

(e) "Cross-Connection" is an unprotected actual or potential connection between a potable water system used to supply water for drinking purposes and any source or system containing unapproved water or a substance that is not or cannot be approved as safe, wholesome, and potable. By-pass arrangements, jumper connections, removable sections, swivel or changeover devices, or other devices through which backflow could occur, shall be considered to be cross-connections.

(f) "Double Check Valve Assembly (DC)" is an assembly of at least two independently acting check valves including tightly closing shut-off valves on each side of the check valve assembly and test cocks available for testing the water tightness of each check valve.

(g) "Health Agency" means the county or city health authority.

(h) "Local Health Agency" means the county or city health authority.

(i) "Reclaimed Water" is a wastewater which as a result of treatment is suitable for used other than potable use.

(j) "Reduced Pressure Principle Backflow Prevention Device (RP)" is a backflow preventer incorporating less than two check valves, an automatically operated differential relief valve located between the two check valves, a tightly closing shut-off valve on each side of the check valve assembly, and equipped with necessary test cocks for testing.

(k) "User Connection" is the point of connection of a user's piping to the water supplier's facilities.

(l) "Water Supplier" is the person who owns or operates the public water system.

(m) "Water User" is any person obtaining water from a public water supply.

7584. Responsibility and Scope Program.

The water supplier shall protect the public water supply from contamination by implementation of a cross-connection control program. The program or any portion thereof, may be implemented directly by the water supplier or by means of a contract with the local health agency, or with another agency approved by the health agency. The water supplier's cross-connection control program shall for the purpose of addressing the requirements of Sections 7585 through 7605 include, but not be limited to, the following elements:

(a) The adoption of operating rules or ordinances to implement the cross-connection program.

(b) The conducting of surveys to identify water user premises where cross-connections are likely to occur,

(c) The provisions of backflow protection by the water user at the user's connection or within the user's premises or both,

(d) The provision of at least one person trained in cross-connection control to carry out the cross-connection program,

(e) The establishment of a procedure or system for testing backflow preventers and,

(f) The maintenance of records of locations, tests, and repairs of backflow preventers.

7585. Evaluation of Hazard.

The water supplier shall evaluate the degree of potential health hazard to the public water supply which may be created as a result of conditions existing on a user's premises. The water supplier, however, shall not be responsible for abatement of cross-connections which may exist within a user's premises. As a minimum, the evaluation should consider: the existence of cross-connections, the nature of materials handled on the property, the probability of a backflow occurring, the degree of piping system complexity and the potential for piping system modification. Special consideration shall be given to the premises of the following types of water users:

(a) Premises where substances harmful to health are handled under pressure in a manner which could permit their entry into the public water system. This includes chemical or biological process waters and water from public water supplies which have deteriorated in sanitary quality.

(b) Premises having an auxiliary water supply, unless the auxiliary supply is accepted as an additional source by the water supplier and is approved by the health agency.

(c) Premises that have internal cross-connections that are not abated to the satisfaction of the water supplier or the health agency.

(d) Premises where cross-connections are likely to occur and entry is restricted so that cross-connections inspections cannot be made with sufficient frequency or at sufficiently short notice to assure that cross-connections do not exist.

(e) Premises having a repeated history of cross-connections do not exist.

(f) Premises having a repeated history of cross-connections being established or re-established.

NOTE: Authority cited: Sections 208 and 4026, Health and Safety Code.
Reference: Section 4026, Health and Safety Code.

7586. User Supervisor

The health agency and water supplier may, at their discretion, require an industrial water user to designate a user supervisor when the water user's premises has a multipiping system that convey [sic] various types of fluids, some of which may be hazardous and where changes in the piping system are frequently made. The user supervisor shall be responsible for the avoidance of cross-connections during the installation, operation and maintenance of the water user's pipelines and equipment.

NOTE: Authority cited: Sections 208 and 2406, Health and Safety Code.
Reference: Section 4026, Health and Safety Code.

Article 2. Protection of Water System

7601. Approval of Backflow Preventers.

Backflow preventers required by this Chapter shall have passed laboratory and field evaluation tests performed by a recognized testing organization which has demonstrated their competency to perform such test to the Department.

NOTE: Authority cited: Section 4026, Health and Safety Code.

HISTORY:

1. New section filed 5-26-87; operative 6-25-87 (Register 87, No. 23).

7602. Construction of Backflow Preventers.

(a) Air-gap Separation. An Air-gap separation (AG) shall be at least double the diameter of the supply pipe, measured vertically from the flood rim of the receiving vessel to the supply pipe; however, in no case shall this separation be less than one inch.

(b) Double Check Valve Assembly. A required double check valve assembly (DC) shall, as a minimum, conform to the AWWA Standard C506-78 (R83) adopted on January 28, 1978 for Double Check Valve Type Backflow Preventive [sic] Devices which is herein incorporated by reference.

(c) Reduced Pressure Principle Backflow Prevention Device. A required reduced pressure principle backflow prevention device (RP) shall, as a minimum, conform to the AWWA Standard C506-78 (R83) adopted on January 28, 1978 for Reduced Pressure Principle Type Backflow Prevention Devices which is herein incorporated by reference.

NOTE: Authority cited: Sections 208 and 4026, Health and Safety Code.
Reference: Section 4026, Health and Safety Code.

7603. Location of Backflow Preventers.

(a) Air-gap Separation. An air-gap separation shall be located as close as practical to the user's connection and all piping between the user's connection and the receiving tank shall be entirely visible unless otherwise approved in writing by the water supplier and the health agency.

(b) Double Check Valve Assembly. A double check valve assembly shall be located as close as practical to the user's connection and shall be installed above grade, if possible, and in a manner where it is readily accessible for testing and maintenance.

(c) Reduced Pressure Principle Backflow Prevention Device. A reduced pressure principle backflow prevention device shall be located as close as practical to the user's connection and shall be installed a minimum of twelve inches (12") side clearance.

NOTE: Authority cited: Sections 208 and 4026, Health and Safety Code.
Reference: Section 4026, Health and Safety Code.

7604. Type of Protection Required.

The type of protection that shall be provided to prevent backflow into the public water supply shall be commensurate with the degree of hazard that exists on the consumer's premises. The type of protective device that may be required (listed in an increasing level of protection) includes: Double Check Valve Assembly – (DC), Reduced Pressure Principle Backflow Prevention Device – (RP), and an Air-gap Separation – (AG). The water user may choose a higher level of protection than required by the supplier. The minimum types of backflow protection required to protect the public water supply, at the water user's connection to premises with various degrees of hazard are given in Table 1. Situations which are not covered in Table 1 shall be evaluated on a case-by-case basis and the appropriate backflow protection shall be determined by the water supplier or health agency.

TABLE 1
TYPE OF BACKFLOW PROTECTION REQUIRED

Degree of Hazard	*Minimum Type of Backflow Prevention*
(a) Sewage and Hazardous Substances	
(1) Premises where the public water system is used to supplement the reclaimed water supply.	AG
(2) Premises where there are wastewater pumping and/or treatment plants and there is no interconnection with the potable water system. This does not include a single-family residence that has a sewage lift pump. An RP may be provided in lieu of an AG if approved by the health agency and water supplier.	AG
(3) Premises where reclaimed water is used and there is no interconnection with the potable water system. An RP may be provided in lieu of an AG if approved by the health agency and water supplier.	AG
(4) Premises where hazardous substances are handled in any manner in which the substances may enter the potable water system. This does not include a single-family residence that has a sewage lift pump. An RP may be provided in lieu of on AG if approved by the health agency and water supplier.	AG
(5) Premises where there are irrigation systems into which fertilizers, herbicides, or pesticides are, or can be, injected.	RP
(b) Auxiliary Water Supplies	
(1) Premises where there is an unapproved auxiliary water supply which is interconnected with the public water system. An RP or DC may be provided in lieu of an AG if approved by the health agency and water supplier.	AG
(2) Premises where there is an unapproved auxiliary water supply and there are no interconnections with the public water system. A DC may be provided in lieu of a an RP if approved by the health agency and water supplier.	RP
(c) Fire Protection Systems	

(1) Premises where the fire system is directly supplied from the public water system and there is an unapproved auxiliary water supply on or to the premises (not interconnected). DC

(2) Premises where the fire system is supplied from the public water system and interconnected with an unapproved auxiliary water supply. An RP may be provided in lieu of an AG if approved by the health agency and water supplier. AG

(3) Premises where the fire system is supplied from the public water system and where either elevated storage tanks or fire pumps which take suction from private reservoirs or tanks are used. DC

(d) Dockside Watering Points and Marine Facilities DC

(1) Pier hydrants for supplying water to vessels for any purpose. RP

(2) Premises where there are marine facilities. RP

(e) Premises where entry is restricted so that inspections for cross-connections cannot be made with sufficient frequency or at sufficiently short notice to assure that cross-connections do not exist. RP

(f) Premises where there as a repeated history of cross-connections being established or re-established. RP

7605. Testing and Maintenance of Backflow Preventers.

(a) The water supplier shall assure that adequate maintenance and periodic testing are provided by the water user to ensure their proper operation.

(b) Backflow preventers shall be tested by persons who have demonstrated their competency in testing of these devices to the water supplier or health agency.

(c) Backflow preventers shall be tested at least annually or more frequently if determined to be necessary by the health agency or water supplier. When devices are found to be defective, they shall be repaired or replaced in accordance with the provisions of this Chapter.

(d) Backflow preventers shall be tested immediately after they are installed, relocated or repaired and not placed in service unless they are functioning as required.

(e) The water supplier shall notify the water user when testing of backflow preventers is needed. The notice shall contain the date when the test must be completed.

(f) Reports of testing and maintenance shall be maintained by the water supplier for a minimum of three years.

NOTE: Authority cited: Sections 208 and 4026, Health and Safety Code.
Reference: Section 4026, Health and Safety Code.

Appendix F. Sample Pesticide Labels

NEMACUR 3

Specimen Label

RESTRICTED USE PESTICIDE

Due to High Acute Toxicity and Toxicity to Wildlife

For Sale to and Use Only by a Certified Applicator for Uses Authorized by His Certification, or By Persons Under His Direct Supervision.

NEMACUR® 3

Emulsifiable Systemic Insecticide-Nematicide

For effective control of nematodes in certain field, fruit and vegetable crops.

ACTIVE INGREDIENT:
Ethyl 3-methyl-4-(methylthio)phenyl (1-methylethyl)phosphoramidate 35%
INERT INGREDIENTS: 65%
100%

Do not store below 32°F. Keep in a cool dry place. Do not use or store near heat or open flame.
Contains Aromatic Petroleum Distillates
Contains 3 pounds of ethyl 3-methyl-4-(methylthio)phenyl(1-methylethyl)phosphoramidate per gallon.

Two 2.5-Gallon Jugs per Case;
30-Gallon Drum;
5-Gallon Multi-trip Container

EPA Reg. No. 3125-283

STOP - Read The Label Before Use
KEEP OUT OF THE REACH OF CHILDREN

DANGER  **POISON**

PELIGRO

Si usted no entiende la etiqueta, busque a alguien para que se la explique a usted en detalle.

(If you do not understand the label, find someone to explain it to you in detail.)

PRECAUTIONARY STATEMENTS

HAZARDS TO HUMANS AND DOMESTIC ANIMALS

Fatal if swallowed or absorbed through the skin. Causes irreversible eye damage. May be fatal if inhaled. Do not get in eyes, on skin or on clothing. Do not breathe vapor or spray mist. Do not contaminate feed or food.

Personal Protective Equipment: Some materials that are chemical-resistant to this product are listed below. If you want more options, follow the instructions for category G on an EPA chemical-resistance category selection chart.

Applicators and other handlers must wear:
- Chemical-resistant protective suit
- Chemical-resistant gloves, such as barrier laminate or viton
- Chemical-resistant footwear plus socks
- Protective eyewear
- Chemical-resistant headgear for overhead exposure
- For exposures in enclosed areas, a respirator with either an organic vapor-removing cartridge with a prefilter approved for pesticides (MSHA/NIOSH approval number prefix TC-23C), or a canister approved for pesticides (MSHA/NIOSH approval number prefix TC-14G)

Specimen Label

- For exposures outdoors, dust/mist filtering respirator (MSHA/NIOSH approval number prefix TC-21C)

Discard clothing and other absorbent materials that have been drenched or heavily contaminated with this products concentrate. Do not reuse them. Follow manufacturer's instructions for cleaning/maintaining PPE. If no such instructions for washables, use detergent and hot water. Keep and wash PPE separately from other laundry.

Engineering controls statements: When handlers use closed systems or enclosed cabs in a manner that meets the requirements listed in the Worker Protection Standard (WPS) for agricultural pesticides [40 CFR 170.240(d)(4-6)], the handler PPE requirements may be reduced or modified as specified in the WPS.

User Safety Recommendations:
User should:

- Wash hands before eating, drinking, chewing gum, using tobacco or using the toilet.
- Remove clothing immediately if pesticide gets inside. Then wash thoroughly and put on clean clothing.
- Remove PPE immediately after handling this product. Wash the outside of gloves before removing. As soon as possible, wash thoroughly and change into clean clothing.

SYMPTOMS OF POISONING: A sense of "tightness" in the chest. Sweating. Contracted pupils. Stomach pains. Vomiting and diarrhea.

STATEMENTS OF PRACTICAL TREATMENT

Organophosphate

In case of poisoning, call a physician immediately. Have patient lie down and keep quiet. **If swallowed, vomiting should be induced**. Administer water freely and induce vomiting by giving one dose (1/2 oz or 15 mL) of syrup of ipecac. If vomiting does not occur within 10 to 20 minutes, administer second dose. If syrup of ipecac is not available, induce vomiting by sticking finger down throat. Repeat until vomit fluid is clear. Never give anything by mouth to an unconscious person. Professional medical assistance should be secured immediately. **If on skin,** remove contaminated clothing and wash skin immediately with soap and warm water. **If eyes are contaminated,** hold eyelids open and flush with a steady, gentle stream of water for at least 15 minutes. **If inhaled,** remove victim to fresh air. If not breathing, give artificial respiration, preferably mouth to mouth.

NOTE TO PHYSICIAN: ANTIDOTE - Administer atropine sulfate in large therapeutic doses. Repeat as necessary to the point of tolerance. 2-PAM is also antidotal and may be administered in conjunction with atropine.

Compound inhibits cholinesterase resulting in stimulation of the central nervous system, the parasympathetic nervous system, and the somatic motor nerves. Do not give morphine. Watch for pulmonary edema, which may develop in serious cases of poisoning even after 12 hours. At first sign of pulmonary edema, the patient should be placed in an oxygen tent and treated symptomatically.

ENVIRONMENTAL HAZARDS

This pesticide is toxic to fish and wildlife. Drift and runoff from treated areas may be hazardous to aquatic organisms in neighboring areas. Do not apply directly to water, or to areas where surface water is present or to intertidal areas below the mean high water mark. Do not contaminate water when disposing of equipment washwaters.

Do not use mist sprayers. Use only coarse sprays directed at soil to eliminate spray drift.

Aerial application of this product is prohibited.

Ground Water Advisory: Fenamiphos is known to leach through soil and has been found in ground water as a result of agriculture use. Users are advised not to apply in areas where soils are permeable, particularly where ground water is used for drinking water. Consult with the pesticide state lead agency for information regarding soil permeability and aquifer vulnerability in your area.

DIRECTIONS FOR USE

It is a violation of Federal law to use this product in a manner inconsistent with its labeling. Do not apply this product in a way that will contact workers or other persons, either directly or through drift. Only protected handlers may be in the area during application. For any requirements specific to your State or Tribe, consult the agency responsible for pesticide regulation.

AGRICULTURAL USE REQUIREMENTS
Use this product only in accordance with its labeling and with the Worker Protection Standard, 40 CFR part 170. This Standard contains requirements for the protection of agricultural workers on farms, forests, nurseries, and greenhouses, and handlers of agricultural pesticides. It contains requirements for training, decontamination, notification, and emergency assistance. It also contains

Page 2

Specimen Label

AGRICULTURAL USE REQUIREMENTS (Continued)
specific instructions and exceptions pertaining to the statements on this label about personal protective equipment (PPE) and restricted-entry interval. The requirements in this box only apply to uses of thisproduct that are covered by the Worker Protection Standard.

Do not enter or allow worker entry into treated areas during the restricted entry interval (REI) of 48 hours.

Exception: If the product is soil-injected or soil-incorporated, the Worker Protection Standard, under certain circumstances, allows workers to enter the treated area if there will be no contact with anything that has been treated.

PPE required for early entry to treated areas that is permitted under the Worker Protection Standard and that involves contact with anything that has been treated, such as plants, soil, or water, is:

- Chemical-resistant protective suit
- Chemical-resistant gloves, such as barrier laminate or viton
- Chemical-resistant footwear plus socks
- Protective eyewear
- Chemical-resistant headgear for overhead exposure

IMPORTANT: Read these entire Directions and Conditions of Sale before using NEMACUR 3 emulsifiable systemic insecticide-nematicide.

CONDITIONS OF SALE: THE DIRECTIONS ON THIS LABEL WERE DETERMINED THROUGH RESEARCH TO BE THE DIRECTIONS FOR CORRECT USE OF THIS PRODUCT. THIS PRODUCT HAS BEEN TESTED FOR A RANGE OF WEATHER CONDITIONS SIMILAR TO THOSE WEATHER CONDITIONS THAT ARE ORDINARY AND CUSTOMARY IN THE GEOGRAPHIC AREA WHERE THE PRODUCT IS USED. INSUFFICIENT CONTROL OF PESTS AND/OR INJURY TO THE CROP TO WHICH THE PRODUCT IS APPLIED MAY RESULT FROM THE OCCURRENCE OF EXTRAORDINARY OR UNUSUAL WEATHER, OR FROM FAILURE TO FOLLOW LABEL DIRECTIONS. IN ADDITION, FAILURE TO FOLLOW LABEL DIRECTIONS MAY CAUSE INJURY TO OTHER CROPS, ANIMALS, MAN, OR THE ENVIRONMENT. MILES OFFERS, AND THE BUYER ACCEPTS AND USES, THIS PRODUCT SUBJECT TO THE CONDITIONS THAT EXTRAORDINARY OR UNUSUAL WEATHER, OR FAILURE TO FOLLOW LABEL DIRECTIONS ARE BEYOND THE CONTROL OF MILES AND ARE, THEREFORE, THE RESPONSIBILITY OF THE BUYER.

NEMACUR 3 may be applied by ground equipment or low pressure irrigation (apple, cherry, grape, nectarine, peach, kiwifruit, citrus and pineapple (Hawaii only). Do not apply by mist sprayers or aerial spray equipment.

Ground application equipment: Unless otherwise specified, apply NEMACUR 3 in a minimum of 20 gallons of water per **treated** acre. Maintain sufficient agitation during application to insure a uniform mixture. Incorporate NEMACUR 3 immediately after application for best results.

Mixing: NEMACUR 3 mixes readily with water. No special premixing should be required if used in spray equipment equipped with a recirculation system or agitation. If the spray mixture is allowed to sit overnight, redispersion is readily obtained by agitation or recirculation.

Tank Mixing: NEMACUR 3 can be tank mixed with registered fungicides, herbicides and insecticides. Observe all cautions and limitations on labeling of all products used in tank mixtures.

Compatibility: NEMACUR 3 is physically compatible with many registered pesticides and liquid fertilizers. To determine the compatibility of NEMACUR 3 with other products, the following procedure should be followed: Pour the recommended proportions of the products into a suitable container of water, mix thoroughly and allow to stand at least five (5) minutes. If the combination remains mixed or can be re-mixed readily, the mixture is considered physically compatible. It is recommended that combinations not be allowed to sit overnight due to possible crystal growth, clabbering, etc. For further information contact your local Miles representative.

Consult your State Agricultural Experiment Station, or Extension Service Specialist or Miles representative for guidance on the selection of dosage rates and timing of applications appropriate for your specific crop. State regulations may contain additional restrictions or requirements. Check the remarks and restrictions sections of this label for additional directions.

CHEMIGATION

Apply NEMACUR 3 only through low pressure irrigation equipment (including drip, mini-sprinklers, strip tubing and jets). Do not apply NEMACUR 3 through any other type of irrigation system.

The system must contain a functional check valve, vacuum relief valve and low pressure drain appropriately located on the irrigation pipeline to prevent water source contamination from backflow.

The pesticide injection pipeline must contain a functional, automatic, quick-closing check valve to prevent the flow of fluid back toward the injection pump.

The pesticide injection pipeline must also contain a functional, normally closed, solenoid-operated valve located on the intake side of the injection pump and connected to the system interlock to prevent fluid from being withdrawn from the supply tank when the irrigation system is either automatically or manually shut down.

The system must contain functional interlocking controls to automatically shut off the pesticide injection pump when the water pump motor stops.

Page 3

Specimen Label

The irrigation line or water pump must include a functional pressure switch which will stop the water pump motor when the water pressure decreases to the point where pesticide distribution is adversely affected.

Systems must use a metering pump such as a positive displacement injection pump (e.g., diaphragm pump) effectively designed and constructed of materials that are compatible with pesticides and capable of being fitted with a system interlock.

Do not apply when wind speed favors drift beyond the areas intended for treatment.

A pesticide supply tank is not necessary for premixing since NEMACUR 3 mixes well with water in the irrigation line. If a pesticide supply tank is used, add NEMACUR 3 to water in supply tank and maintain sufficient agitation during application to insure a uniform mixture.

Begin irrigation and check emitters to insure all systems are operating normally. Make NEMACUR 3 injection in the first 1/3 to 1/2 of the irrigation with an injection time of at least 30 minutes. Maintain the system's operating pressure low enough to prevent fogging and/or misting during application. Inject NEMACUR 3 into the irrigation system after the filter(s). Continue irrigation until system's lines are flushed.

Crop injury, lack of effectiveness, or illegal pesticide residues in the crop can result from nonuniform distribution of treated water.

If you have questions about calibration, you should contact State Extension Service specialists, equipment manufacturers or other experts.

Do not connect an irrigation system (including greenhouse systems) used for pesticide application to a public water system unless the pesticide label-prescribed safety devices for public water systems are in place.

A person knowledgeable of the chemigation system and responsible for its operation, or under the supervision of the responsible person, shall shut the system down and make necessary adjustments should the need arise.

Posting: This sign is in addition to any sign posted to comply with the Worker Protection Standard. Posting of areas to be chemigated is required when 1) any part of a treated area is within 300 feet of sensitive areas such as residential areas, labor camps, businesses, day care centers, hospitals, inpatient clinics, nursing homes or any public areas such as schools, parks, playgrounds, or other public facilities not including public roads; or 2) when the chemigated area is open to the public such as golf courses or retail greenhouses.

Posting must conform to the following requirements: Treated areas shall be posted with signs at all usual points of entry and along likely routes of approach from the listed sensitive areas. When there are no usual points of entry, signs must be posted in the corners of the treated areas and in any other location affording maximum visibility to sensitive areas. The printed side of the sign should face away from the treated area towards the sensitive area. The signs shall be printed in English. Signs must be posted prior to application and must remain posted until foliage has dried and soil surface water has disappeared. Signs may remain in place indefinitely as long as they are composed of materials to prevent deterioration and maintain legibility for the duration of the posting period.

All words shall consist of letters at least 2-1/2 inches tall, and all letters and the symbol shall be a color which sharply contrasts with their immediate background. At the top of the sign shall be the words KEEP OUT, followed by an octagonal stop sign symbol at least 8 inches in diameter containing the word STOP. Below the symbol shall be the words PESTICIDES IN IRRIGATION WATER.

RECOMMENDED APPLICATIONS

CROP	PEST	DOSAGE NEMACUR 3	REMARKS
FIELD CROPS Cotton	Thrips (early season reduction)	2.4 to 3.3 fl oz per 1000 ft of row for any row spacing (1 to 1-1/3 qt per acre based on 40 inch rows)	In furrow/covering soil: Apply specified dosage per 1000 ft of row as a spray in furrow or band in covering soil behind the seed drop and in front of covering devices.
	Nematodes Thrips	3.3 to 7.1 fl oz per 1000 ft of row for any row spacing (1-1/3 to 2.9 qt per acre based on 40 inch rows)	**In furrow/covering soil:** Apply specified dosage per 1000 ft of row as a spray in furrow or band in covering soil behind the seed drop and in front of covering devices. **Band:** Apply specified dosage per 1000 ft of row as a spray in a 6- to 12-inch band over the row. Incorporate thoroughly to insure uniform distribution.[1] **Suggested use rates of NEMACUR 3 in fluid oz/1000 ft of row based on nematode infestation levels:*** **Bandwidth / Low / Moderate / High*** In furrow 3.3 4.0 4.8 In covering soil 3.3 4.0 4.8 6 inch band 3.3 4.0 4.8 8 inch band 4.0 4.8 5.7 10 inch band 4.4 5.2 6.4 12 inch band 4.8 5.7 7.1 *Use the highest recommended rates in fields with a history of nematode problems.

Page 4

Specimen Label

RECOMMENDED APPLICATIONS

CROP	PEST	DOSAGE NEMACUR 3	REMARKS
FIELD CROPS (Continued) Cotton (Continued)	Nematodes Thrips	9.8 fl oz per 1000 ft of row or 4.0 qt per acre based on 40 inch rows. Do not exceed 4 qt per acre regardless of row spacing.	**SOIL INJECTION APPLICATIONS:** (California only) Apply specified dosage in a water emulsion or with liquid fertilizer through 2 or more injection shanks centered on the seed row covering an 18" band. Plant crop in usual manner.
	Nematodes Thrips (early season reduction)	3.9 to 8.9 fl oz per 1000 ft of row for any row spacing	**Tank Mix Recommendations:** Apply specified dosage as a water emulsion spray in combination with Treflan 4 EC and/or liquid fertilizer in a 12- to 18-inch band over the row prior to planting. Refer to table below for specific band width application rates for NEMACUR 3. Use sufficient water and incorporate thoroughly to insure uniform distribution. Use the high rate in fields with high populations of nematodes or in fields having a history of severe nematode damage. Consult federally approved product labels for special precautionary statements, restrictions, and directions for use. Suggested use rates of NEMACUR 3 in fluid oz/1000 ft of row based on nematode infestation levels: Band Width / Low / Mod to High 12" / 3.9 / 5.9 18" / 5.9 / 8.9
IMPORTANT: Do not overlap bands on narrow row crops. Do not feed or graze cotton foliage. Under very heavy thrip pressure, additional control measures may be needed. [1] The action of some planters during planting operation may provide adequate incorporation. When using planters which do not disturb enough soil for incorporation or under dry conditions, additional incorporation can be achieved with scratchers, paddlewheels, ground driven rototillers or similar equipment behind the planter.			
Peanuts	Nematodes Thrips (early season reduction)	4.5 to 7.3 fl oz for any row spacing (or 2 to 3 qt per acre on 36-inch rows)	Apply specified dosage per 1000 ft of row as a water emulsion spray in a 12-inch band over the row at planting. Use sufficient water and incorporate into the soil. ON NARROW ROW CROPS, DO NOT ALLOW BANDS TO OVERLAP. Use the high rate in fields with high populations of nematodes, in fields where root-knot species other than southern root-knot, *Meloidogyne incognita*, are present or in fields having a history of serious nematode damage. Do not feed or graze green peanut vines or peanut vine hay. Do not hog down treated peanut fields.
Tobacco (Not for use on shade-grown tobacco)	Nematodes	1-1/3 to 2 gal	Uniformly broadcast the specified dosage in a minimum of 20 gallons of water per acre over the soil surface. Incorporate to a depth of 2 to 4 inches by disking or tilling. Use the high rate in fields with high populations of nematodes, in fields where tobacco cyst and/or root-knot species other than southern root-knot, *Meloidogyne incognita*, are present or in fields having a history of serious nematode damage. Plant crop in the usual manner.
	Tobacco cyst nematodes Aphids (suppression)	2 gal	
	Tank Mixes: The following tank mix recommendations are made for the control of additional pests on tobacco. Before mixing NEMACUR 3 with any other pesticide, read and carefully observe the precautionary statements, directions for use, and other information appearing on the EPA registered product labels.		
	Additional Pests Controlled	Product Dosage/Acre	NEMACUR 3 Dosage/Acre
	Wireworms Flea beetles Cutworms	Lorsban 4EC 2 qt	1 to 2* gal
	Wireworms Cutworms Flea beetle larvae Mole crickets	Mocap EC 2 to 4 qt	1 to 2* gal
	* For the control of tobacco cyst nematode, do not apply less than two gal of NEMACUR 3 per acre.		

Page 5

Specimen Label

RECOMMENDED APPLICATIONS

CROP	PEST	DOSAGE NEMACUR 3	REMARKS
FRUIT Apple Cherry Nectarine Peach	Nematodes	1-2/3 to 3-1/3 gal	**BAND:** Apply specified dosage in not less than 10 gallons of solution per acre to the soil with ground equipment and incorporate immediately.* Center the treated band on the tree row using a band width of 50% of the row spacing and covering the feeder root system of the plant. Do not apply more than 3-1/3 gallons per acre per year per planting site.
		2 qt to 1-1/2 gal	**LOW PRESSURE IRRIGATION:** Apply specified dosage per acre to the soil in 1 to 6 applications with at least 14 days between applications using a minimum of 1-1/2 gallons to a maximum of 3 gallons of NEMACUR 3 per acre per season.
	* Incorporate this product into the soil mechanically or supply sufficient irrigation (not to exceed the depth of the root zone). Control of nematodes is best obtained when there is adequate rainfall or irrigation after application. Do not apply within 72 days of harvest of apples or within 45 days of harvest of peaches, nectarines and cherries. Do not graze livestock in treated orchards. Do not feed cover crops grown in treated orchards to livestock. **Special Precautions**: Avoid contacting tree foliage with the spray mixture. Do not add oils or surfactants to the spray mixture. Do not allow spray or drift to contact operator. On non-bearing apple, cherry, nectarine and peach make initial applications when trees have recovered from transplanting shock and have initiated new growth, provided that one or more irrigations have been applied and soil has settled around the roots. Phytotoxicity may occur if applications are made before soil has settled and plants are growing.		
Grape	Nematodes	3 gal	**Band:** Apply specified dosage in not less than 10 gallons of solution per acre to the soil with ground equipment and incorporate immediately.* Center the treated band on the vine row using a band width of 50% of the row spacing.
		1 qt to 1 gal	**LOW PRESSURE IRRIGATION:** Apply the specified dosage per acre to the soil in one or more applications using a minimum of 1 gallon and a maximum of 2 gallons of NEMACUR 3 per acre per season.
	* Incorporate this product into the soil mechanically or supply sufficient irrigation (not to exceed the depth of the root zone). **RESTRICTIONS:** On bearing grapes, the last application may be made up to 2 days of harvest. Do not apply more than 3 gallons of NEMACUR 3 (9 lb active ingredient) per acre in a 50% band or 2 gallons per acre by low pressure irrigation in one season. Do not graze or feed treated crop to livestock. On non-bearing grapes, make initial applications when vines have recovered from transplanting shock and have initiated new growth, provided that one or more irrigations have been applied and soil has settled around the roots. Phytotoxicity may occur if applications are made before soil has settled and plants are growing.		
Kiwifruit (California only)	Nematodes	2 qt to 1 gal	**LOW PRESSURE IRRIGATION:** Apply specified dosage per acre to the soil in 2 to 6 applications using a minimum of 2 gallons and a maximum of 4 gallons of NEMACUR 3 per acre per year. Start NEMACUR applications when the soil temperature in the spring reaches 55°F or above at the 12-inch depth. To maintain control, an application may be made 5 days after harvest if the soil temperature at the 12-inch depth is 55°F or above. A maximum of 3 gallons of NEMACUR 3 may be applied preharvest, and a maximum of 1 gallon may be applied postharvest. Do not apply NEMACUR 3 to newly established kiwifruit vineyards which are less than one year old. Do not apply within 31 days of harvest. Do not graze livestock in treated vineyards. Do not feed cover crops grown in treated vineyards to livestock.

Page 6

Specimen Label

RECOMMENDED APPLICATIONS

CROP	PEST	DOSAGE NEMACUR 3	REMARKS
Fruit (Continued) Citrus (In California do not apply to Kumquat, Tangelo, or Citrus hybrids)	Nematodes Citrus root weevil complex including Fuller rose beetle (suppression)	1-2/3 to 3-1/3 gal	**Band:** Apply specified dosage in not less than 10 gallons of solution per acre to the soil with ground equipment and incorporate immediately.* Two applications may be made per season, not to exceed 3-1/3 gallons per acre. Center the treated band on the tree row using a band width of 50% of the row spacing and covering the feeder root system of the plant.
		2 qt to 1-1/2 gal	**LOW PRESSURE IRRIGATION:** Apply specified dosage per acre to the soil in 1 to 6 applications with at least 14 days between applications using a minimum of 1-1/2 gallons to a maximum of 3 gallons of NEMACUR 3 per acre per season.
	* Incorporate this product into the soil mechanically or supply sufficient irrigation (not to exceed the depth of the root zone). Control of nematodes is best obtained when there is adequate rainfall or irrigation after application. Under dry conditions, follow with irrigation. Do not apply within 30 days of harvest. Do not exceed 10 pounds active ingredient (3-1/3 gallons of NEMACUR 3) per acre in a 50% band or 3 gallons of NEMACUR 3 (9 pounds active ingredient) per acre by low pressure irrigation per season. Do not graze livestock in treated areas. On non-bearing citrus make initial applications when trees have recovered from transplanting shock and have initiated new growth, provided that one or more irrigations have been applied and soil has settled around the roots. Phytotoxicity may occur if applications are made before soil has settled and plants are growing.		
Pineapple (Hawaii only)	Nematodes (*Rotylenchulus* and *Meloidogyne* species)	6-2/3 gal	**PREPLANT APPLICATION:** Apply the specified dosage per acre as a water emulsion spray over the entire area to be treated. Incorporate thoroughly to insure uniform distribution. See "NOTE" below. Additional postplant applications may be made to the plant crop at 1 to 3 month intervals by broadcast spray or drip irrigation equipment as indicated below. However, do not apply more than a total of 13-1/3 gallons of NEMACUR 3 (40 pounds of active ingredient) per acre per plant crop.
		1-1/3 to 8 pt	**POSTPLANT APPLICATION: PLANT CROP**. Apply specified dosage in 50 to 250 gallons of water per acre as a broadcast spray or through drip irrigation. Applications may begin immediately after planting. Make additional applications at intervals of 1 to 3 months as needed. Do not apply within 30 days before harvest. See "NOTE" below. Do not apply more than a total of 13-1/3 gallons of NEMACUR 3 per acre per plant crop. **RATOON CROP.** Apply specified dosage in 50 to 250 gallons of water per acre as a broadcast spray or through drip irrigation. The first application may be made immediately following crop harvest. Make additional applications at intervals of 1 to 3 months as needed. Do not apply within 30 days before harvest. See "NOTE" below. Do not apply more than a total of 6-2/3 gallons of NEMACUR 3 to each ratoon crop. NOTE: Do not use green forage or green fodder (cannery waste, such as cull fruit, fruit skin or shells, crowns, cores and basal ends of fruit may be fed) for animal feed.

Page 7

Specimen Label

RECOMMENDED APPLICATIONS

CROP	PEST	DOSAGE NEMACUR 3	REMARKS
Pineapple* (Puerto Rico only)	Nematodes	1-2/3 to 3-1/3 gal	**POSTPLANT BROADCAST APPLICATION: PLANT CROP** Apply specified dosage (5 to 10 lb active ingredient) in 50 to 250 gallons of water per acre as a broadcast spray. Begin applications 1 to 3 months after planting. When used in conjunction with a preplant soil application of NEMACUR 15% Granular, postplant applications should be initiated 3 to 6 months after the preplant soil application. Make additional applications at intervals of 3 to 6 months as needed. Do not apply within 7-1/2 months before harvest. **FIRST RATOON CROP:** Apply specified dosage in 50 to 250 gallons of water per acre as a broadcast spray. Make the first application immediately following crop harvest. Make additional applications at intervals of 3 to 6 months as needed. Do not apply within 7-1/2 months before harvest.
	Do not apply more than a total of 20 lb active ingredient per acre per crop season regardless of the formulation or method of application. Do not use forage or fodder for animal feed. * NEMACUR 3 emulsifiable nematicide has shown excellent activity against the major genera of nematodes infesting pineapple plants. When applied as a broadcast spray after planting, NEMACUR protects young developing root systems. For best results, applications should be made when plants are actively growing.		
Raspberry (Except California)	Nematodes	1 to 2 gal	**Band:** Apply specified dosage in not less than 10 gallons of solution per acre to the soil with ground equipment and incorporate immediately using mechanical incorporation or irrigation. Center the treated band on the row using a band width of 50% of the row spacing and covering the feeder root system of the plant. Apply during the period of October 1 to December 31 when adequate rainfall can be expected. Do not apply within six months of harvest. Do not apply more than once per year. Use the high rate in fields with high populations of nematodes or in fields having a history of serious nematode damage.
Strawberries	Nematodes	5.9 to 8.8 fl oz for any row spacing (or 2.4 to 3.6 qt per acre on 40-inch rows)	Apply specified dosage per 1000 ft of row in a 12- to 18-inch band over the row and incorporate immediately by cultivation or by sprinkler irrigation prior to transplanting. Do not apply more than one application. Do not apply within 110 days of harvest. Apply by ground equipment only. **BAND WIDTH:** 12-INCH: 5.9 fl oz/1000 ft OR 2.4 qt/acre (40-inch rows); 18-INCH: 8.8 fl oz/1000 ft OR 3.6 qt/acre (40-inch rows) Use the maximum band width and rate in fields with high populations of nematodes or in fields having a history of serious nematode damage.
VEGETABLES Asparagus (Connecticut Delaware Maine Maryland Massachusetts New Hampshire New Jersey New York Pennsylvania and Rhode Island only)	Nematodes	5-1/3 pt	**BAND APPLICATION** **Nursery:** Apply the specified dosage per acre to the soil and immediately incorporate to a depth of 2 to 6 inches by disking or tilling. Center the treated band on the row using a band width of 50% of the row spacing. Plant crop in the usual manner. **Production Field:** **Newly Planted Crowns:** Apply the specified dosage per acre to the soil before planting crowns but after making furrows. Center the treated band on the row using a band width of 50% of the row spacing. Plant crowns and cover with soil. The following Spring apply before spear emergence and immediately incorporate by reshaping beds. Do not apply within 9 months of harvest. **Post Harvest:** Apply after the last harvest and all spears have been removed. Apply the specified dosage per acre to the soil and immediately incorporate by reshaping beds. Center the treated band on the row using a band width of 50% of the row spacing. Do not apply within 9 months of harvest. Do not apply more than once per year.

Page 8

Specimen Label

RECOMMENDED APPLICATIONS

CROP	PEST	DOSAGE NEMACUR 3	REMARKS
Eggplant	Nematodes	5.9 fl oz (or 2-2/3 qt per acre on 36-inch rows)	Apply specified dosage per 1000 ft of row in a 12-inch band over the row at transplanting and immediately incorporate into the soil. ON NARROW ROW CROPS, DO NOT ALLOW BANDS TO OVERLAP.
Table beets (Illinois, Indiana, Michigan, New York, Ohio and Pennsylvania only)	Cyst nematode	4 to 6 fl oz per 1000 ft of row for any row spacing	Apply specified dosage per 1000 ft of row in a band over the row before or at seeding. Center the treated band on the row using a band width of 33% to 50% of the row spacing. See the table below for dosages for specific band widths. Incorporate immediately by shallow cultivation or by sprinkler irrigation. Do not make more than one application per season. Do not apply within 90 days of harvest. Apply by ground equipment only. **Dosage for Various Band Widths:** Band Width (in.) / Fl oz per 1000 ft of row: 8 – 4; 9 – 4.5; 10 – 5; 11 – 5.5; 12 – 6

RESTRICTIONS

Do not use on other crops. Use only according to label directions.

Any food crop not specified on this label may be planted into treated areas 120 days after the last application. Any cover crops that are planted during the 120 day period must be plowed under and not grazed.

STORAGE AND DISPOSAL

Pesticide Disposal: Do not contaminate water, food or feed by storage or disposal.

Pesticide wastes are acutely hazardous. Improper disposal of excess pesticide, spray mixture, or rinsate is a violation of Federal Law. If these wastes cannot be disposed of by use according to label instructions, contact your State Pesticide or Environmental Control Agency, or the Hazardous Waste representative at the nearest EPA Regional Office for guidance.

Container Disposal: Non-refillable Metal: Triple rinse (or equivalent). Then offer for recycling or reconditioning, or puncture and dispose of in a sanitary landfill, or by other procedures approved by state and local authorities. **Non-refillable Plastic:** Triple rinse (or equivalent). Then offer for recycling or reconditioning, or puncture and dispose of in a sanitary landfill, or incineration, or, if allowed by state and local authorities, by burning. If burned, stay out of smoke. **Returnable Refillable Sealed Containers:** Do not rinse container. Do not break seals. Replace the secondary containment cover cap and return container intact to point of purchase.

Date of Draft: 12/10/93
Supersedes Draft Dated: 5/24/93
Reason to Issue: To delete non-bearing fruit.

Store in a cool, dry place and in such a manner as to prevent cross contamination with other pesticides, fertilizers, food, and feed. Do not store below 32°F. Do not store near heat or open flame. Store in original container and out of the reach of children, preferably in a locked storage area.

Handle and open container in a manner as to prevent spillage. If container is leaking, invert to prevent leakage. If the container is leaking or material spilled for any reason or cause, carefully dam up spilled material to prevent runoff. Refer to Precautionary Statements on label for hazards associated with the handling of this material. Do not walk through spilled material. Absorb spilled material with absorbing type compounds and dispose of as directed for pesticides above. In spill or leak incidents, keep unauthorized people away. You may contact the Miles Emergency Response Team for decontamination procedures or any other assistance that may be necessary. The Miles Kansas City Emergency Response Telephone No. is 816-242-2582, or contact Chemtrec at 800-424-9300.

Page 9

Specimen Label

SPECIAL LOCAL NEED REGISTRATIONS [SECTION 24(c)] FOR NEMACUR 3

CROP	PEST		DOSAGE NEMACUR 3	STATE / EPA SLN NUMBER
Cabbage, transplanted	Nematodes		5 Fl Oz / 1000 Ft Row	Florida FL-840019
Bananas	Nematodes Burrowing nematodes	Banana root borer Root-knot nematodes	2-1/5 to 6-23/ Qt / Acre	Hawaii HI-920001
Tobacco	Nematodes Wireworms	Flea beetles Cutworms	1 - 2 Gal / Acre	North Carolina NC-880005
Iris Lily bulbs	Nematodes		2 - 4 Gal / Acre	Oregon OR-800063
Pineapple	Nematodes		3-1/3 to 6-2/3 Gal / Acre	Puerto Rico PR-910035
Cotton	Nematodes	Thrips	2.6 - 5.3 Fl Oz / 1000 Ft	Texas TX-790035
Iris Narcissus bulbs	Nematodes		2 - 4 Gal / Acre	Washington WA-760034

For distribution and use only within the state specified. These Special Local Need registrations are approved as of date of this publication are, however, subject to cancellation by the state. Contact your nearest Miles Inc. Agriculture Division Office for the current status and a copy of the complete individual supplemental Special Local Need labeling. This labeling must be in the possession of the user at time of pesticide application.

Miles Inc.
Crop Protection Products
Box 4913
Kansas City, MO 64120-0013
N 9501
Printed in U.S.A. on recycled paper.

IMPORTANT
Before using this product, read and carefully observe directions, cautionary statements and other information appearing on the product packaging label. This product is sold subject to the Conditions of Sale set forth on the container label.

Page 10

RESTRICTED USE PESTICIDE

Due to Acute Toxicity And Toxicity to Birds and Mammals.

For retail sale to and use only by Certified Applicators or persons under their direct supervision and only for those uses covered by the Certified Applicator's certification.

INSECTICIDE \ NEMATICIDE

WATER SOLUBLE LIQUID

1 GALLON CONTAINS 2 LBS. ACTIVE INGREDIENT PER GALLON.

ACTIVE INGREDIENT	**BY WEIGHT**
Oxamyl [Methyl N'N'-dimethyl-N-[(methylcarbamoyl)oxy]-1-thiooxamimidate]	24%
INERT INGREDIENTS	76%
TOTAL	100%

EPA Reg. No. 352-372

KEEP OUT OF REACH OF CHILDREN

DANGER **POISON**

PELIGRO

Si usted no entiende la etiqueta, busque a alguien para que se la explique a usted en detalle. (If you do not understand this label, find someone to explain it to you in detail.)

STATEMENT OF PRACTICAL TREATMENT

This product is an N-Methyl Carbamate insecticide.

If swallowed: Call a physician or Poison Control Center. Drink 1 or 2 glasses of water and induce vomiting by touching back of throat with finger. Do not induce vomiting or give anything by mouth to an unconscious person.

If inhaled: Remove from exposure and have patient lie down and keep quiet. If patient is not breathing, start artificial respiration immediately. Never give anything by mouth to an unconscious person.

In case of contact, wash skin with plenty of soap and water; for eyes, flush with water for 15 minutes and get medical attention; remove and wash contaminated clothing before re-use.

ATROPINE IS AN ANTIDOTE -- SEEK MEDICAL ATTENTION AT ONCE IN ALL CASES OF SUSPECTED POISONING

If warning symptoms appear (see WARNING SYMPTOMS), get medical attention.

For medical emergencies involving this product, call toll free 1-800-441-3637.

PRECAUTIONARY STATEMENTS

HAZARDS TO HUMANS AND DOMESTIC ANIMALS

DANGER-POISON! CONTAINS METHANOL. MAY BE FATAL OR CAUSE BLINDNESS IF SWALLOWED. MAY BE FATAL IF ABSORBED THROUGH SKIN OR INHALED.

Do not breathe vapors or spray mist. Do not get in eyes, on skin, or on clothing. Pilot should not assist in the mixing and loading operation.

WARNING SYMPTOMS--Oxamyl poisoning produces effects associated with anticholinesterase activity which may include weakness, blurred vision, headache, nausea, abdominal cramps, discomfort in the chest, constriction of pupils, sweating, slow pulse, muscle tremors.

NOTE TO PHYSICIAN

TREATMENT--Atropine sulfate should be used for treatment. Administer repeated doses, 1.2 to 2.0 mg intravenously every 10 to 30 minutes until full atropinization is achieved. Maintain atropinization until the patient recovers. Artificial respiration or oxygen may be necessary. Allow no further exposure to any cholinesterase inhibitor until recovery is assured.

Do not use 2-PAM for exposure to "Vydate" L alone. However, for exposure to combinations of "Vydate" L and organophosphorous insecticides, 2-PAM may be used as required to supplement the atropine sulfate treatment. Do not use morphine.

For medical emergencies involving this product, call toll free 1-800-441-3637.

PRECAUTIONARY STATEMENTS (continued on next page)

1

PRECAUTIONARY STATEMENTS

HAZARDS TO HUMANS AND DOMESTIC ANIMALS

PERSONAL PROTECTIVE EQUIPMENT

Some materials that are chemical-resistant to this product are listed below. If you want more options, follow the instructions for category C on an EPA chemical-resistance category selection chart.

Applicators and other handlers must wear:

Coveralls over short-sleeved shirt and short pants.

Chemical-resistant gloves, such as barrier laminate or butyl rubber or neoprene rubber or polyvinyl chloride (PVC) or viton or nitrile rubber.

Chemical-resistant footwear plus socks.

Protective eyewear.

Chemical-resistant headgear for overhead exposure.

Chemical-resistant apron when cleaning equipment, mixing, or loading.

A respirator with either an organic vapor-removing cartridge with a prefilter approved for pesticides (MSHA/NIOSH approval number prefix TC-23C), or a canister approved for pesticides (MSHA/NIOSH approval number prefix TC-14G).

Discard clothing or other absorbent materials that have been drenched or heavily contaminated with this product's concentrate. Do not reuse them. Follow manufacturer's instructions for cleaning/maintaining PPE. If no such instructions for washables, use detergent and hot water. Keep and wash PPE separately from other laundry.

ENGINEERING CONTROL STATEMENTS

Human flaggers must be in enclosed cabs.

When handlers use closed systems, enclosed cabs, or aircraft in a manner that meets the requirements listed in the Worker Protection Standard (WPS) for agricultural pesticides [40 CFR part 170.240 (d)(4-6)], the handler PPE requirements may be reduced or modified as specified in the WPS.

The enclosed cabs must be used in a manner that meets the requirements listed in the Worker Protection Standard (WPS) for agricultural pesticides [40 CFR part 170.240 (d)(4-6)]. The handler PPE requirements may be reduced or modified as specified in the WPS.

USER SAFETY RECOMMENDATIONS

USERS SHOULD: Wash hands before eating, drinking, chewing gum, using tobacco or using the toilet.

Remove clothing immediately if pesticide gets inside. Then wash thoroughly and put on clean clothing.

Remove personal protective equipment immediately after handling this product. Wash the outside of gloves before removing. As soon as possible, wash thoroughly and change into clean clothing.

ENVIRONMENTAL HAZARDS

This pesticide is toxic to aquatic organisms(fish and invertebrates) and extremely toxic to birds. Cover or disc spill areas. Birds in treated areas may be killed. Do not apply directly to water, or to areas where surface water is present, or to intertidal areas below the mean high water mark. Drift and runoff may be hazardous to aquatic organisms in neighboring areas. Do not contaminate water when disposing of equipment washwaters.

This product is highly toxic to bees exposed to direct treatment or residues on blooming crops or weeds. Do not apply this product or allow to drift to blooming crops or weeds if bees are visiting the treatment area.

GROUND WATER ADVISORY--Residues of "Vydate" L can seep or leach through soil and can contaminate ground water which may be used for drinking. Users are advised not to apply "Vydate" L where the water table is close to the surface and where soils are very permeable, i.e., well-drained soils such as loamy sands. Local agricultural Agencies can provide information on the soil type in your area and the location of the ground water.

PHYSICAL AND CHEMICAL HAZARDS

Flammable. Keep away from heat, sparks, and open flame. Keep container closed. Use with adequate ventilation.

2

DIRECTIONS FOR USE

It is a violation of federal law to use this product in a manner inconsistent with its labeling.

Do not apply this product in a way that will contact workers or other persons, either directly or through drift. Only protected handlers may be in the area during application. For any requirements specific to your State or Tribe, consult the agency responsible for pesticide regulation.

AGRICULTURAL USE REQUIREMENTS

Use this product only in accordance with its labeling and with the Worker Protection Standard, 40 CFR part 170. This Standard contains requirements for the protection of agricultural workers on farms, forests, nurseries, and greenhouses, and handlers of agricultural pesticides. It contains requirements for training, decontamination, notification, and emergency assistance. It also contains specific instructions and exceptions pertaining to the statements on this label about personal protective equipment(PPE) and restricted-entry interval. The requirements in this box only apply to uses of this product that are covered by the Worker Protection Standard.

Do not enter or allow worker entry into treated areas during the restricted entry interval (REI) of 48 hours.

PPE required for early entry to treated areas that is permitted under the Worker Protection Standard and that involves contact with anything that has been treated, such as plants, soil, or water, is:

Coveralls over short-sleeved shirt and short pants.
Chemical-resistant gloves, such as barrier laminate or butyl rubber or neoprene rubber or polyvinyl chloride (PVC) or viton or nitrile rubber.
Chemical-resistant footwear plus socks.
Protective eyewear.
Chemical-resistant headgear for overhead exposure.

Du Pont "Vydate" L Insecticide/Nematicide should be used only in accordance with recommendations on this label or in separate Du Pont recommendations available through local dealers.

Du Pont will not be responsible for losses or damages resulting from use of this product in any manner not specifically recommended by Du Pont. User assumes all risks associated with such non-recommended use.

"Vydate" L is a water soluble liquid to be diluted with water for application. Use only in commercial and farm plantings; do not use in home plantings. Do not plant crops other than those with registered "Vydate" L uses within 6 months after the last application.

Do not use in Suffolk and Nassau Counties, Long Island, New York.

Spray Preparation--Fill spray tank 1/4 to 1/2 full of water. Add "Vydate" L directly to tank. Mix thoroughly while adding remaining water. No further agitation is necessary. Spray mix should not be stored overnight in spray tank.

Chemigation--Refer to supplemental labeling entitled, "Vydate" L Use in Chemigation Systems" for use directions for chemigation. Do not apply this product through any irrigation system unless the supplemental labeling on chemigation is followed.

FRUITS

APPLES

ROSY APPLE APHID

Apply 4 to 8 pts. "Vydate" L per acre at pink (before bloom - no petals open) when aphids are present in significant numbers.

APPLE APHID

Apply 4 to 8 pts. "Vydate" L per acre when 50% of terminals are infested.

SPOTTED TENTIFORM LEAFMINERS

Apply 2 to 4 pts. Vydate L per acre (1-2 pts/100 gal. dilute, not to exceed 4pts/acre) as follows:

For control of 1st Brood Leafminer: Make a single application at 1/2" green stage to early pink stage of bud development. Do not apply Vydate L after the blossom clusters have separated.

For control of 2nd Brood Leafminer: Make a single application when there are an average of two or more larvae per leaf in the sap-feeding stage. For best results, the application should be made before the larvae enter into the tissue-feeding stage. Make a second application, if needed, at 7 to 14 days after the first application.

EUROPEAN RED MITE AND TWOSPOTTED SPIDER MITE

Apply 2 to 4 pts. Vydate L per acre (1-2 pts/100 gal. dilute, not to exceed 4 pts/acre) when mite populations reach 2 to 4 mites/leaf and repeat at 7 to 14 day intervals as needed.

WHITE APPLE LEAFHOPPERS

Apply 2 to 4 pts. Vydate L per acre (1-2 pts/100 gals. dilute, not to exceed 4 pts/acre) and repeat at 10 to 14 day intervals as needed.

NOTE: APPLY VYDATE L IN SUFFICIENT WATER TO OBTAIN THOROUGH COVERAGE BUT NOT IN EXCESS OF 400 GALS NOR LESS THAN 50 GALS PER ACRE.

DO NOT APPLY AT BLOOM OR WITHIN 30 DAYS AFTER BLOOM, OR FRUIT THINNING MAY OCCUR.

DO NOT APPLY MORE THAN 8 PTS. OF VYDATE L PER ACRE PER GROWING SEASON.

DO NOT APPLY TO APPLES WITHIN 14 DAYS OF HARVEST.

DO NOT GRAZE LIVESTOCK IN TREATED ORCHARDS.

BANANAS, PLANTAINS (PUERTO RICO ONLY)

For control of nematodes (Radopholus similis, and species of Pratylenchus, Meloidogyne, Rotylenchulus, Helicotylenchus), and the banana corm borer (Cosmopolites sordidus).

Apply only with the specially designed "Vydate" L spotgun applicator with a coarse spray nozzle.

Rate and Frequency of Application

1. At planting time - apply undiluted "Vydate" L at the rate of 5 to 10 ml. per corm or "seed" in the planting hole. Cover the treated corm with soil.

3

2. Post-planting applications:

a) Make a second application at the same rate 2 to 3 months after planting. If the developing pseudostem is 1 foot or less in height, the treatment should be made directly over the top, wetting the leaves and leaf axils. When pseudostem is higher, direct treatment to the soil, applying product as close as possible to the developing pseudostem in a semicircular pattern.

b) Proceed with other applications to the soil at the same rate, but at 3 to 4 month intervals following instructions as previously indicated in 2a.

c) When a sucker or "follower" has been selected for the production of the ratoon crop, switch the application from around the mother plant toward the selected sucker at the same rate and frequency.

Use the higher rate and shorter frequency of application when nematodes or corm borer population levels are high.

"Vydate" L is most effective when ground applications are made at the beginning of the rainy season, or when the soil moisture is adequate.

For ground applications, remove weeds and leaf trash from the treated area and apply to the bare soil.

Note: Do not harvest for 1 day after last application. Do not permit animals to graze or forage in treated areas. Do not apply more than 2 gals per acre per year.

CITRUS

CITRUS RUST MITE

Apply 1/4 to 1 pt. "Vydate" L per 100 gals. water. Spray to run-off using up to 400 gals. of water per acre. Do not apply more than 2 qts. of product per acre.

Begin applications as soon as significant infestations are found and continue on an as-needed basis. Under light to moderate pressure, apply on 4- to 6-week intervals or as needed; whereas under moderate to heavy pressure, apply every 2 to 3 weeks as long as the infestation continues.

DO NOT MAKE MORE THAN 6 APPLICATIONS PER YEAR.

CITRUS THRIPS

Apply 2 to 4 pts of "Vydate" L per acre in sufficient water (100 to 500 gals. per acre) to give good coverage. Apply in the early spring before bloom when the new growth is about 3 to 4 inches long. Make additional applications as needed. Applications at petal fall may be critical to prevent fruit scarring. Applications are also recommended during mid-summer to protect new growth on young trees.

Note: Do not apply to citrus within 7 days of harvest. Do not graze livestock in treated orchards. Do not handle plants or come in contact with foliage within 3 days of application. This product is toxic to bees and should not be applied when bees are actively visiting the area. Treatment during bloom can be made provided application is during the period from one hour before sunset until one hour after sunrise or when the ambient temperature is below 55° F. Do not apply more than 24 pts(3 gals) "Vydate" L per acre per year.

NON-BEARING FRUITS

"Vydate" L is recommended for use on the following non-bearing fruit trees for control of insects, mites, and nematodes:

Apple* Citrus* Pear* Cherry*
Peach* Strawberry(Nurseries)* **

* Non-bearing trees and strawberries that will not bear fruit within 12 months after application.

** Note: Certain varieties of strawberries are sensitive to "Vydate" L exhibiting severe phytotoxic symptoms. Known sensitive varieties that should not be treated with "Vydate" L include Bounty, Canoga (NY 1362), Cavalier, Darrow, Dunlop, Earlidawn, Earliglow, Empire, Holiday, Marlate, Midland, Midway, Mic Mac, NY 1409, Ogallalla, Redglow, Redstar, Sequoia, Stark Red, Sunrise, and Viestar. However, other varieties may also be sensitive. Therefore, "Vydate" L should be used initially only on a small scale to determine crop sensitivity prior to large scale field application.

INSECTS AND MITES including aphids, fungus fly, fungus gnats, Japanese beetle, leafhoppers, leafminers, mealybugs, royal palm bug, scales, thrips, white fly, and mites:

Foliar Treatment--Mix 2 to 4 pts. "Vydate" L per 100 gals. and apply up to 200 gals. of this mixture per acre or apply 4 to 8 pts. "Vydate" L per acre in a maximum of 600 gals. water per acre. For small areas, mix 3 to 6 tablespoonfuls "Vydate" L per 5 to 14 gals. water and apply to 1,000 sq. ft. Apply as needed to maintain control.

NEMATODES including root knot (except Javanese), sting, lesion, and burrowing nematodes:

"Vydate" L may be used as a soil treatment. When transplanting, plants should be set into soil within 24 hours after application. It may also be used as a foliar treatment alone or as a supplement to the above "Vydate" L or other nematicide treatments.

Preplant Soil Incorporation Treatment--Apply 3 to 4 gals. "Vydate" L in a minimum of 20 gals. of water per acre. Thoroughly incorporate with a rotary tiller to a depth of 4 to 8 inches immediately after application. If applied in a band, use proportionately less material.

Soil Mix Treatment--Mix 2-3/4 fl. ozs. of "Vydate" L in 10 gals. of water. Spray 2-1/2 to 3-1/3 gals. of the dilute mix on to 1 cubic yard of soil while tumbling in a soil mixer (use 1 to 1-1/3 pts. for 1 bushel of soil).\

Foliar Treatment--As a foliar treatment, mix 2 to 4 pts. "Vydate" L per 100 gals. of water and apply as a dilute spray, not to exceed 8 pts. per acre. Apply on a 2- to 3-week schedule for 4 applications. Apply first spray at first full leaf or when plant is in active growth phase. Do not apply to plants under water stress or to plants not actively growing. Include a spreader sticker.

Note: Use only on commercial plantings; do not use on home plantings. Since varieties are numerous, continually changing, and may respond differently to "Vydate" L, test "Vydate" L on a small scale before proceeding to large-scale application. Varietal response may also vary if "Vydate" L is mixed with other products. Do not apply more than 32 pts (4 gals) per acre per season.

4

PEARS

Use not registered in California

EUROPEAN RED MITE, McDANIEL MITE, TWOSPOTTED SPIDER MITE, PEAR RUST MITE

Apply 6 to 8 pts "Vydate" L per acre in 100 to 600 gals. of water by ground. For best results use a dilute application.

Make application when mites first appear and repeat as needed to maintain control. Use lower rate on light infestations and higher rate on heavy infestations.

Note: Do not apply at bloom or within 30 days after full bloom, or fruit thinning may occur. Do not apply within 14 days of harvest. This product has been tested on Bartlett and d'Anjou varieties of pears without russeting. Its use on other varieties should be on a small scale until the possibility of russeting has been evaluated. Do not graze livestock in treated orchards.

Do not apply more than 8 pts (1 gal) "Vydate" L per acre per season.

PINEAPPLE

Use not registered in California

RENIFORM AND ROOT KNOT NEMATODES

Preplant or Planting Treatment--Apply 2 gals. "Vydate" L per acre broadcast within 1 week prior to planting and incorporate 4 to 6 inches into the soil. Alternatively, apply 1 to 2 gals. "Vydate" L per acre broadcast, or apply 1/2 to 2 gal. per acre directly to the soil via drip irrigation injection, within 1 week after planting.

Subsequent Treatments--Apply 1/2 to 1 gal. "Vydate" L per acre by ground equipment as a foliar spray at 2 to 4 week intervals in sufficient water to obtain thorough coverage. Begin applications when pineapple roots begin to grow following planting. Alternatively, apply 1/4 to 1 gal. "Vydate" L per acre directly to the soil via drip irrigation injection at 2, 4, or 8 week intervals. For best results, subsequent applications should be made after preplant or planting application of "Vydate" L or a standard fumigant. Best results occur under optimum soil moisture conditions.

Note: Do not apply more than 6 gals. "Vydate" L per acre per year. Do not harvest or graze treated fields within 30 days of application.

VEGETABLES

CARROTS

Use not registered in California

ROOT KNOT (EXCEPT JAVANESE), LESION, STING, SPIRAL AND STUNT NEMATODES

Preplant Treatment -- Apply 2 to 4 gals. of "Vydate" L per acre in a minimum of 20 gals. water per acre as a broadcast treatment of the soil. Apply within one week of planting. Thoroughly incorporate to a depth of 4 to 6 inches in the soil.

In-Furrow Treatment -- Alternatively, apply 1 to 2 gals. of "Vydate" L per acre in a minimum of 20 gals. water per acre in the seed furrow during the planting operation.

Carrot Weevil

Apply 2 to 4 pts. "Vydate" L per acre as a directed spray in 20 gals. water per acre. Apply up to 3 treatments, 2 to 3 weeks apart beginning when insects appear in damaging numbers. Do not apply within 14 days of harvest.

Note: Do not apply more than 32 pts(4 gals) "Vydate" L per acre per season.

CELERY (FL)

SERPENTINE LEAFMINERS (EXCEPT LIRIOMYZA TRIFOLII)

Apply 2 to 4 pts. "Vydate" L per acre as a foliar spray using sufficient water (minimum 5 gals per acre by air) to obtain uniform coverage. Treat when insect first appears and repeat at 5- to 7-day intervals or as needed. Use the low rates for light infestations, the intermediate rates for heavier infestations, and the highest rates for severe infestations.

Note: Do not apply within 14 days of harvest. Do not apply more than 24 pts(3 gals) "Vydate" L per acre per season.

ROOT KNOT NEMATODE (MELOIDOGYNE HAPLA) AND PIN NEMATODE

Transplant Treatment--Apply 1/2 to 1 gal. "Vydate" L per acre immediately after transplanting celery seedlings in the field. Use a minimum of 100 gals. of water per acre.

Foliar Treatment--Make two foliar applications. Apply 1 gal. "Vydate" L per acre in a minimum of 100 gals. of water as a directed spray. Apply first spray 3 weeks after transplanting. Apply second spray 3 weeks after first foliar spray.

Preplant Row Treatment--Apply 2 gallons per acre. Treat a band 8 to 16 inches wide on the row using 20 gallons of water per acre. Incorporate with a rotary tiller to a depth of 4 inches in the soil.

Foliar Treatment as an Extension of Preplant Treatment--Apply 4 pints per acre. Apply as a directed spray in a minimum of 20 gallons of water per acre. Apply two treatments 2 to 3 weeks apart beginning 2 to 3 weeks following transplanting.

CARROT WEEVIL

Foliar Treatment Alone or as an Extension of Preplant Nematode Treatment--Apply 4 pints per acre. Apply as a directed spray in a minimum of 20 gallons of water per acre. Apply three treatments 2 to 3 weeks apart beginning 2 to 3 weeks following transplanting. As an extension of the preplant treatment do not make more than 2 foliar applications.

Note: Do not apply within 14 days of harvest. Do not apply more than 24 pts(3 gals) "Vydate" L per acre per season.

5

CUCUMBER, CANTALOUPE, HONEYDEW MELON, WATERMELON, SQUASH, PUMPKIN

ROOT KNOT (EXCEPT JAVANESE), LESION, RING, STING AND STUNT NEMATODES

Preplant and Planting Treatment -- Apply 1 to 2 gals. "Vydate" L per acre (broadcast); for band treatment, use proportionately less. Following application and before planting, incorporate "Vydate" L 2 to 4 inches into the soil. Use the low rate for light infestations.

Foliar Treatment -- Apply 2 to 4 pts. "Vydate" L per acre in sufficient water to obtain uniform coverage of foliage. Make the first application 2 to 4 weeks after planting and repeat 2 to 3 weeks later. Use the lower rate on light infestations. The best results follow usage of "Vydate" L as a soil treatment as described above.

LEAFMINERS (Liriomyza spp.), APHIDS

Apply 2 to 4 pts. "Vydate" L per acre in sufficient water to obtain uniform coverage of foliage. Make applications when insects first appear and repeat weekly or as needed. Apply the low rate for light infestations of insects and the high rate for severe infestations.

Where leafminer infestations occur annually, initiate treatment schedule 2 to 4 weeks after planting and repeat weekly or as needed.

Note: Do not apply within 1 day of harvest. Do not apply more than 24 pts(3 gals) "Vydate" L per acre per season.

EGGPLANT

APHIDS, COLORADO POTATO BEETLE, LEAFMINERS, AND MITES

Apply 1 to 2 qts. "Vydate" L per acre by ground equipment as a foliar spray when insects first appear. Repeat at 1 to 3 week intervals as needed.

NEMATODES (EXCEPT IN CALIFORNIA)

Apply 1 gal. "Vydate" L per acre as a band treatment 2 to 3 weeks after transplanting and again 4 weeks later. Two to four weeks following these soil treatments, make 2 foliar applications of 2 qts. per acre by ground equipment at 1 to 2 week intervals.

Note: Do not apply within 1 day of harvest when using only foliar applications or within 7 days of harvest using soil with foliar applications. Do not apply more than 24 pts(3 gals) "Vydate" L per acre per season.

ONIONS (DRY BULB ONLY) (MI, OR, WA, TX)

Onion Thrips, Western Flower Thrips

Apply 1 to 2 pints "Vydate" L per acre in sufficient water (minimum 5 gallons) to obtain uniform coverage. Make applications when insects first appear in significant numbers and repeat at 14-day intervals if needed. For light infestations of insects, use the lower rates, increasing rates as the infestation increases.

Nematodes

For stubby root, stem and bulb nematodes, apply 3/4 to 1 gal. of "Vydate" L per acre at planting as an in-furrow drench using 100 to 150 gals. of water per acre, or 1 1/2 to 2 gals. per acre as an in-furrow band spray using 20 to 50 gals. of water per acre.

Note: Do not apply within 14 days of harvest. Do not harvest tops of treated onions. Do not apply more than 24 pts(3 gals) "Vydate" L per acre per season. Do not use on green onions.

PEPPERS

Use not registered in California

ROOT KNOT (EXCEPT JAVANESE), STING, RING, STUBBY ROOT, AND STUNT NEMATODES --

Transplant Water Treatment: Supplemental Foliar Treatments -- Apply 2 pts. "Vydate" L per acre. Add the "Vydate" L to the transplant water, and use a minimum of 200 gals. per acre. Under very high nematode populations, begin with a transplant water treatment and supplement with foliar sprays. For the best results, make the first foliar application 14 days after transplanting and continue in accordance with the foliar treatment described below.

Note: Do not apply as a transplant water treatment during periods of slow plant growth such as when temperatures fall below 45oF, or crop injury may result.

GREEN PEACH APHID, VEGETABLE LEAFMINER, AND PEPPER WEEVIL --

Foliar Treatments -- Apply 2 to 4 pts. "Vydate" L per acre in sufficient water to obtain uniform coverage. Make applications when insects first appear and repeat at 1- to 2-week intervals or as needed. Apply low rates for light infestations of insects. Use highest recommended rates and shorter intervals for controlling severe infestations.

Note: Do not apply within 7 days of harvest. Do not apply more than 24 pts(3 gals) "Vydate" L per acre per season.

POTATOES

Use not registered in California

ROOT KNOT (EXCEPT JAVANESE) STING, LESION AND RING NEMATODES (EXCEPT WEST OF THE ROCKIES); APHIDS, COLORADO POTATO BEETLE, FLEA BEETLES, POTATO LEAFHOPPER, AND TARNISHED PLANT BUG --

Preplant Treatment -- For control of nematodes and early season control of the above insects, apply either one of the following:

In-Furrow Treatment -- Apply 1 to 2 gals. of "Vydate" L per acre in a minimum of 20 gals. water per acre in the seed furrow during the planting operation.

Broadcast Treatment -- Apply 2 to 4 gals. of "Vydate" L per acre in a minimum of 20 gals. of water per acre as a band or broadcast treatment of the soil within one week of planting. Thoroughly incorporate to a depth of 4 to 6 inches in the soil.

Scout fields to determine when early season control diminishes and use registered foliar insecticides including "Vydate" L to retain control.

Foliar Treatments -- For control of the above pests, apply 2 to 4 pts. "Vydate" L per acre, except in the Midwest and Northeast where 1 to 4 pts. "Vydate" L per acre may be used for Colorado potato beetle. Use sufficient water (minimum 4 gals. per acre by air) to obtain uniform coverage. Make applications when pests first appear and

6

repeat at 5- to 7-day intervals or as needed. Apply the low rates for light infestations. Use the higher rates for severe infestations.

Note: Do not apply within 7 days of harvest. Do not apply more than 36 pts(4 1/2 gals) "Vydate" L per acre per season.

SWEET POTATOES

Use not registered in California

ROOT KNOT (EXCEPT JAVANESE) AND SPIRAL NEMATODES --

Preplant Treatment -- Apply 2 to 3 gals. of "Vydate" L per acre in a minimum of 20 gals. of water per acre as a broadcast treatment of the soil; for band treatment, use proportionately less. Apply within one week of planting. Thoroughly incorporate to a depth of 4 to 6 inches in the soil.

In-Furrow Treatment -- As an alternate to the above, apply 1 to 2 gals. of "Vydate" L per acre in a minimum of 200 gals. of water per acre in the transplant water during the planting of slips.

Note: Do not apply as a transplant water treatment during periods of slow plant growth such as when temperatures fall below 45oF, or crop injury may result. Do not apply more than 3 gals(24 pts) "Vydate" L per acre per season.

TOMATOES

SERPENTINE LEAFMINERS (EXCEPT LIRIOMYZA TRIFOLII), COLORADO POTATO BEETLE, APHIDS

Apply 2 to 4 pts. "Vydate" L per acre in sufficient water (minimum 4 gal per acre by air) to obtain uniform coverage. Make application when insect first appears and repeat at 5- to 7-day intervals or as needed.

Apply the low rates for light infestations of insects. Use intermediate rates for heavier infestations of insects. Use the highest recommended rates for controlling severe infestation.

NEMATODES

For control of root knot, sting, stubby root, stunt and reniform nematodes, apply 2 to 4 pts. "Vydate" L per acre in sufficient water (minimum 100 gals.) to obtain uniform coverage, starting as soon as plants become established and repeat at 1- to 2-week intervals throughout the season. Use higher rate and greater frequency for high nematode pressure and larger plants.

Under very high nematode populations, other effective soil treatments at or before planting may be necessary, followed by foliar sprays of "Vydate" L to extend and maintain control. Supplemental sprays of "Vydate" L should begin when nematode populations start to recover. This will depend on the longevity of protection offered by the treatment applied to the soil.

Note: Do not apply more than 48 pts(6 gals) "Vydate" L per acre per season. Do not apply within 1 day of harvest.

YAMS (DIOSCOREA) -- (PUERTO RICO ONLY)

For control of nematodes only on commercial farming operations.

Seed Piece Dip -- Dip seed pieces into solution containing 2 to 4 qts "Vydate" L in 100 gals of water (1,200 to 2,400 PPM active ingredient) for 15 minutes. Allow to dry for 24 hrs, then plant.

Note: Protective clothing must be worn throughout the dipping and planting period. Do not handle treated seed pieces without hand protection.

Foliar Sprays -- For best results, as a supplement to the seed piece dip, apply "Vydate" at the rate of 2 pts per acre at 2-week intervals. Use sufficient water to obtain complete coverage of the foliage (minimum 25 gals per acre). Begin applications when adequate foliage is present to absorb the "Vydate" L - approximately two months after planting - and continue for a maximum of twelve applications.

Note: Make foliar applications wearing full protective clothing during nonwindy conditions. Do not allow "Vydate" L spray to drift away from the target crop.

Note: Do not apply within 60 days of harvest. Do not apply more than 24 qts (6 gals) "Vydate" L per acre per season.

FIELD CROPS

COTTON

COTTON LEAFPERFORATOR

Apply 1 to 4 pts. "Vydate" L per acre in sufficient water to obtain thorough coverage. Make initial applications when damaging populations begin to build, and continue on a 6 to 8 day schedule for up to 4 applications. Do not apply more than 2 gals "Vydate" L per acre per growing season

BOLL WEEVIL, COTTON FLEAHOPPER, AND TARNISHED PLANT BUG

Apply 1/2 to 1 pt. "Vydate" L per acre in sufficient water to obtain thorough coverage. Begin applications when damaging populations appear and repeat as needed to maintain control. Do not apply more than 1 1/4 gals "Vydate" L per acre per growing season

Note: Do not apply within 21 days of harvest. Do not graze or feed treated cotton to livestock.

PEANUTS

Use not registered in California

ROOT KNOT (EXCEPT JAVANESE), STING, RING, AND LESION NEMATODES, AND THRIPS--

Soil Treatment--Apply 2 to 6 qts. of "Vydate" L in a 7" to 12" band using a minimum of 10 gals. of water per acre. Thoroughly incorporate with a rotary tiller to a depth of 3" to 5" immediately after application. Use the higher rate and wider band for severe infestations. Peanuts should be planted within 24 hours after application. Alternatively, "Vydate" L may be applied in a 7" band immediately behind planting. Use 2 to 6 qts. in a minimum of 10 gals. water per acre.

7

Foliar Treatment--Foliar treatment is to be used only following soil fumigation, or following preplant or at planting soil application of "Vydate" L or other contact nematicides. Make 2 foliar applications of "Vydate" L at 2 to 4 pts. per acre in 20 to 40 gals. of water. Apply first spray 3 weeks after emergence and the second spray 6 weeks after emergence. For best results, concentrate the spray on the row using 3 cone-type nozzles positioned over and to each side of the row. Thorough spray coverage is important.

Note: Do not apply more than 20 pts(2 1/2 gals) "Vydate" L per acre per season.

SOYBEANS

Use not registered in California

ROOT KNOT (EXCEPT JAVANESE), LESION, RING, STUNT, LANCE, AND CYST NEMATODES--

Infurrow Band Treatment--Apply 2 to 4 pts. "Vydate" L in 10 to 20 gals. of water per acre (based on 36" row spacing) at planting. Spray over open drill row at the juncture where the seed is covered with soil to assure mixing of "Vydate" L with the soil around the seed.

Incorporated Band Treatment--Apply 2 to 4 pts. "Vydate" L in 10 to 20 gals. water per acre to a 7" to 10" band in the drill area (based on 36" row spacing) as a preplant treatment (up to 1 week before planting) or as a treatment at planting. Incorporate the "Vydate" L 2" to 4" into the soil in the seed zone.

Incorporated Broadcast Treatment--Apply 1 to 2 gals. of "Vydate" L in 10 to 20 gals. of water per acre as a preplant treatment (up to 1 week before planting) or as a treatment at planting. Incorporate the "Vydate" L 2" to 4" into the soil in the seed zone.

Note: Do not cut for hay or feed treated forage to livestock. Do not apply "Vydate" L to areas with severe infestations of nematodes such as where injury to plants is manifested by severe stunting and chlorosis, as "Vydate" L will not control severe infestations.

Do not apply more than 16 pts (2 gals) "Vydate" L per acre per season

TOBACCO

ROOT KNOT (EXCEPT JAVANESE) AND LESION NEMATODES AND FLEA BEETLES

Soil Treatment--"Vydate" L may be applied to the soil as a band treatment or it may be broadcast, disced, and bedded. For best results, the tobacco should be transplanted within 24 hours after soil treatment.

Row Treatment--Apply 1 gal. of "Vydate" L in an 18- to 24-inch band in a minimum of 20 gals. of water per acre of tobacco (12,000 row feet). Thoroughly incorporate with a rotary tiller to a depth of 4 to 6 inches.

Broadcast and Bed Treatment--Apply a broadcast spray of 1 gal. per acre in a minimum of 40 gals. of water. Incorporate thoroughly to a depth of 4 to 6 inches and bed the field in such a manner that only treated soil is used to form the beds.

Note: Do not apply more than 1 gal. "Vydate" L per acre per growing season.

STORAGE AND DISPOSAL

STORAGE: Do not subject to temperatures below 32 degrees F. Store product in original container only. Do not contaminate water, other pesticides, fertilizer, food or feed in storage.. Not for use or storage in or around the home.

PESTICIDE DISPOSAL: Do not contaminate water, food, or feed by storage or disposal. Pesticide wastes are acutely hazardous. Improper disposal of excess pesticide, spray mixture, or rinsate is a violation of Federal Law. If these wastes cannot be disposed of by use according to label instructions, contact your State Pesticide or Environmental Control Agency, or the Hazardous Waste representative at the nearest EPA Regional Office for guidance.

CONTAINER DISPOSAL: Triple rinse (or equivalent) the container. Then offer for recycling or reconditioning, or puncture and dispose of in a sanitary landfill, or, if allowed by state and local authorities, by burning. If burned, stay out of smoke.

NOTICE TO BUYER--Purchase of this material does not confer any rights under patents of countries outside of the United States

NOTICE OF WARRANTY

Du Pont warrants that this product conforms to the chemical description on the label thereof and is reasonably fit for purposes stated on such label only when used in accordance with directions under normal use conditions. It is impossible to eliminate all risks inherently associated with the use of this product. Crop injury, ineffectiveness, or other unintended consequences may result because of such factors as weather conditions, presence of other materials, or the manner of use or application, all of which are beyond the control of Du Pont. In no case shall Du Pont be liable for consequential, special or indirect damages resulting from the use or handling of this product. All such risks shall be assumed by the buyer. DU PONT MAKES NO WARRANTIES OF MERCHANTABILITY OR FITNESS FOR A PARTICULAR PURPOSE NOR ANY OTHER EXPRESS OR IMPLIED WARRANTY EXCEPT AS STATED ABOVE.

AG - 7732 9034 1/14/94

8

Code 279

RESTRICTED USE PESTICIDE

Due to acute oral and inhalation toxicity. For retail sale to and application only by certified applicators or personnel under their direct supervision.

Net Contents

4 F Insecticide— Nematicide

EPA Reg. No. 279-2876 ZC **EPA Est. 279-**

Active Ingredient:

*Carbofuran	44.0%
Inert Ingredients:	56.0%
	100.0%

*2,3-Dihydro-2,2-dimethyl-7-benzofuranyl methylcarbamate.

This product contains 4 lbs. of carbofuran per gallon.

KEEP OUT OF REACH OF CHILDREN

☠ DANGER-POISON ☠

PELIGRO

See Other Panels for Additional Precautionary Information.

Si usted no entiende la etiqueta, busque a alguien para que se la explique a usted en detalle. (If you do not understand the label, find someone to explain it to you in detail.)

FMC Corporation
Agricultural Chemical Group
Philadelphia PA 19103

3/94

STATEMENT OF PRACTICAL TREATMENT

If swallowed: Drink 1 or 2 glasses of water and induce vomiting by touching back of throat with finger. Do not induce vomiting or give anything by mouth to an unconscious person. Get medical attention.

If inhaled: Remove to fresh air. Call a physician immediately.

If in eyes: Flush with plenty of water for at least 15 minutes. Get medical attention.

If on skin: Wash skin immediately with soap and water.

Antidote

Note to Physician: Carbofuran is an N-methyl carbamate and a reversible cholinesterase inhibitor. Do not use oximes such as 2-PAM. Start by giving 2 mg. atropine intramuscularly. According to clinical response, continue until signs of atropinization occur (dry mouth or dilated pupils). If in eye, instill one drop of homatropine.

For Emergency Assistance Call (800) 331-3148.

PRECAUTIONARY STATEMENTS

Hazards to Humans (and Domestic Animals)

Danger

Poisonous if swallowed or inhaled. May be fatal or harmful as a result of skin or eye contact or by breathing spray mist. Causes cholinesterase inhibition. Warning symptoms of poisoning include weakness, headache, sweating, nausea, vomiting diarrhea, tightness in chest blurred vision, pinpoint eye pupils, abnormal flow of saliva, abdominal cramps, and unconsciousness. Atropine sulfate is antidotal.

Personal Protective Equipment:

Some materials that are chemical-resistant to this product are listed below. If you want more options, follow the instructions for category C on an EPA chemical resistance category selection chart.

Applicators and other handlers must wear: Long-sleeved shirt and long pants; Chemical-resistant gloves; such as Barrier Laminate or Butyl Rubber, or Nitrile Rubber or Neoprene Rubber or Polyvinyl Chloride or Viton; Shoes plus socks; Protective eyewear when mixing or loading; For exposure in enclosed areas: A respirator with either an organic vapor-removing cartridge with a prefilter approved for pesticides (MSHA/NIOSH approval number prefix TC-23C), or a canister approved for pesticides (MSHA/NIOSH approval number prefix TC-14G); For exposures outdoors: Dust/mist filtering respirator (MSHA/NIOSH approval number prefix TC-21C).

Discard clothing and other absorbent materials that have been drenched or heavily contaminated with this product's concentrate. Do not reuse them. Follow manufacturer's instructions for cleaning/maintaining PPE. If no such instructions for washables, use detergent and hot water. Keep and wash PPE separately from other laundry.

When handlers use closed systems, enclosed cabs, or aircraft in a manner that meets the requirements listed in the Worker Protection Standard (WPS) for agricultural pesticides [40 CFR 170.240 (d) (4-6)], the handler PPE requirements may be reduced or modified as specified in the WPS.

the handler PPE requirements may be reduced or modified as specified in the WPS.

User Safety Recommendations:
Users should:
- Wash hands before eating, drinking, chewing gum, using tobacco or using the toilet.
- Remove clothing immediately if pesticide gets inside. Then wash thoroughly and put on clean clothing.
- Remove PPE immediately after handling this product. Wash outside of gloves before removing. As soon as possible, wash thoroughly and change into clean clothing.

Environmental Hazards

This product is toxic to fish, birds and other wildlife. Birds feeding on treated areas may be killed. For waterfowl protection, do not apply immediately before or during irrigation, or on fields in proximity of waterfowl nesting areas, or on fields where waterfowl are known to repeatedly feed. Drift and runoff from treated areas may be hazardous to fish in neighboring areas. Do not apply directly to water.

Notice: It is a federal offense to use any pesticide in a manner that results in the death of a member of an endangered species.

The use of Furadan 4 F may pose a hazard to the following Federally designated endangered/threatened species known to be found in certain areas within the named locations.

Attwater's Greater Prairie Chicken—Texas counties including: Aransas, Austin, Brazoria, Colorado, Galveston, Goliad, Harris, Refugio and Victoria

Aleutian Canada Goose—California counties including: Colusa, Merced, Stanislaus and Sutter

Kern Primrose Sphinx Moth—Walker Basin of Kern County, California

This product may not be used in areas where adverse impact on the Federally designated endangered/threatened species noted above is likely. Prior to making applications, the user of this product must determine that no such species are located in or immediately adjacent to the area to be treated. If the user is in doubt whether or not the above named endangered species may be affected, he should contact either the regional U.S. Fish and Wildlife Service office (Endangered Species Specialist) or personnel of the State Fish and Game office.

This product is highly toxic to bees exposed to direct treatment or residues on crops. Do not apply this product or allow it to drift to blooming crops or weeds if bees are visiting the treatment area. Protective information may be obtained from your Cooperative Agricultural Extension Service.

Carbofuran is a chemical which can travel (seep or leach) through soil and can contaminate groundwater which may be used as drinking water. Carbofuran has been found in groundwater as a result of agricultural use. Users are advised not to apply carbofuran where the water table (groundwater) is close to the surface and where the soils are very permeable, i.e., well-drained soils such as loamy sands. Your local agricultural agencies can provide further information on the type of soil in your area and the location of groundwater.

DIRECTIONS FOR USE

It is a violation of Federal law to use this product in a manner inconsistent with its labeling.

Do not apply this product in a way that will contact workers or other persons, either directly or through drift. Only protected handlers may be in the area during application. For any requirements specific to your State or Tribe, consult the agency responsible for pesticide regulation.

Do not apply this product through any type of irrigation system.

Do not use this product on Long Island, N.Y.

Shake Well Before Using

AGRICULTURAL USE REQUIREMENTS

Use this product only in accordance with its labeling and with the Worker Protection Standard, 40 CFR part 170. This Standard contains requirements for the protection of agricultural workers on farms, forests, nurseries, and greenhouses, and handlers of agricultural pesticides. It contains requirements for training, decontamination, notification, and emergency assistance. It also contains specific instructions and exceptions pertaining to the statements on this label about personal protective equipment (PPE), notification to workers, and restricted-entry interval. The requirements in this box only apply to uses of this product that are covered by the Worker Protection Standard.

Do not enter or allow worker entry into treated areas during the restricted entry interval (REI) of 48 hours (except for foliar applications to corn, sunflowers, or sorghum).
After foliar applications on corn, sunflowers, and sorghum, do not enter or allow worker entry into treated areas during the restricted entry interval (REI) of 14 days. Exception: for the last 12 days of the REI, workers may enter the treated area to perform hand labor or other tasks involving contact with anything that has been treated, such as plants, soil, or water, without time limit, if they wear the early entry personal protective equipment listed below.

PPE required for early entry to treated areas that is permitted under the Worker Protection Standard and that involves contact with anything that has been treated, such as plants, soil, or water, is: Coveralls, Chemical-resistant gloves, such as Barrier Laminate or Butyl Rubber, or Nitrile Rubber or Neoprene Rubber or Polyvinyl Chloride or Viton, and Shoes plus socks.

Notify workers of the application by warning them orally and by posting warning signs at entrances to treated areas.

STORAGE AND DISPOSAL

Pesticide Storage
Not for use or storage in or around the house.

Do not store below 35°F, (2°C).

Keep out of reach of children and animals. Store in original containers only. Store in a cool, dry place and avoid excess heat. Carefully open containers. After partial use, replace lids and close tightly. Do not put concentrate or dilute material into food or drink containers. Do not contaminate other pesticides, fertilizers, water, food, or feed by storage or disposal.

Bulk holding containers used for Furadan 4F, and associated equipment (e.g., pumps, meters, hoses, etc.), should be thoroughly rinsed to avoid cross contamination with materials subsequently introduced into the bulk system. Begin by draining any remaining product from tanks and equipment and storing in approved appropriately labeled containers. Thoroughly rinse down the inside of the tank and cycle the rinsate through the pump and metering system, repeating this rinsing and cycling procedure at least three times. Refer to the Pesticide Disposal statement for directions for disposal of rinsate.

In case of spill, avoid contact, isolate area and keep out animals and unprotected persons. Confine spills. Call FMC: (800) 331-3148

To confine spill: If liquid, dike surrounding area or absorb with sand, cat litter or commercial clay. If dry material, cover to prevent dispersal. Place damaged package in a holding container. Identify contents.

Pesticide Disposal
Pesticide wastes are acutely hazardous. Improper disposal of excess pesticide, spray mixture, or rinsate is a violation of Federal law. If these wastes cannot be disposed of by use according to label instructions, contact your State Pesticide or Environmental Control Agency, or the Hazardous Waste representative at the nearest EPA Regional Office for guidance.

Container Disposal
Metal Containers: Triple rinse (or equivalent). Then offer for recycling or reconditioning, or puncture and dispose of in a sanitary landfill, or by other procedures approved by state and local authorities. Do not cut or weld metal containers.

Plastic Containers: Triple rinse (or equivalent). Then offer for recycling or reconditioning, or puncture and dispose of in a sanitary landfill, or incineration, or, if allowed by state and local authorities, by burning. If burned, stay out of smoke.

Returnable/Refillable Sealed Container: Do not rinse container. Do not empty remaining formulated product. Do not break seals. Return intact to point of purchase.

GENERAL INSTRUCTIONS

Do not plant with any crop other than alfalfa, artichokes, bananas, barley, coffee, corn (field, pop or sweet), cotton, cranberries, cucurbits (cucumbers, melons, pumpkins, squash), flax, grapes, non-bearing fruit, oats, ornamentals, peanuts, peppers, potatoes, rice, seed crops (Bermudagrass, spinach), sorghum, strawberries, soybeans, sugar beets, sugarcane, sunflowers, tobacco, or wheat for at least 10 months following use of this product.

Do not rotate with any crop on soil treated at greater than 3.0 pounds active ingredient per acre for at least 10 months.

Alfalfa: Alfalfa Weevil Larvae, Egyptian Alfalfa Weevil, Pea Aphid and in N.Y. State for Snout Beetle control—Apply the amount of Furadan 4 F, indicated in the chart, when feeding is noticed or when insects appear. Alfalfa Weevil Adult—Apply 1 to 2 pints per acre when insects appear. Alfalfa Blotch Leafminer and Potato Leafhopper. Apply 1 to 2 pints per acre when insects appear. Lygus Bugs—Apply 2 pints per acre prior to bloom. Grasshoppers—Apply ¼ to ½ pint per acre when grasshopper feeding is noticed. For control of Blue Alfalfa Aphid (nymphs and wingless adults)—Apply Furadan 4 F at ½ to 1 pint per acre when insect feeding is noticed or when insects appear. Do not apply more than twice per season. Do not apply more than once per cutting. Do not use more than 1 pint per acre in the second application. Apply only to fields planted to pure stands of alfalfa. Do not move bees to alfalfa fields within 7 days of application. Observe the indicated number of days after application before cutting or grazing.

Pints of Furadan 4 F Per Acre	Do not cut or graze within
½	**7 days**
1	**14 days**
2	**28 days**

Minimum gallonage requirements. Ten gallons of finished spray per acre with ground equipment, two gallons per acre with aircraft, except Blue Alfalfa Aphid use 30 gallons of finished spray per acre with ground equipment, 5 gallons per acre with aircraft.

For waterfowl protection do not apply on fields in proximity of waterfowl nesting areas and/or fields where waterfowl are known to repeatedly feed.

Cotton: Thrips—Use Furadan 4 F at 2.5 fluid ounces per 1,000 linear feet of row (1 quart per acre with 40 inch row spacing). Apply in the seed furrow at planting. Furadan 4 F may be mixed with water or liquid fertilizer. When Furadan 4 F is used with liquid fertilizers, premix one part of Furadan 4 F with two parts of water. Check physical compatibility before mixing large quantities. Add this premix to the tank of fertilizer along with the rinsings from the premixing container. Maintain agitation in the tank after mixing and during application. Do not mix until ready to use. Do not feed cotton forage.

Field Corn, Popcorn, Sweet Corn—At Planting: Corn Rootworms, Flea Beetles and to aid in the control of first generation European Corn Borer—Use 2.5 fluid ounces of Furadan 4 F per 1,000 linear feet of row (1 quart per acre with 40 inch row spacing). Apply at planting, as a 7 inch band over the row or inject on each side of the row by mixing with water or liquid fertilizers.

Corn Rootworms, Flea Beetles, Seedcorn Maggot, Wireworms, and to aid in the control of first generation European Corn Borer and Armyworm for approximately 4 to 6 weeks after planting—Use 2.5 fluid ounces of Furadan® 4 F insecticide-nematicide per 1,000 linear feet of row (1 quart per acre with 40 inch row spacing). Apply at planting directly into the seed furrow.

Furadan 4 F may be mixed with water or liquid fertilizers. If Furadan 4 F is used with liquid fertilizers, premix one part of Furadan 4 F with two parts of water. Check physical compatibility before mixing large quantities. Add this premix to the tank of fertilizer along with the rinsings from the premixing container. Maintain agitation in the tank after mixing and during application. Do not mix until ready to use. Do not feed forage within 30 days of last application.

Field Corn, Popcorn—Post Plant: Corn rootworms (northern, southern and western)—use 2.5 fluid ounces of Furadan 4F per 1,000 linear feet of row (1 quart per acre with 40 inch row spacing). Apply as a post emergent spray by banding over the row, or by side dressing or basal spraying both sides of the row after corn emerges. Control will generally be improved if the treatment is cultivated into the soil. Do not feed forage within 30 days of last application.

Field Corn—Foliar Application

European Corn Borer—Use 1½ to 2 pints of Furadan 4 F per acre as a foliar spray when corn borer eggs begin to hatch. Use the higher rate for heavier pest infestations. For treatment with aerial equipment apply as a broadcast spray using a minimum of 1 gallon of finished spray per acre. For treatment with ground equipment, direct the spray into the corn whorl for first brood and into the ear zone for second brood, using a minimum of 10 gallons of finished spray per acre. Repeat if necessary. Observe all precautions listed below.

Southwestern Corn Borer—Use Furadan 4 F at 1 to 2 pints per acre as a broadcast foliar spray when eggs begin to hatch. Rate used will depend on desired residual activity. If infestation continues, retreat in 7 days after a 1 pint application and within 14 days after a 2 pint application. Apply in sufficient water for thorough coverage using a minimum of 1 gallon of finished spray per acre with air equipment or 10 gallons of finished spray per acre with ground equipment. Observe all precautions listed below.

Banks Grass Mites (suppression)—Furadan 4 F when applied at 2 pints per acre for the control of European or Southwestern Corn Borers will also suppress Banks Grass Mites.

Grasshoppers—Use Furadan 4 F as a foliar spray at ¼ to ½ pint per acre when Insects appear or feeding is noticed.
—Use the ¼ pint rate under light to moderate population levels (0 to 14 grasshoppers per sq. yd.).
—Use the ½ pint rate under more severe population levels (15 or more grasshoppers per sq. yd.).
Apply in sufficient water for thorough coverage using a minimum of 2 gallons of finished spray per acre with air equipment or 10 gallons of finished spray per acre with ground equipment. Observe all precautions listed below.

Do not make more than two applications per season at the 1½ to 2 pint use rate. Do not make more than four applications per season at the 1 pint use rate. Do not forage cut or harvest within 30 days of last application. Do not apply on seed corn less than 14 days prior to detasseling or rogueing.

Sweet Corn—Foliar Application (Machine Harvested Only): European Corn Borer—For control of second generation borers apply 1 pint of Furadan 4 F per acre. Make first application just prior to first silking and repeat at weekly intervals. Do not make more than four (4) applications per season. Do not apply within seven (7) days of harvest. If prolonged, intimate contact will result do not reenter treated field within 14 days of application without wearing proper protective clothing. Do not graze or harvest stalks within 21 days of last application.

Do not make a foliar application if more than 12 ozs. of Furadan 10 G per 1,000 linear feet of row (10 lbs. per acre with 40″ row spacing) or 8 ozs. of Furadan 15 G per 1 000 linear feet of row (6.7 lbs. per acre with 40″ row spacing) or 2.5 fluid ounces of Furadan 4 F per 1,000 linear feet of row (1 quart per acre with 40 inch row spacing) were used in an at-planting application.

Minimum gallonage requirements: 10 gallons of finished spray per acre with ground equipment, 2 gallons per acre with aircraft.

Grapes (For use in California): Nematodes (Root Knot and Dagger) and Grape Phylloxera—Apply Furadan® 4 F insecticide-nematicide at 2½ gallons per acre as a broadcast treatment only to the soil surface between the vine rows and immediately incorporate by mechanical means. Do not apply within 200 days of harvest. Remove dense weed growth prior to treatment. Do not use on soils of pH 8.0 or greater. Do not make more than one application per crop year.

Ornamentals—Container Grown
Root Weevil Larvae—Prepare a stock solution by thoroughly mixing 1 to 2 fluid ounces of Furadan 4 F per 100 gallons of water. Apply as a soil drench in sufficient volume to saturate the entire soil profile within each container. The following guidelines give approximate amounts to use on various sized containers. Make a single application when larvae are present (usually from July to mid-October in outdoor growing areas). Later application may be less effective due to lower temperatures and/or the presence of more mature larvae.

Container Size	Amount of Stock Solution Per Container
6 inches diameter	1 pint
8 inches diameter	1 to 2 pints
10 inches diameter	2 to 4 pints
12 inches diameter	3 to 6 pints
16 inches diameter	4 to 8 pints

Early marginal necrosis or leaf drop may occur on Hydrangea or Birch. Not all species or varieties of ornamentals have been tested. Before treating large numbers of plants of a particular variety, treat a few plants and observe prior to full scale application.

Application of Furadan 4 F through overhead sprinkler equipment is prohibited .

Pine Seedlings: Pales Weevil and Pitch-Eating Weevil in pine plantations—Apply a 1% (W/W) active Furadan clay slurry (see following for preparation) to the roots of pine seedlings prior to transplanting. Treat seedlings by dipping roots or use any other suitable means which allows thorough coating. Keep roots moist until transplanted. Prepare the slurry as follows: Add 1.6 ounces (2½ tablespoons) of Furadan 4 F to ½ gallon of water. Mix thoroughly. Add 2 pounds of pulverized kaolin clay (pH 4.5)to this suspension. Mix thoroughly. This is sufficient to treat the roots of 150 to 200 seedlings. Adequate ventilation is required for indoor treatment.

Potatoes: Potato Tuberworm (Virginia only); Colorado Potato Beetle, European Corn Borer, Potato Flea Beetle, Potato Leafhopper—Use 1 to 2 pints of Furadan 4 F in sufficient water to treat one acre. Apply when insects first appear and repeat as necessary to maintain control. Do not make more than 8 foliar applications per season. Do not apply more than 6 pints to the foliage if either Furadan 10 G or Furadan 15 G was used at planting. Do not apply more than 2 pints per application. Do not apply within 14 days of harvest. Minimum gallonage requirements: 10 gallons of finished spray per acre with ground equipment and 3 gallons per acre with aircraft.

Do not use this product on Long Island, New York.

For waterfowl protection do not apply:
—immediately before or during furrow irrigation
—on fields in proximity of waterfowl nesting areas
—on fields where waterfowl are known to repeatedly feed

Small Grains (Wheat, Oats, Barley): Grasshoppers—Use Furadan 4 F at ¼ to ½ pint per acre when insects appear. Cereal Leaf Beetle—Use ½ pint per acre when insects appear. Apply before heads emerge from boot. Do not make more than two applications per season. Do not feed treated forage to livestock.

Minimum gallonage requirements: Ten gallons of finished spray per acre with ground equipment two gallons per acre with aircraft.
For waterfowl protection do not apply on fields in proximity of waterfowl nesting areas and/or on fields where waterfowl are known to repeatedly feed.

Soybeans—At Planting Application: Nematodes (Root Knot Cyst, Stunt, Ring Sting Spiral Lesion, Lance Dagger, Stubby Root)—Use 3.75 to 5 fluid ounces per 1,000 linear feet of row (3 to 4 pints per acre with 40-inch row spacing). Apply at planting as a 12-inch surface band. Soybeans may be grazed or cut for forage 30 days or later following an at planting application.

Soybeans—Foliar Application: Mexican Bean Beetle—Use Furadan 4F at ½ to 1 pint per acre when insects appear. Grasshoppers—Use Furadan 4F at ¼ to ½ pint per acre when insects appear. Do not use Furadan 4F as a foliar application if Furadan 10G, Furadan 15G, Furadan 4F was applied to soybeans at planting time. Do not make more than 2 foliar applications per season. Do not apply within 21 days of harvest. Do not graze or feed foliar-treated forage to livestock or cut for silage or hay. Minimum gallonage requirements: 20 gallons of finished spray per acre with ground equipment, 1½ gallons per acre with aircraft.

Strawberry: (For use in Washington & Oregon) Root Weevils—Use Furadan 4 F at 2 to 4 pints per acre (2.6 to 5.1 fluid ounces per 1,000 linear feet of row for 42 inch row spacing). Apply as a 10 to 12 inch band over the row after last harvest but before October 1st. Do not make more than one application per season. Do not apply if berries are present.

Sugarcane: Sugarcane Borer—Apply 1 to 1½ pints Furadan 4 F per acre using ground or aerial equipment. Check sugarcane fields weekly, beginning in early June and continuing through August. Make first application only after visible joints form and 5% or more of the plants are infested with young larvae feeding in or under the leaf sheath and which have not bored into the stalks. Repeat whenever field checks indicate the infestation exceeds 5 percent. Do not apply within 17 days of harvest. Do not use in Hawaii.

Sunflowers (Confectionary and Oil)—At Planting
Sunflower Stem Weevil, Sunflower Beetle, Grasshoppers—Use 2.5 to 5.0 fluid ounces of Furadan 4 F per 1000 linear feet of row (1.4 to 2.8 quarts per acre with 30 inch row spacing). Apply directly into the seed furrow—OR—Apply in a 7 inch band. Furadan 4 F may be mixed with water or liquid fertilizer. When Furadan 4 F is used with liquid fertilizers, premix one part of Furadan 4 F with two parts of water. Check physical compatibility before mixing large quantities. Add this premix to the tank of fertilizer along with the rinsings from the premixing container. Maintain agitation in the tank after mixing and during application. Do not mix until ready to use.

Use the higher rate when heavier insect infestations are anticipated.

Sunflowers (Confectionary and Oil)—Foliar Application
Sunflower Moth, Banded Sunflower Moth, Stem Weevils, Seed Weevils—Use Furadan 4 F at 1 pint per acre. Grasshoppers—Use Furadan 4 F at ¼ to 1 pint per acre. Sunflower Beetle—Use Furadan 4 F at ¼ to ½ pint per acre. Apply as a foliar spray when insects appear. When a rate range is indicated, use higher rate for heavier insect infestations. Repeat applications as necessary but not more than four times per season. Apply a minimum of two gallons of finished spray per acre with aircraft and 10 gallons by ground equipment. Do not harvest crop within 28 days of last application.

Tobacco—Burley: Flea Beetles—Use 1 gallon of Furadan 4 F per acre. Apply as a broadcast spray over the soil surface prior to transplanting and incorporate into the top 3 inches of soil with a suitable device.

Tobacco—Flue-Cured: Flea Beetles, Wireworms, and to aid in the control of Budworms and Root Knot Nematodes—Use 1½ gallons of Furadan 4 F per acre—OR—Use 1 gallon of Furadan 4 F per acre for control of Flea Beetles only. Apply as a broadcast spray over the soil surface prior to forming beds. Incorporate into the top 3 inches of soil. Form beds and plant as usual. This product may induce flecking of the bottom or lower leaves.

Dealers Should Sell in Original Packages Only.

Terms of Sale or Use: On purchase of this product buyer and user agree to the following conditions:

Warranty: FMC makes no warranty, expressed or implied, concerning the use of this product other than indicated on the label. Except as so warranted, the product is sold as is. Buyer and user assume all risk of use and/or handling and/or storage of this material when such use and/ or handling and/or storage is contrary to label instructions.

Directions and Recommendations: Follow directions carefully. Timing and method of application, weather and crop conditions, mixture with other chemicals not specifically recommended and other influencing factors in the use of this product are beyond the control of the seller and are assumed by the buyer at his own risk.

Use of Product: FMC's recommendations for the use of this product are based upon tests believed to be reliable. The use of this product being beyond the control of the manufacturer, no guarantee, expressed or implied, is made as to the effects of such or the results to be obtained if not used in accordance with directions or established safe practice.

Damages: Buyer's or user's exclusive remedy for damages for breach of warranty or negligence shall be limited to direct damages not exceeding the purchase price paid and shall not include incidental or consequential damages

Furadan and FMC—FMC Trademarks (279-9/29/92-A)

Metam 426

A SOIL FUMIGANT SOLUTION FOR ALL CROPS
MAY BE APPLIED BY SPRINKLER, SOIL INJECTION, OR CHEMIGATION FOR CONTROL OF SOIL-BORNE PESTS THAT ATTACK ORNAMENTALS, FOOD AND FIBER CROPS.
Controls Weeds such as Annual Bluegrass, Bermudagrass, Chickweed, Dandelion, Ragweed, Henbit, Lambsquarter, Amaranthus species, Watergrass, Johnsongrass, Nutgrass, Wild Morningglory and Purslane, Nematodes and Symphylids. Soil-Borne diseases such as Rhizoctonia, Pythium, Phytophthora, Verticillium, Sclerotinia, Oak Root Fungus and Club Root of Crucifers.

ACTIVE INGREDIENT: Sodium methyldithiocarbamate (anhydrous) . . . 42.0%
INERT INGREDIENTS: 58.0%
TOTAL 100.0%

Contains 4.26 lbs. METAM SODIUM per gallon

KEEP OUT OF REACH OF CHILDREN

DANGER PELIGRO

Si usted no entiende la etiqueta, busque a alquien para que se la explique a usted en detalle.
(If you do not understand the label, find someone to explain it to you in detail.)

For Emergencies call 24 hours a day:
Transportation: Chemtrec 1-800-424-9300
Medical: Hazard Information Services 1-800-228-5635, Ext. 169

STATEMENT OF PRACTICAL TREATMENT

Immediately start the procedures below and contact a Poison Control Center, a physician or the nearest hospital. Describe the type and extent of exposure, the victim's symptoms, and follow the advice given.

IF ON SKIN: Immediately flush skin with large amounts of running water for at least 15 minutes while removing contaminated clothing and shoes. Get medical attention immediately. **IF IN EYES:** Immediately flush eyes with large amounts of water for at least 15 minutes. Hold eye lids apart to ensure rinsing of the entire surface of the eye and lids with water. Get medical attention immediately. **IF INHALED:** Remove to fresh air. If not breathing, clear the victim's airway and start mouth to mouth artificial respiration. If breathing is difficult, give oxygen, preferably with a physician's advice. Get medical attention immediately. **IF SWALLOWED:** Immediately give several glasses of water but do not induce vomiting. If vomiting occurs, give fluids again. Have a physician determine if condition of patient will permit induction of vomiting or evacuation of stomach. Do not give anything by mouth to an unconscious or convulsing person.

PRECAUTIONARY STATEMENT

HAZARDS TO HUMANS AND DOMESTIC ANIMALS

DANGER: Corrosive: Causes skin damage. May be fatal if absorbed through the skin. Do not get on skin or clothing. Prolonged or frequently repeated skin contact may cause allergic reactions in some individuals. Harmful if swallowed. Harmful if inhaled. Irritating to eyes, nose, and throat. Avoid breathing vapor or spray mist. Irritating to eyes. Do not get in eyes.

PERSONAL PROTECTIVE EQUIPMENT (PPE)

(1) Handlers Performing Direct-Contact Tasks. Direct-contact tasks include:
- mixing, loading, or fumigant transfer with or without dry-disconnect fittings
- equipment calibration or adjustment
- equipment cleanup and repair
- product sampling
- application or soil-sealing outside an enclosed cab
- any activity less than 6 feet from an unshielded pressurized hose containing this product
- spill cleanup
- removal of tarp or plastic film
- rinsate disposal
- cleanup of small spills
- preparing containers for aeration
- any other handling task not otherwise listed in (2) or (3) below

Applicators and other handlers performing direct-contact activities must wear:
- Coveralls over long-sleeved shirt and long pants,
- Waterproof gloves,
- Chemical-resistant footwear plus socks,
- Chemical-resistant headgear for overhead exposure,
- Chemical-resistant apron when cleaning equipment, or when mixing, loading, or transferring without dry-disconnect fittings,
- Face-sealing goggles, unless full-face respirator is worn,
- A respirator with either an organic-vapor-removing cartridge with a prefilter approved for pesticides (MSHA/NIOSH approval number prefix TC-23C), or a canister approved for pesticides (MSHA/NIOSH approval number prefix TC-14G)

(2) Handlers in Enclosed Cabs. Applicators and other handlers in enclosed cabs must wear:
- Coveralls,
- Shoes and socks

Plus, if pungent, rotten-egg odor of this product can be detected inside the enclosed cab, the handlers in the cab must wear:
- Face-sealing goggles, unless full-face respirator is worn
- A respirator with either an organic-vapor-removing cartridge with a prefilter approved for pesticides (MSHA/NIOSH approval number prefix TC-23C), or a canister approved for pesticides (MSHA/NIOSH approval number prefix TC-14G)

In addition, the PPE specified in (1) for direct-contact activities must be immediately available in the enclosed cab and must be worn if the handler leaves the enclosed cab to perform any direct-contact activity.

The enclosed cab must meet the requirements listed in the Worker Protection Standard (WPS) for agricultural pesticides—40CFR 170.240(d)(5).

(3) Handlers in Treated Areas While Entry Is Restricted.
While entry is restricted (see "Entry Restrictions" in the Agricultural Use Requirements box elsewhere in this labeling), only the following handling tasks may be performed in a treated area outdoors or in the entire greenhouse (or other enclosed structure) in which a treatment took place:
- Assessing/adjusting the soil seal
- Assessing pest control, application technique, or application efficacy
- Operating ventilation equipment
- Sampling air or soil for this product

All other tasks are prohibited until the entry restriction is over.

Handlers performing the above tasks must wear:
- Coveralls over long-sleeved shirt and long pants,
- Waterproof gloves,
- Chemical-resistant footwear and socks

Plus: Handlers must wear (1) in a treated greenhouse before ventilation criteria have been met OR (2) if pungent, rotten egg odor of this product can be detected outdoors or in a treated greenhouse after ventilation criteria have been met:
- Face-sealing goggles (unless full-face respirator is worn) and
- A respirator with either an organic-vapor-removing cartridge with a prefilter approved for pesticides (MSHA/NIOSH approval number prefix TC-23C), or a canister approved for pesticides (MSHA/NIOSH approval number prefix TC-14G).

NET CONTENTS 5 GALLONS

EPA REG. NO. 5481-423 EPA EST. NO. 5481-CA-1

AMVAC CHEMICAL CORPORATION
4100 East Washington Blvd. Los Angeles, CA 90023 U.S.A.

1/94 WPS

PRODUCT INFORMATION

METAM 426 is a water soluble liquid. When applied to properly prepared soil, the liquid is converted into a gaseous fumigant. After sufficient interval of time, the gas dissipates leaving the soil ready for planting.

WHEN TO USE MAXIMUM AND MINIMUM RATES
The application rate of METAM 426 is dependent on the soil type to be treated and the position in the soil of the pest to be suppressed or controlled. Generally a light sandy soil requires a lower application rate than a heavier mineral soil. In addition, if the pest is in the upper portion of the soil profile (annual weeds) a lower application rate is generally required than if the pest is deeper in the soil profile and deeper penetration is desired (perennial weed seeds). When a range of application rates is given in this label consult your local agricultural extension service for more specific information.

METAM 426 is recommended for the control of the following soil-borne pests that attack ornamental, food and fiber crops: Weeds and germinating weed seeds; Annual Bluegrass, Bermudagrass, Chickweed, Dandelion, Ragweed, Henbit, Lambsquarter, Amaranthus spp. (Pigweed & Careless Weed), Watergrass, Johnsongrass, Nutgrass, Wild Morningglory and Purslane; Nematodes and Symphylids (Garden Centipede) and Soil-borne Diseases such as Rhizoctonia, Pythium, Phytophthora, Verticillium, Sclerotinia, Oak Root Fungus and Club Root of Crucifers.

USE PRECAUTIONS: Keep children and pets out of treated area. Keep off desirable lawns and plants. Do not apply within 3 feet of the drip line of desirable plants, shrubs or trees. Do not use in confined areas or when fumes may enter nearby houses. Do not use in greenhouses. Keep container tightly closed when not in use. Do not store near feed or food.

CULTIVATION AND PLANTING AFTER APPLICATION
On well drained soil of light to medium texture which is not excessively wet or cold following application, planting may take place 14 to 21 days after treatment. If soils are heavy or especially high in organic matter or remain wet and/or cold (below 60°F) following application of METAM 426 an extended interval should be observed. Where dosages are greater than 75 gals. per acre, wait at least 60 days. On heavy, wet soils, light surface cultivation to break up crusting and promote drying of the soil should be done 5 to 7 days after application. This cultivation may be repeated as necessary. To avoid reinfesting treated soils, cultural practices should be such that untreated soils are not mixed with treated soils. SPECIAL INSTRUCTIONS: When treating heavier soils, including soil high in clay or organic matter, it is essentially important that the soils be allowed to aerate and dry thoroughly after using METAM 426. During cold and/or wet weather, frequent shallow cultivations may aid the escape of METAM 426 from the soil. If in doubt transplant a seedling plant and examine for injury before planting crop.

FIELD APPLICATION-WHERE ENTIRE AREA IS BEING TREATED

SOIL INJECTION: Space thin injection shanks 5 inches apart and inject METAM 426, 4 inches deep into well prepared soil. Follow immediately with a roller to smooth and compact the soil surface. Light watering or a tarp after rolling helps to prevent gas escape. For field use, 30 to 75 gallons of METAM 426 per acre is recommended.

ROTARY TILLER or POWER MULCHER Spray METAM 426 immediately in front of the tiller or mulcher. Use 37.5 to 75 gallons per acre. Follow immediately with a roller to seal soil surface. Light watering or a tarp after rolling may be used to help prevent gas escape.

SPRINKLER SYSTEM: Use only those sprinkler systems which give large water droplets to prevent excessive loss. Use 56.25 to 75 gals. METAM 426 per acre. For control of shallow pests (top 1 ft. of soil), run sprinklers 5 to 10 minutes. In next 10 to 20 minutes inject all METAM 426 needed for the area covered. On very light soils keep surface moist by sprinkling for 2 to 3 days. For control of pests deeper than 18 inches in the soil, divide METAM 426 into 3 or more equal parts and apply at intervals during the sprinkling period.

CHECK OR FLOOD IRRIGATION: Meter METAM 426 at a steady rate into water during irrigation. Apply 37.5 to 75 gals. per acre depending upon the kind of pest and depth desired, in 3 inches to 18 inches of water per acre.

FIELD APPLICATION TO BEDS OR ROWS

SOIL INJECTION: METAM 426 may be injected into pre-formed plant beds following the directions given above under soil injection. If a wider treated band is desired space 2 or more shanks at intervals of 5 inches to cover the desired treating width. Use thin injection shank(s) and inject METAM 426, 4 inches deep into well prepared soil. Follow immediately with a roller to seal chisel channel(s). Light watering or a tarp after rolling helps to prevent gas escape. Apply at the rate of 37 to 92 fluid ounces/1,000 linear feet of bed per chisel (30 to 75 gals./treated acre). Space shanks 5 inches apart to cover the desired treating width. If METAM 426 is injected into established plant beds through plastic tarps to terminate growth of a previous crop, and to fumigate the bed in preparation of planting a subsequent crop, the terminated crop must not be used for any food or feed purposes after METAM 426 has been applied.

SOIL COVERING METHOD: (Bed-over methods): METAM 426 may be sprayed or dripped in a bed wide band onto the soil immediately ahead of bedshaping equipment. Cover the METAM 426 with soil to a depth of 3 to 6 inches. The soil should be rolled and compacted immediately. Apply at the rate of 37.5 to 75 gallons per acre of treated soil or 11 to 22 fluid ounces per 100 linear feet of row (12-inch wide bed.) If a narrower or wider bed is to be treated, adjust the fluid ounces per 100 linear feet of row to reflect the actual treated acres.

ROTARY TILLER or POWER MULCHER Spray METAM 426 immediately in front of the tiller or mulcher. Use 37.5 to 75 gallons per treated acre. Follow immediately with a roller or bed shaper to seal soil surface. Light watering or a tarp after rolling may be used to help prevent gas escape.

METHOD OF DETERMINING FLUID OUNCES PER 100 FEET OF LINEAR ROW

1) Determine width of bed in feet by dividing width of bed in inches by 12. Example: 8 inches bed = divided by 12 = 0.666 feet
2) Determine square feet in 100 linear feet of bed by multiplying the width of the bed by 100. Example: 0.666 feet x 100 feet = 66.66 square feet
3) Determine the treated acres per 100 linear feet of bed by dividing the square feet by 43,560 (square feet in acre). Example 66.66 square ft divided by 43,560 = 0.0015 acre.
4) To determine the fluid ounces per 100 linear feet
   a) 1 gallon = 128 fl. oz; 50 gallons = 6400 fl. oz; 100 gallons = 12,800 fl. oz.
   b) multiply fluid ounces by acres. Example: 50 gallons = 6400 fl. oz. x 0.0015 = 9.6 fl. oz. per 100 linear feet row.

DRIP IRRIGATION SYSTEM

METAM 426 must be applied through a drip irrigation system to wet the soil thoroughly in the area being treated. Meter 19 to 56 gallons METAM 426 per treated acre into the drip system during the entire irrigation period. APPLICATION MUST BE CONTINUOUSLY SUPERVISED. THIS IS VERY IMPORTANT: WEED ELIMINATION WILL NOT BE SATSIFACTORY IF TOO MUCH WATER IS APPLIED. An adequate concentration of METAM 426 must be present at the time of weed seed germination in order to be effective. Further directions for use are as follows:

1) Ground must be in seed-bed condition, no clods larger than ½ inch in diameter.
2) Beds must be listed, shaped and ready for planting.
3) Soil moisture must be 50% of field capacity in the top 2-3 inches at time of application.

TOBACCO PLANT BEDS

Fall applications are recommended whenever possible. Read and follow directions carefully. Treatment in the South should generally be made before November 30th. Prepare the bed 5 to 7 days before application to insure best conditions for weed seed germination and fumigant action. The bed should be free of clods, level and in good tilth. Cover the bed immediately after treatment with plastic cover. Keep covered no less than one day but no more than two days. The cover need not be tented but should be secure to prevent wind from uncovering the treated area. Seven days after date of application loosen the treated soil to a depth of 2 inches. Do not seed tobacco earlier than 21 days after application.

TARP METHOD: Apply ¾ to 1 gallon of METAM 426 in a minimum of 40 gallons of water per 100 square yards. Apply uniformly over the entire bed. Follow GENERAL DIRECTIONS given above.

DRENCH METHOD: Apply 2 gallons of METAM 426 in 150 to 200 gallons of water per 100 square yards. Application may be made with sprinklers, sprayers with nozzles or any suitable equipment. Follow GENERAL INSTRUCTIONS given above.

POTATOES

For suppression of potato pests such as nematodes, weed seeds and Verticillium dahliae;

SOIL INJECTION: Apply 30 gallons per acre of METAM 426. Follow directions for FIELD APPLICATION WHERE ENTIRE AREA IS TREATED.

SPRINKLER SYSTEM PRE-PLANT: Apply 37.5 to 75 gallons of METAM 426 per acre. Inject into a sprinkler system that can deliver an even water distribution for the area being treated. Soil temperatures should be in the range of 40°F to 90°F in the treatment zone. On very light soils, keep surface area most by sprinkling periodically for 2 to 3 days. Apply in a minimum of 1 acre inch of water. METAM 426 may be applied where crop stubble or vegetation exist without prior tillage, provided there is adequate soil penetration of METAM 426.

PEPPERMINT

Verticillium Wilt: When infestation is limited to small spots in a field, the spread of Verticillium can be reduced by treating the infected spots. Apply 75 gallons per acre of METAM 426 using injector blade or thin shank injector rig. Follow directions for "FIELD APPLICATION—WHERE ENTIRE AREA IS BEING TREATED."

USER SAFETY REQUIREMENTS:

1. Respirator Requirements: When a respirator is required for use with this product, the following criteria must be met:
   a. Cartridges or canisters must be replaced daily or when odor or irritation from this product becomes apparent, whichever is sooner.
   b. Respirators must be fit-tested and fit-checked using a program that conforms to OSHA's requirements (described in 29 CFR Part 1910.134).
2. Dispose of Contaminated Clothing: Discard clothing and other absorbent materials that have been drenched or heavily contaminated with liquid from this product. Do not reuse them.
3. Clean and Maintain PPE: Follow manufacturer's instructions for cleaning/maintaining PPE. If no such instructions for washables, use detergent and hot water. Keep and wash PPE separately from other laundry. Wash PPE after each day's use.

User Safety Recommendations:
Users should:
- Wash hands before eating, drinking, chewing gum, using tobacco, or using the toilet.
- Remove clothing immediately if pesticide gets inside. Then wash thoroughly and put on clean clothing.
- Remove PPE immediately after handling this product. Wash the outside of gloves before removing. As soon as possible, wash thoroughly and change into clean clothing.

ENVIRONMENTAL HAZARDS

Do not apply directly to water, to areas where surface water is present, or to intertidal areas below the mean high water mark. Do not contaminate water when disposing of equipment washwaters. Do not contaminate irrigation ditches or water used for irrigation or domestic purposes. Do not apply when conditions favor drift from treated areas. Do not use in a greenhouse.

DIRECTIONS FOR USE

It is a violation of Federal law to use this product in a manner inconsistent with its labeling. Do not apply this product in a way that will contact workers or other persons, either directly or through drift. Only protected handlers may be in the area during application. For any requirements specific to your State or Tribe, consult the agency responsible for pesticide regulation. Use this product only in accordance with its labeling and with the Worker Protection Standard, 40 CFR 170. Refer to supplemental labeling under "Agricultural Use Requirements" in the Direction for Use section for information about this standard.

CALIFORNIA ONLY: Application must be in compliance with Technical Information Bulletin-California: "Metam Sodium Guidelines for all Application Methods for Metam Sodium in California." This information bulletin may be obtained from your local pesticide dealer or a Metam 426 registrant.

AGRICULTURAL USE REQUIREMENTS

Use this product only in accordance with its labeling and with the Worker Protection Standard, 40 CFR 170. This Standard contains requirements for the protection of agricultural workers on farms, forests, nurseries, and greenhouses, and handlers of agricultural pesticides. It contains requirements for training, decontamination, notification, and emergency assistance. It also contains specific instructions and exceptions pertaining to the statements in this labeling about personal protective equipment, restricted-entry intervals, and notification to workers. The requirements in this box only apply to uses of this product that are covered by the Worker Protection Standard (WPS).

ENTRY RESTRICTIONS:

Outdoors: Entry (including early entry that would otherwise be permitted under the WPS) by any person—other than a correctly trained and equipped handler who is performing a handling task permitted on this labeling—is PROHIBITED from the start of application until 48 hours after application. In addition, if tarps are used for the application, non-handler entry is prohibited while tarps are being removed.

Greenhouses: Entry (including early entry that would otherwise be permitted under the WPS) by any person—other than a correctly trained and equipped handler who is performing a handling task permitted on this labeling—is PROHIBITED in the entire greenhouse (entire enclosed structure) from the start of application until 48 hours after application AND until one of the WPS ventilation criteria for air exchanges, mechanical ventilation, or passive ventilation has been met. In addition, if tarps are used for the application, non-handler entry is prohibited during tarp removal and until one of the WPS ventilation criteria has been met.

NOTIFICATION: Notify workers of the application by warning them orally and by posting fumigant warning signs. The signs must bear the skull and crossbones symbol and state: (1) "DANGER/PELIGRO," (2) "Area under fumigation, DO NOT ENTER/NO ENTRE," (3) the date and time of fumigation, (4) "Metam 426 Fumigant in use," and (5) "name, address, and telephone number of the applicator." Post the fumigant warning sign instead of the WPS sign for this application, but follow all WPS requirements pertaining to location, legibility, size, and timing of posting and removal.

Greenhouses: Post the fumigant warning signs outside all entrances to the greenhouse.

Outdoors: Post the fumigant warning signs at entrances to treated areas.

PPE FOR ENTRY DURING THE RESTRICTED PERIOD: PPE for entry that is permitted by this labeling is listed in the "Hazards to Humans and Domestic Animals" section of this labeling.

GENERAL PRECAUTIONS FOR IRRIGATION SYSTEMS:

Posting of areas to be chemigated is required when 1) any part of a treated area is within 300 feet of sensitive areas such as residential areas, labor camps, businesses, day care centers, hospitals, in-patient clinics, nursing homes or any public areas such as schools, parks, playgrounds, or other public facilities not including public roads, or 2) when chemigated area is open to the public such as golf courses or retail greenhouses.

Posting must conform to the following requirements. Treated areas shall be posted with signs at all usual points of entry and along likely routes of approach from the listed sensitive areas. When there are no usual points of entry, signs must be posted in the corners of the treated areas and in any other location affording maximum visibility to sensitive areas. The printed side of the sign should face away from the treated area towards the sensitive area. The signs shall be printed in English. Signs must be posted prior to application and must remain posted until foliage has dried and soil surface water has disappeared. Signs may remain in place indefinitely as long as they are composed of materials to prevent deterioration and maintain legibility for the duration of the posting period.

All words shall consist of letters of at least 2½ inches tall, and all letters and the symbol shall be a color which sharply contrasts with their immediate background. At the top of the sign shall be the words KEEP OUT, followed by an octagonal stop sign symbol of at least 8 inches in diameter containing the word STOP. Below the symbol shall be the words PESTICIDES IN IRRIGATION WATER.

This sign is in addition to any sign posted to comply with the Worker Protection Standard.

STORAGE AND DISPOSAL

Do not contaminate water, food or feed by storage or disposal.

STORAGE: Store product in a cool, dry, locked place out of reach of children. Do not store below 0°F. Product crystalizes at lower temperatures. If exposed, warm or store at higher temperatures and mix to redissolve crystals and assure uniformity before use.

PESTICIDE DISPOSAL: Pesticide wastes are toxic. Improper disposal of excess pesticide, spray mixture, or rinsate is a violation of Federal law. If these wastes cannot be disposed of according to label instructions, contact your State Pesticide or Environmental Control Agency, or the Hazardous Waste Representative at the nearest EPA regional office for guidance.

CONTAINER DISPOSAL: (METAL) Triple rinse or equivalent. Then offer for recycling or reconditioning, or puncture and dispose of in a sanitary landfill, or by other procedures approved by State and local authorities. (PLASTIC) Triple rinse or equivalent. Puncture and dispose of in a sanitary landfill, or by incineration, or if allowed by State and local authorities by burning. If burned stay out of smoke.

GENERAL INSTRUCTIONS

Before applying this product always thoroughly cultivate the area to be treated, breaking up clods and loosening soil deeply and thoroughly. A week before treatment, moisten soil after cultivation to the desired depth; sprinkle or flood irrigate. This step is essential for all methods of use. Immediately before application, cultivate lightly if the soil has crusted. See POTATOES section for specific directions on the application of METAM 426 to potato fields where no-till stubble or cover crop exist. To prevent loss from evaporation, use only at times when air temperature is moderate and there is little wind movement. Soil temperature must be from 40 to 90°F in the treated zone. Treated zone is defined as the depth of treatment that METAM 426 achieves at the time of application. For other conditions, see section, "CULTIVATION AND PLANTING AFTER APPLICATION". Do not apply to soil surface, as in the sprinkler method, when air temperature is over 90°F or when low humidity or high winds would cause loss of METAM 426 before it can be drenched into the soil with additional water. If fumes become unpleasant during treatment, apply more water to seal the fumes into the soil where they should be confined to achieve maximum fumigation benefit.

The activity of METAM 426 is increased by the use of tarp (plastic, paper or fabric) spread loosely over the treated areas and secured to prevent removal by wind. Keep covered for a minimum period of 48 hours. Seven days after treatment cultivate areas to depth of 2 inches to aerate the soil. Do not seed earlier than 21 days after application when tarping method is used. Use promptly after mixing with water. Do not allow solution to stand. Flush equipment with water after each day's use. Disassemble valves and clean carefully.

TREATMENT OF TREE REPLANT SITES IN COMMERCIAL ORCHARDS

After removing dead or diseased trees and as much of the root system as possible, make a shallow basin over the planting site. Add METAM 426 to the stream of water while filling the basin. Use ¾ quart of METAM 426 per 100 sq. ft. in sufficient water (depending on the soil type) to penetrate at least 6 ft. For control of Oak Root Fungus, use a basin of at least 20 ft. square. Increase dosage to 1½ quarts per 100 sq. ft. in sufficient water to penetrate to the depth of the root system. If water is tanked to the planting site, add METAM 426 to the water and mix before filling the basin. Tarping of replant sites is required when near (½ mile) to populated areas such as schools, hospitals, commercial or office buildings, factories, residential areas, etc. Tarping is not required if treatment if further than ½ mile from such populated areas.

SYMPHYLID CONTROL

Soil should be in good bed condition to a depth of 8 to 10 inches. Maintain adequate soil moisture during the Spring season to bring symphylids to the upper soil surface. Treat during July-August when symphylids are in the upper soil surface. Apply as a broadcast treatment at 15 gallons METAM 426 per acre using blade or chisel injectors. Inject below level of symphylid concentration, usually 6 to 8 inches. Pack soil with a roller immediately after application.

STATEMENTS CONCERNING CHEMIGATION OF METAM 426

Apply this product only through (choose one or more of the following types of systems: sprinkler including center pivot, lateral move, end tow, side (wheel) roll, traveler, big gun, solid set, or hand move: flood (basin); furrow; border; or drip (trickle); irrigation system(s).

Do not apply this product through any other type of irrigation system. Crop injury, lack of effectiveness, or illegal pesticide residues in the crop can result from nonuniform distribution of treated water. If you have questions about calibration, you should contact State Extension Service specialists, equipment manufacturers or other experts. Do not connect an irrigation system (including greenhouse systems) used for pesticide application to a public water system unless the pesticide label prescribed safety devices for public water systems are in place. A person knowledgeable of the chemigation system and responsible for its operation or under the supervision of the responsible person, shall shut the system down and make necessary adjustments should the need arise.

OBSERVE THE FOLLOWING PRECAUTIONS IF YOUR CHEMIGATION SYSTEM IS CONNECTED TO A PUBLIC WATER SYSTEM

Public water system is defined as a system for the provision to the public of piped water for human consumption if such system has at least 15 service connections or regularly serves an average of at least 25 individuals daily at least 60 days out of a year.

Chemigation systems connected to public water systems must contain a functional, reduced-pressure zone, backflow preventer (RPZ) or the functional equivalent in the water supply line upstream from the point of pesticide introduction. As an option to the RPZ the water from the public water system should be discharged into a reservoir tank prior to pesticide introduction.

There shall be a complete physical break (air gap) between the outlet end of the fill pipe and the top of overflow rim of the reservoir tank of at least twice the inside diameter of the fill pipe.

The pesticide injection pipeline must contain a functional, automatic, quick-closing check valve to prevent the flow of fluid toward the injection pump.

The pesticide injection pipeline must also contain a functional, normally closed, solenoid-operated valve located on the intake side of the injection pump and connected to the system interlock to prevent fluid from being withdrawn from the supply tank when the irrigation system is either automatically or manually shut down.

The system must contain functional interlocking controls to automatically shut off the pesticide injection pump when the water pump motor stops, or in the cases where there is no water pump, when the water pressure decreases to the point where pesticide distribution is adversely affected.

Systems must use a metering pump, such as a positive displacement injection pump (e.g. diaphragm pump) effectively designed and constructed of materials that are compatible with pesticides and capable of being fitted with a system interlock. Do not apply when wind speed favors drift beyond the area intended for treatment.

SPRINKLER CHEMIGATION SYSTEMS

Use only those sprinkler systems which give large water droplets to prevent excessive loss. Use 56.25 to 75 gallons of METAM 426 per acre. For control of shallow pests (top 1 ft. or less of soil), run sprinklers 5 to 10 minutes, in the next 10 to 20 minutes inject into the sprinkler system all METAM 426 needed for the area covered. On very light soils keep surface moist by sprinkling for 2 or 3 days. For control of pests deeper than 18 inches in the soil, divide METAM 426 in 3 or more equal parts and apply at intervals during the sprinkling period.

STATEMENTS CONCERNING THE OPERATION OF SPRINKLER CHEMIGATION; DRIP (TRICKLE); OR OTHER APPROVED SYSTEMS UTILIZING A PRESSURIZED WATER AND PESTICIDE INJECTION SYSTEM

The system must contain a functional check valve, vacuum relief valve, and low pressure drain appropriately located on the irrigation pipeline to prevent water source contamination from backflow. The pesticide injection pipeline must contain a functional, automatic, quick-closing check valve to prevent the flow of fluid toward the injection pump. The pesticide injection pipeline must also contain a functional, normally closed, solenoid-operated valve located on the intake side of the injection pump and connected to the system interlock to prevent fluid from being withdrawn from the supply tank when the irrigation system is either automatically or manually shut down.

The system must contain functional interlocking controls to automatically shut off the pesticide injection pump when the water pump motor stops. The irrigation line or water pump must include a functional pressure switch which will stop the water pump motor when the water pressure decreases to the point where pesticide distribution is adversely affected.

Systems must use a metering pump, such as a positive displacement injection pump (e.g. diaphragm pump) effectively designed and constructed of materials that are compatible with pesticides and capable of being fitted with a system interlock. Do not apply when wind speed favors drift beyond the area intended for treatment.

STATEMENTS CONCERNING FLOOD (BASIN), FURROW AND BORDER CHEMIGATION

Systems using a gravity flow pesticide dispersing system must meter the pesticide into the water at the head of the field and downstream of a hydraulic discontinuity such as a drop structure or weir box to decrease potential for water source contamination from backflow if water flow stops.

Do not dilute in supply tanks. Agitation of supply tank recommended after freezing.

IMPORTANT: PLEASE READ BEFORE USE

By using this product, the user accepts the following:

LIMITED WARRANTY: AMVAC warrants that (a) this product conforms to the chemical description on its label; (b) this product is reasonably fit for the purposes stated on its label, subject to the inherent risks referred to herein, when used in accordance with its directions; and (c) that the directions, warnings, cautions, and other statements on this label are based upon responsible experts evaluations of reasonable tests of effectiveness, of toxicity to laboratory animals and plants, of residues on food crops, and upon reports of field experience. Testing has not been performed on all varieties of food crops and plants, in all states, or under all application, weather, and crop conditions.

There are no express warranties other than those set forth herein. Amvac neither makes nor intends, nor does it authorize any agent or representative to make, any other warranty, express or implied. Amvac expressly excludes and disclaims all implied warranties of merchantability, fitness for a particular purpose, or any other warranty of quality or performance. This warranty does not extend to, and the user shall be solely responsible for, any loss or damage which results from the use of this product in any manner which is inconsistent with this label's directions, warnings, or cautions. User's exclusive remedy and Amvac's or seller's exclusive liability for any claim, loss, damage, or injury resulting from the use or handling of this product, whether or not based in contract, negligence, strict liability in tort, or otherwise, shall be limited, at Amvac's option, to replacement of, or repayment of the purchase price for, the quantity of product with respect to which damages are claimed. In no event shall Amvac or seller be liable for special, indirect, or consequential damages resulting from the use or handling of this product.

REFERENCES

Abbott, J.S. 1985. "Emitter Clogging – Causes and Prevention." *ICID Bulletin.* July.

Anonymous. Undated. *Efficient Fertilizer Use.* International Minerals and Chemical Corporation. Mundelein, IL.

Anonymous. 1982a. "Chemicals Today." *Irrigation Age.* October.

Anonymous. 1982b. "Drip Line Pest Control Useful in Tomatoes." *Agrichemical Age.* August/ September.

Anonymous. 1983a. "Research Shows Vapam Potential for Application with Sprinklers." *Agrichemical Age.* July.

Anonymous. 1983b. "Drip Applied Chemical Kills Citrus Nematodes." *California Farmer.* October 15.

Anonymous. 1985a. "New Fumigation Method Ups Strawberry Profits." *California Farmer.* September 21, pp 34 – 35.

Anonymous. 1985b. "Drip Nematicides Shield Table Grapes." *California Farmer.* June 15, p 41.

Anonymous. 1986a. "Fall Chemigation in Grapes." *Agrichemical Age.* September.

Anonymous. 1986b. "Chemigation Guide." *Irrigation Age.* March, pp 24 – 25.

Anonymous. 1988. "To Figure Injection Rate." *Irrigation Journal.* May/June, p 35.

Anonymous. 1990. "Micro Type Irrigation Systems Perform Well When Applying Herbicides to Pistachio Orchards." *California-Arizona Farm Press.* October 20, p 23.

Arceneaux, W. 1974. "Operation and Maintenance of Wells." *Journal of AWWA.* March, pp 199 – 204.

Ayers, R.S. and D.W. Westcot. 1989. "Water Quality for Agriculture." Irrigation and Drainage Paper 29, Rev. 1. Food and Agriculture Organization of the United Nations. Rome, Italy.

Barber, S.A. 1974. "Influence of the Plant Root on Ion Movement in Soil." E.W. Carson. (Ed.) *The Plant Root and Its Environment.* University Press of Virginia, Charlottesville, VA. pp 525 – 564.

Barber, S.A. and R.A. Olson. 1968. "Fertilizer Use on Corn." L.B. Nelson, M.H. McVickar, R.D. Munson, L.F. Seatz, S.L. Tisdale and W.C. White (Eds.). *Changing Patterns in Fertilizer Use.* American Society of Agronomy, Inc., Madison, WI. pp 163 – 188.

Bean, B. W. 1990. "Herbigation: Application of Herbicides with Center Pivots." *Chemigation Workbook.* Texas A & M. College Station, Texas.

Beverley, R.B., J.C. Stark, J.C. Ojala, T.W. Embleton. 1984. "Nutrient Diagnosis of Valencia Oranges by DRIS." *Journal American Society of Horticultural Science* 109:649 – 654.

Boffardi, B.P. 1992. "The Chemistry of Polyphosphate." Proceedings Water Quality Technology Conference. Advances in Water Analysis and Treatment, Part II. Orlando, FL. pp 1293 – 1304.

Boffardi, B.P. 1984. "Corrosion Control of Industrial Cooling Water Systems." Paper No. 274, Corrosion/84. National Association of Corrosion Engineers.

Bouma, D. 1983. "Diagnosis of Mineral Deficiencies Using Plant Tests." *Encyclopedia of Plant Physiology, 15A: Inorganic Plant Nutrition*, A. Lauchli, and R.L. Bieleski. (Eds.). Springer-Verlag. Berlin.

Boswell, M. J. 1990. *Micro-Irrigation Design Manual.* Hardie Irrigation. El Cajon, CA.

Bowsmith. 1995. *Gypsum Use in Micro-Irrigation Systems.* Bowsmith, Inc. Exeter, CA.

Bryant, D. 1993. "Soil Bath." *California Farmer.* December, pp 8 – 9.

Bucks, D.A. and F.S. Nakayama. 1980. "Injection of Fertilizer and Other Chemicals for Drip Irrigation." Proceedings Agri-Turf Irrigation Conference, Houston, Texas. The Irrigation Association, Silver Spring, MD. pp 166 – 180.

Burt, C.M., 1993. *Irrigation Water Management.* Agricultural Engineering Department, California Polytechnic State University, San Luis Obispo, CA.

Burt, C.M. and S.W. Styles. 1994. *Drip/Microirrigation Design and Management.* Irrigation Training and Research Center, Agricultural Engineering Department, California Polytechnic State University, San Luis Obispo, CA.

Burt, C.M., R.E. Walker, and S.W. Styles. 1995. *Irrigation Evaluation Manual.* Irrigation Training and Research Center, Agricultural Engineering Department, California Polytechnic State University, San Luis Obispo, CA.

California Department of Food and Agriculture. 1993. "Fertilizing Materials." *Tonnage Report.* July/December.

California Fertilizer Association. 1980. *Western Fertilizer Handbook*, Seventh Edition. The Interstate Printers and Publishers. IL.

Caron, J. and L.E. Parent. 1989. "Derivation and Assessment of DRIS Norms for Greenhouse Tomatoes." *Canadian Journal of Plant Science* 69:1027 – 1035.

Christiansen, P. 1989. *Vineyard Tissue Sampling Guide for Plant Analysis.* University of California Cooperative Extension Service.

Cooper, K.G., L.G. Hanlon, G.M. Smart, and R.E. Talbot. 1979. "25 Years Experience in the

Development and Application of Scale Inhibitors. *Desalination.* pp 243 – 254.

Curley, S. (Ed.), Not dated. "Fertilizer and Chemical Applications through Irrigation Systems." A & L Agricultural Labs, Modesto, CA. pp 6 – 7.

Dart, F.J. 1983. "Recent Developments in Iron and Manganese Control." AWWA seminar on control of inorganic contaminants. pp 51 – 62.

Daudet, G. 1993. "Low-Flow Irrigation: System Maintenance from the Inside Out." *Irrigation Journal* November/December pp 18, 20 – 22.

Doerge, T.A., R.L. Roth and B.R. Gardner. 1991. *Nitrogen Fertilizer Management in Arizona.* The University of Arizona. Tucson, AZ.

Dickinson, B. 1995. "Irrigation and Nutrient Management: A Conference and Trade Fair." February 2. Salinas, CA.

Dollar, F. 1992. "Polyphosphates Eliminate Rusty Water Complaints." American Water Works Association. *Opflow.* 18(6): 1 – 3.

Driscoll, F. G. 1986. *Groundwater and Wells.* Johnson Division. St. Paul, MN.

Everts, C.J. and R.S. Kanwar. 1990. "Estimating Preferential Flow to a Subsurface Drain with Tracers." Transcript ASAE 33(2):451 – 457.

Follett, R.H., L.S. Murphy, and R.L. Donahue. 1981. *Fertilizers and Soil Amendments.* Prentice-Hall, Inc., Englewood Cliffs, NJ.

Goldberg, D., B. Gornat, and Y. Bar. 1971. "The Distribution of Roots, Water, and Minerals as a Result of Trickle Irrigation." *Journal of the American Society of Horticultural Science.* 96:645 – 648.

Hageman, R.H. 1984. "Ammonium vs. Nitrate Nutrition of Higher Plants. Roland D. Hauck (Ed.). *Nitrogen in Crop Production.* American Society of Agronomy, Inc., Madison, WI. pp 67 – 85.

Hallmark, W.B., J.F. Adams, H.F. Morris, J.D. Fontenot. 1986. "Effect of Plant Growth Stage on Detection of Soybean Nutrient Deficiencies." *Journal of Fertilizer Issues.* 3(3): 66 – 71.

Hartz, T.K., R.F. Smith, K.F. Schulbach, M. LeStrange. 1994. "On-Farm Nitrogen Tests Improve Fertilizer Efficiency, Protect Groundwater." *California Agriculture.* July/August, pp 29 – 32.

Hartz, T.K. 1994. *Drip Irrigation and Fertigation Management of Vegetable Crops.* California Department of Food and Agriculture. Sacramento, CA.

Holm, T.R. and M.R. Schock. 1991. "Potential Effects of Polyphosphate Products on Lead Solubility in Plumbing Systems." *Journal of the AWWA.* July, pp. 76 – 82.

Hornack, R.S. 1994a. *The Story of Polyphosphates.* Chem Craft Corp. Oklahoma City, OK.

Hornack, R.S. 1994b. Personal communication. Chem Craft Corp. Oklahoma City, OK.

Hornung, C. 1994. Personal communication. Soil Solutions Corp. Visalia, CA.

Huntmacher, R.B., C.J. Phene, R.M. Mead, K.R. Davis, and S.S. Vail. 1994. "Fertigation

Management with Subsurface Drip Irrigation: Principles and Precautions." Proceedings of the Microirrigation Workshop. Santa Maria, CA.

Johnson, A.. W., J.R. Young, E.D. Threadgill, C.C. Dowler, and D.R. Sumner. 1987. "Chemigation's Potential." *Agrichemical Age.* March, pp 6 – 8.

Jones, U.S. 1982. *Fertilizers and Soil Fertility.* Second Edition, Reston Publishing Co., Reston, VA.

Jones, J.B., Jr., B. Wolf, and H.A. Mills. 1991. *Plant Analysis Handbook – A Practical Sampling, Preparation, Analysis, and Interpretation Guide.* Micro-Macro Publishing, Inc., Athens, GA.

Jungk, A. and S.A. Barber. 1975. "Plant Age and the Phosphorus Uptake Characteristics of Trimmed and Untrimmed Corn Root Systems. *Plant and Soil.* 42:227 – 239.

Kaufman, H.W. and T. Lee. 1990 "Fungigation: Sprinkler Application of Fungicides for Disease Control." *Chemigation Workbook.* Texas A & M. College Station, TX.

Kelling, K.A. and E.E. Schulte. 1986. "DRIS as a Part of a Routine Plant Analysis Program." *Journal of Fertilizer Issues.* 3(3): 107 – 112.

Kirkby, E.A. 1968. "Influence of Ammonium and Nitrate Nutrition on the Cation-Anion Balance and Nitrogen and Carbohydrate Metabolism of White Mustard Plants Grown in Dilute Nutrient Solutions." *Soil Science.* 105:133-144.

Knutson, A. and Carl D. Patrick. 1990. "Insectigation: Application of Insecticides with Center Pivot Irrigation Systems." *Chemigation Workbook.* Texas A & M. College Station, TX.

Kranz, W.L. and D.E. Eisenhauer. 1990. "Calibration Accuracy of Chemical Injection Devices." ASAE Paper No. 90-1597 presented at the 1990 International Winter Meeting, Chicago, IL.

Lauer, D.A. 1988. "Vertical Distribution in Soil of Sprinkler-Applied Phosphorus." *Soil Science Society of America Journal.* 52:862 – 868.

Lorenz, O.A. and K.B. Tyler. 1983. "Plant Tissue Analysis of Vegetable Crops" in "Soil and Plant Tissue Testing in California." *University of California Bulletin.* 1879.

Martin, W.E. 1953. Proceedings of the Sprinkler Irrigation Conference, Davis, CA.

Mikkelsen, R.L. 1989. "Phosphorus Fertilization through Drip Irrigation." *Journal of Production Agriculture.* 2:279 – 286.

Mikkelsen, R.L. and W.M. Jarrell. 1987. "Application of Urea Phosphate and Urea Sulfate to Drip-Irrigated Tomatoes Grown in Calcareous Soil." *Soil Science Society of America Journal* 51:464 – 468.

Miller, R.O. and J. Kotuby-Amacher. 1995. "A Proficiency Testing Program for the Agricultural Laboratory Industry." Draft Report. Utah State University, Logan, UT.

Miller, R.W., and R.L. Donahue. 1990. *Soils: An Introduction to Soils and Plant Growth.* Simon and Schuster, NJ. p 768.

Mortvedt, J.J. 1972. *Micronutrients In Agriculture.* Soil Science Society of America, Madison, WI.

Mullinier, H.R. 1974. "Applying Anhydrous Ammonia in Irrigation Water." *University of Nebraska Guide.* G74-129.

Overeem, E. 1992. "Additives to Increase Fertilizer Efficiency." Agricultural water conservation and nutrient management conference. Salinas, CA.

New, L.L. 1993. "Current Chemigation Technology with Center Pivots." Conference and Exposition Proceedings of the Irrigation Association. pp 12 – 16.

New, L.L., H.W. Kaufman, A. Knutson, T. Lee, et al. 1990. *Chemigation Workbook.* Texas Agriculture Extension Service.

Purdy, S. 1994. Personal communication. Crop Production Services. Stockton, CA.

Rain Bird International. 1990. *Low Volume Irrigation System Maintenance Manual.* Glendora, CA.

Rauschkolb, R.S., D.E. Rolston, R.J. Miller, A.B. Carlton, and R.G. Burau. 1976. "Phosphorus Fertilizer with Drip Irrigation." *Soil Science Society of America Journal.* 40:68 – 72.

Rehm, G. and G. Hergert. 1978. "Applying Zinc in Fluid Fertilizer." *Agrichemical Age.* September/ October, pp 24A – 24B.

Rivers, J. 1993. Personal communication. Growers Testing Services. Visalia, CA.

Robinson, R.B., G.D. Reed, D. Christodos, B. Frazier, and V. Chidambariah. 1990. *Sequestering Methods of Iron and Manganese Treatment.* Department of Civil Engineering, University of Tennessee. Knoxville, TN. p 296.

Rubeiz, I.G. 1984. "Subsurface Drip Irrigation and Urea Phosphate Use in Vegetables in Calcareous Soils." Ph. D. dissertation, University of Arizona Library, Tucson, AZ, Dissertation Abstracts 84-24933.

Sanchez, C.A., G.H. Snyder, and H.W. Burdine. 1991. "DRIS Evaluation of the Nutritional Status of Crisphead Lettuce." *Horticulture Science.* 26(3): 274 – 276.

Schwankl, L. J. and T.L. Prichard. 1990. "Clogging of Buried Drip Irrigation Systems." *California Agriculture.* 44(1): 16 – 17.

Schwankl, L., B. Hanson, and T.L. Prichard. 1994. *Micro-Irrigation of Trees and Vines.* Draft Copy. University of California, Davis, CA

Schwankl, L. 1993. "Microirrigation System Maintenance." Proceedings of the Microirrigation Workshop and Trade Show. Cachuma Resource Conservation District. Santa Maria, CA.

Selna. G. 1994. Personal communication. Crop Production Services. Stockton, CA.

Soil Science. 1994. A series of 10 articles on polyacrylamides. *Soil Science.* 158(4):233 – 300.

Sumner, M.E. and F.C. Boswell. 1981. "Alleviating Nutrient Stress." G. G. Arkin and H.M. Taylor (Eds.), *Modifying the Root Environment to Reduce Crop Stress.* Monograph 4. American Society of Agronomy. Madison, WI. pp 99 – 137.

Sumner, M.E. 1990. "Advances in the Use and Application of Plant Analysis." *Communication in Soil Science Plant Analysis.* 21: 13 – 16.

Tanji, K. (Ed.). 1990. "Agricultural Salinity Assessment and Management." *ASCE Manual and Report on Engineering Practices No. 71.* American Society of Civil Engineers. New York, NY

Tisdale, S. L., W. L. Nelson, and J. D. Beaton. 1985. *Soil Fertility and Fertilizers.* Fourth Edition. Macmillan Publishing Company, New York, NY.

Trimmer, W.L., T.W. Ley, G. Clough, and D. Larsen. 1992. "Chemigation in the Pacific Northwest." Pacific Northwest Extension Publication 360. Oregon State University, Corvallis, OR.

Unocal Corporation. 1993. *N-pHURIC® Reference Manual.* Unocal Corporation. Sacramento, CA.

Uriu, K., R.M. Carlson, and D.W. Henderson. 1977. "Application of Potassium Fertilizer to Prunes through a Drip Irrigation System." Seventh International Agricultural Plastics Congress Proceedings, San Diego, CA. pp 211 – 214.

U.S. Department of Commerce, Bureau of the Census. 1986. "1984 Farm and Ranch Irrigation Survey." Special Report Series AG84-SR-1. U.S. Government Printing Office, Washington, DC.

Viets, F.G., R.P. Humbert, and C.E. Nelson. 1967. "Fertilizers in Relation to Irrigation Practice." *Irrigation of Agricultural Lands, Agronomy, No. 11.* Hagan et al. (Eds). American Society of Agronomy, Inc. Madison, WI.

Walworth, J.L. and M.E. Sumner. 1987. "The Diagnosis and Recommendation Integrated System (DRIS)." *Advances in Soil Science.* Vol. 6.

Walworth, J.L. and M.E. Sumner. 1988. "Foliar Diagnosis: A Review." *Advances in Plant Nutrition.* Vol. 3, pp 193 – 241.

Watanabe, F.S., S.R. Olsen, and R.E. Danielson. 1960. "P Availability as Related to Soil Moisture." Transactions of the Seventh International Congress of Soil Science. III:450.

Weijnen, M.P.C. and G.M. van Rosmalen. 1985. "The Influence of Various Polyelectrolytes on the Precipitation of Gypsum." *Desalination.* 54: 239 – 261.

Westcot, D.W. and R.S. Ayers. 1984. "Irrigation Water Quality Criteria." *Irrigation with Reclaimed Municipal Wastewater - A Guidance Manual,* Pettygrove, G.S. and T. Asano (Eds.) Report No. 84-1. California State Water Resources Control Board. Sacramento, CA.

White, J.G. 1986. "Nematicides Come to Life with Drip Tape Systems." *Microirrigation.* May, p 160.

Young, J.R. 1988. "Safe Chemigation of Insecticides." *Irrigation News.* February, p 4.

GLOSSARY

Acid A substance with a pH less than 7.0 which releases H^+.

Acid-forming fertilizer Fertilizer which increases acidity and lowers the soil pH after it is applied and reacts with the soil.

Adjusted Sodium Adsorption Ratio (adj. R_{Na}) An index of permeability problems, based on water quality.

Anion A negatively charged ion (e.g., sulfate, nitrate, phosphate, and chloride).

Attapulgite clay A hydrous magnesium aluminum silicate which is used for suspending fertilizers, lime, gypsum, and other materials. It has high swelling and adsorptive properties and can plug microirrigation systems.

Cation A positively charged ion (e.g., ammonium, calcium, magnesium, sodium, potassium, and hydrogen).

Cation exchange capacity (CEC) The ability of a soil to attract and hold on to cations electrically. It is related to soil texture, clay type, and the amount of organic matter in the soil.

Cation exchange sites The location on the negatively charged clay surfaces where cations are held.

Chelate Certain chemicals which are stable enough to diminish the rate of precipitations with anions. These chemicals have a ring surrounding polyvalent (greater than one charge) metal cations.

Chemigation The process of applying chemicals through the irrigation water.

Compatible Refers to mixtures which do not interact and therefore, remain in stable dispersions; they do not precipitate.

Complete fertilizer A fertilizer containing nitrogen, phosphorus, and potassium.

Controlled release fertilizer A fertilizer which releases the nutrients more slowly than conventional water soluble fertilizers.

Custom fertilizer blend A fertilizer which is specially formulated to have the grade requested by the grower.

Deep percolation The water which flows below the root zone.

Distribution Uniformity (DU) The measure of how uniformly water infiltrates the soil. This is expressed as a percentage.

$$DU = \frac{\text{“ Minimum ” depth infilitrated}}{\text{Average depth infiltrated}} \times 100$$

Electrical Conductivity (EC) A measurement which indicates the total salinity of a sample of soil or water without considering the individual constituents. A high EC occurs with a high salt level. Units are millimhos/centimeter (mmho/cm) or decisiemens/meter (dS/m). There are three EC measurements of concern with water management:

EC_w EC of the irrigation water.

EC_{sw} EC of the soil water solution. This is the salinity that the plant roots “see”. The EC_{sw} increases as the soil dries out, because the same amount of salt is contained in less and less water, resulting in a more concentrated solution.

EC_e EC of the extract of a saturated soil paste. Distilled water is added to a soil sample, mixed up, and then the solution is extracted from the saturated soil sample. The EC of this extract is the EC_e. Estimates of the effects of the EC_{sw} on crop growth are made from this EC_e.

Exchangeable cations Elements held by their electrical charge on the cation exchange sites of the clay.

Exchangeable Sodium Percentage (ESP) The percentage of the cation exchange capacity of a soil which is occupied by sodium.

Fixation The process by which available plant nutrients become unavailable by reacting with another compound in the soil to become insoluble. Sometimes this term refers to nutrients becoming “trapped” in between layers of clays.

Fluid lime A mixture of fine limestone and water.

Foliar fertilization Application of fluid fertilizers directly onto the plants' leaves.

Gypsum Calcium sulfate. Used as a source of calcium and/or sulfur, and for correcting soil permeability problems.

Irrigation efficiency (IE) A measure of the efficiency of applied irrigation water. It is expressed as a percentage.

$$IE = \frac{\text{Irrigation Water Beneficially Used}}{\text{Total Irrigation Water Applied}} \times 100$$

Leaching The movement of soluble constituents below the root zone with water.

Lime The general term for a class of compounds containing calcium or magnesium carbonate, calcium oxide, or calcium hydroxide. These compounds are used to neutralize soil acidity (i.e., raise the pH).

Macronutrients Nutrients required by plants in large amounts (i.e., N, P, K).

Micronutrients Nutrients required by plants in trace amounts (i.e., B, Cl, Cu, Fe, Mo, Mn, Zn).

Nitrification Conversion of ammonium to nitrate by microorganisms. This is an acidifying process.

Nitrification inhibitors Compounds which delay the nitrification process, thus keeping the nitrogen in the ammonium form longer.

Nitrogen fixation The conversion of nitrogen gas to ammonium by soil organisms.

Non-acid forming fertilizer A fertilizer which does not lower the soil pH when applied to the soil.

pH A measure of acidity or alkalinity. Most agricultural plants grow best with a soil pH in the range of 6.5 – 7.5.

Polyphosphate A polyphosphoric acid salt. It contains many phosphates linked together.

Precipitate An insoluble solid material.

Quick test A simple and rapid chemical test of a soil or a plant part which can be done on the farm and gives estimates of plant and soil nutrition.

Secondary nutrients Nutrients required in moderate amounts (e.g., Ca, Mg, S).

Sequestration Refers to the complexing of metallic ions and organic compounds so that plant nutrients are held in solution and do not precipitate. This is what a chelate does in soil.

Sodic (alkaline) soil A soil with an exchangeable sodium percentage (ESP) greater than 15%. Symptoms of a sodic (alkaline) soil include water penetration problems and poor crop growth. Very hard, structureless clods tend to form near the soil surface when the soil dries out. Sodic soils have excess sodium and insufficient exchangeable calcium on the exchange sites.

Soil solution The water held in the soil and the nutrients contained in that water.

Solution fertilizer A liquid fertilizer in which all nutrients have been completely dissolved in water. A clear liquid fertilizer.

Suspension fertilizer A fluid fertilizer containing dissolved and undissolved nutrients. The undissolved nutrients are kept in suspension by special agents.

INDEX

A

acids 12, 20, 96, 117, 164, 165
acidity 67
adjusted sodium adsorption ratio 177–179
advance 47
agitator 192
air-gap separation 10
algae 55, 162, 172
alkaline 58, 86
amino acid 65
Ammonia 68
ammonia 63, 67–69, 70, 96
ammonium 58, 63, 65–69, 73, 75, 84, 121, 188
    ammonium-nitrate solution 107
    ammonium-nitrogen 188
ammonium nitrate 107
ammonium phosphate 108, 118
ammonium polyphosphate 80, 108, 119
ammonium polysulfide 109, 177
ammonium thiosulfate 109, 177
anhydrous ammonia 67, 70, 106, 168
AN-20 107, 199
anion 58, 68, 74, 78
application rate 205, 206
aqua ammonia 67, 68, 69, 107, 109
attapulgite clay 92

B

backflow prevention 10, 192
bacteria 66, 162, 165, 172–173
bell pepper 197–198
bicarbonate 114, 116, 179
biuret 115
blossom end rot 197
border strip sprinkler 43–45, 48, 193
boron 85
broadcasting 60–62
bromide 172
buffering capacity 110, 188

C

cadmium 111
calcareous soil 114, 204, 205
calcium 74, 84–100, 186, 197, 204
    calcium precipitate 175
calcium ammonium nitrate 107, 188
calcium carbonate ($CaCO_3$) 175, 186, 188
calcium polysulfide 110, 177
Calgon® 168
calibration 192, 209
CAN-17 107, 115, 199, 205
carbonate 175, 179, 186
Cardy Meter® 152, 226
California
    considerations 43
    regulations 245
carrier site 77, 78
cation 58, 66, 78, 81, 84, 184, 206
    cation exchange capacity
      59, 88, 110, 111, 126, 187
    cation exchange site 58, 66, 68, 81, 82, 178
    cation-anion balance 73, 74
    cation, monovalent 184
center pivot 1, 43–45, 48, 191, 194, 212
check valve 10
chelates 87, 88, 110, 111, 131, 206
chemical injection 5, 162, 191
chemical travel time 51
chemigation 1, 6, 191–196, 203
chloride 85, 113, 204
chlorine 49, 50, 96, 162–165, 173, 174, 176
    chlorine gas 162, 164
citric acid 165
clear liquid 103, 114
colorimetric 225
compatibility 20, 38, 94–96, 99, 117, 166
competition 73
conditioner 91, 212

constant rate 44, 45, 50
containment structures 12–13
continuous injection 50
controller 16, 20
cooling effect 92
copper 87
copper sulfate 172
corrosion 12, 50, 102, 165, 172
cost comparison 209
critical level 139
crop residue 71, 130

D

DAP 66, 98
deep percolation 47, 53–58, 60, 192
denitrification 66
density 212
diaphragm pump 34, 48
diatomaceous earth 92
differential pressure 21–23, 28, 48
diffusion 77
distribution uniformity (DU) 44–45, 47, 53–59, 128
drip 20, 43, 49–51, 59–60, 67, 79–80, 82, 116, 122, 133, 135, 192, 197
  buried drip 59
  drip systems 161–175
DRIS 149, 150–152, 207
DTPA 89, 131

E

electrical conductivity (EC) 122, 126, 215
  EC_w 177, 180, 184
EDDHA 89
EDTA 89
emitters 48, 50, 55, 67, 79, 82, 108, 116, 135, 162, 172, 175
environmental hazards 192
Environmental Protection Agency (EPA) 7, 229
Enzone® 206
equilibrium 78, 81, 82, 163
evaporation 54, 69
exchangeable sodium percentage 177

F

fertilizers
  liquid 211
  solid 48, 91, 211
  spoon feeding 78
  suspension 104
fertigation 2, 60–63, 66, 113, 115, 122, 135, 137, 202–207
  schedule 200
filter 12, 20, 43, 50, 117, 166, 174
fishmeal 71
fittings 12
float valve 26
flushing 19, 161–175
flow rate 15, 43, 44, 45, 48, 51
foliar application 202
fungicide 2, 193
Furadan® 196, 206
furrow 43–45, 48, 55, 110, 177, 197

G

generators 37
grapes 203–207
green acid 113
groundwater 5, 59, 192
gypsum 36, 96, 110, 114, 180–181, 197, 204, 206
  injection machine 183
  injectors 36

H

hard water 108, 168, 169
hardware 5, 91, 165
heavy metals 186
herbicide 2, 43, 164, 175, 180, 192–195, 207
hoses 11, 172, 173
hydrated silica 92
hydroxide 68, 70
hypochlorite 162–164
hypochlorous acid 163

I

immobilization 65
infiltration 44, 46, 67, 177–184, 204
injection 15–42, 192, 210–213
  injection line check valve 7
  injection rate 43–44, 47–48, 136, 192, 209–210, 213
  non-continuous injection 51
injectors 192
  calibration accuracy 38
  design 24–37
  gypsum injectors 36
  venturi 16, 19, 22, 24
insecticides 2, 12, 194
interaction 204, 207
interlock 7
iron 87, 172–174
iron sulfide 174
iron oxide 162, 174
irrigation 43–52
  check valve 7
  efficiency 57
  event 15
  irrigation set 43–50, 173
  over-irrigation 56

principles 53–57
scheduling 53, 54, 57
surface irrigation 2, 45, 55, 59, 80
systems 44–47
under-irrigation 55
water testing 154

J

jar test 94

L

Label Improvement Program 7, 10, 229
labels, pesticide 253
laminar flow 59, 60
leaching 53, 57–61, 66–68, 121, 203, 206
LEPA 194
lettuce 198–199
lime 55, 70, 85, 87, 110, 177, 186–189
liming effect 97, 169
lime sulfur 110, 177
limestone 67, 186, 187
linear move sprinkler 1, 44, 191
low pressure cutoff 7
low pressure drain 7

M

magnesium 74, 84, 206
magnesium carbonate precipitation 175
magnesium/calcium ratios 132, 177, 183
magnetic water treatment 175
manganese 87, 172, 174
manganese oxide 162, 174
manganese sulfides 162, 174
manure 71, 88
MAP 66, 98
mass flow 77
materials 11–12
hose 11, 172, 173
tanks 12, 16, 20–31, 48, 117, 118
metal chelates (see chelates)
metal micronutrients 86
methyl bromide 195
mica 180
microbe 65, 71
microirrigation 2, 77, 122, 160–161, 172, 180, 206–207, 212
micronutrient 84–88, 97, 100, 110–111, 118, 126, 131, 165, 202
microspray 43
milliequivalents 125, 217
mineral nitrogen 65
minerals 103–120
mineralization 65, 67, 124
miticide 194
mixing 43, 92, 96, 165
mobile unit 49, 50
mobility 59
molybdenum 85

N

N-pHURIC® 116–118, 165, 199, 201, 206, 207
Nemacur® 196, 207
nemagation 194
nitrate 58, 61–68, 73–75, 80, 121, 129, 135, 184
nitrate reductase 75
nutrition 74
nitrate reductase 75
sap test 226
nitrification 58, 65–67, 68, 185
nitrification inhibitor 68, 73
nitrogen 60–61, 63–66
nitrogen cycle 64
nitrogen fertilizer 165
nitrogen fixation 65, 86
nitrogen uptake 73–76, 197
organic nitrogen 67, 71
nitrosomonas 67
non-continuous injection 51
non-uniformity 53, 55
nutrient process 77–90, 154
availability 122
balance 132
concentration 138, 147
interaction 149, 160
management 122
ratio 123, 148

O

organic matter 129, 173
organic nitrogen 67, 71
orthophosphate 80
over-irrigation 56

P

P/Zn ratio 133
perfect timing 56
permanent subsurface drip 175
permeability 110, 161, 178, 180, 183, 206
pesticides 10–12, 43, 49, 59, 164, 191
pesticide labeling 5, 10–12, 253
petiole analysis 147, 200, 205
petiole concentration 137
pH 50, 67–70, 73, 78, 85, 99, 110, 122, 126, 163–165, 175, 206
pH modification 185–190
phosphonates 171
phosphoric acid 80, 97, 113, 116, 118, 165, 175, 206

phosphorus 58, 78–81, 135, 137, 206
phosphorus movement 79
Phyloxxera 206
Phytophthera 197
piston pump 35
plant tissue analysis 122, 136–152
plant uptake 68, 71, 78, 81, 84, 126, 133, 136
plastic 12
plugging 44, 55, 91, 92, 94, 161, 162, 172–176
polyelectrolytes 171, 184
polymers 184
polyphosphate 59, 80, 168, 175
poor timing 57
potassium 58, 74, 81–84, 100, 186, 197
  potassium fixation 81
potassium chloride 113, 204
potassium nitrate 113
potassium phosphate 114
potassium sulfate 114, 204
potassium thiosulfate 114, 204
ppm (parts per million) 216
precipitate 94, 172, 174, 175, 178
preferential flow 59–60
pressure differential 21–23, 55
pressure relief valve 107, 109
proportion control 20
protein 65
PSSD 175
pumps 23, 28, 31–35 166, 174
  diaphragm pump 34, 48
  piston pump 35
PVC 12
pyrophosphate 80, 108

Q

quarter-half-quarter rule 15, 50
quick test 122, 127, 133
  equipment 219–222
  procedures 223–228

R

re-entry intervals 11
record keeping 122
residual nitrogen 135
roots
  interception 77
  intrusion 44, 162, 175
runoff 54

S

safety 5–13, 117, 192
salinity 178, 215–218
sap nitrate 226
secondary nutrients 83–84
sequester 172, 174
slimy bacteria 165, 172
sodic (acidic) soil 289
sodium 180
  sodium ratio 177
sodium hypochlorate 96
soil
  acidity 67, 186, 188, 289
  fertility 122
  nitrate 223 133
  solution 66, 68, 75, 77, 81, 131–137
  solution access tubes 134
  sodic (acidic) soil 289
  structure 204, 206
  testing 121–135, 207
  texture 79, 187, 193, 204
solubility 91–93, 115, 193, 204, 211
spoon feeding 78
sprinklers 47, 48, 51, 55, 70, 80, 82, 192, 193
  border strip 43–45, 48, 193
  center pivot 1, 43–45, 48, 191, 194, 212
  furrow 43–45, 48, 55, 110, 177, 197
  linear move sprinkler 1, 44, 191
  travelers 44
stainless steel 12, 113, 117
sufficiency levels 135, 147
sufficiency range 207
sulfate 74
sulfides 175
sulfur 84, 177
sulfuric acid 109–116, 165, 175, 180
super saturation 82
superchlorination 173
surface irrigation 2, 45, 55, 59, 80
suspension 103, 115
synthetic compounds 167

T

tailwater 44–45, 47
tanks 12, 16, 20–31, 48, 117, 118
test strip 128, 223
Thiobacillus 84, 109, 110, 114
tissue testing 136, 207
titration curve 166
tomatoes 200–202
Treflan® 175
Trifluralan® 207

U

UAN-32 58, 107, 204, 205
under-irrigation 55
uniformity 55, 60, 61, 192, 193, 211
urea 52, 58, 63, 67, 71–73, 115–120, 121, 184, 211, 213
urea ammonium nitrate solution 107

urea phosphate 80, 116
urea sulfuric acid 97, 116–117
urease 71–73

V

vacuum relief 7
Vapam® 195
véraison 205
vine 175, 203, 205, 206, 207
volatilization 58, 66–69, 109, 116, 193
Vydate® 196

W

water
- acidification 164
- infiltration 177
- permeability 177, 180, 184
- pure irrigation 180, 181, 182, 183, 184
- quality 177, 179, 203, 204, 206
- treatment 180
- water-powered injector 16
- well water 174

wetting agents 184
white acid 113

Z

zinc 87, 137, 205